DEBATING CANCER

The Paradox in Cancer Research

DEBATING CANCER

The Paradox in Cancer Research

Henry H. Heng

Wayne State University School of Medicine, USA

World Scientific

NEW JERSEY • LONDON • SINGAPORE • BEIJING • SHANGHAI • HONG KONG • TAIPEI • CHENNAI • TOKYO

Published by

World Scientific Publishing Co. Pte. Ltd.
5 Toh Tuck Link, Singapore 596224
USA office: 27 Warren Street, Suite 401-402, Hackensack, NJ 07601
UK office: 57 Shelton Street, Covent Garden, London WC2H 9HE

Library of Congress Cataloging-in-Publication Data
Heng, Henry H., author.
Debating cancer : the paradox in cancer research / Henry H. Heng.
p. ; cm.
Includes bibliographical references and index.
ISBN 978-9814520843 (hardcover : alk. paper)
I. Title.
[DNLM: 1. Biomedical Research. 2. Neoplasms. 3. Neoplasms--genetics. QZ 206]
RC267
362.19699'400724--dc23
2015022548

British Library Cataloguing-in-Publication Data
A catalogue record for this book is available from the British Library.

Printed in Singapore

Dedication

To Julie, Eric, and Christine

Preface

We are living in interesting times. The explosion of genomic information has generated both excitement and confusion. The paradox of knowing more about cancer's genetic landscape yet understanding less of its common molecular basis represents such an example. It was believed or hoped by many that the cancer genome sequencing project would once and for all solve the mystery of cancer, without anticipating that the powerful technology would further add to the unmanageable complexity of the picture. Following the high hopes of the development and utilization of various large scale — omics technologies, the long expected clear-cut understanding of cancer is actually fading away. "Well, cancer is complicated." This seems to be the general conclusion after the expensive and hard-fought war on cancer. But what is the real problem? Not enough molecular data yet? No suitable model for data analyses? Or on an even more serious note, has there been a wrong conceptual framework all along?

Cancer research is not alone. A few years back, the leading edge section of the journal *Cell* asked a few leading genomic researchers a question: what's been the most surprising regarding the human genome? The answers were: let us remember the chromosome (we need a new platform to synthesize parts information); variation and complexity (DNA holds limited phenotype prediction power); a hidden ecosystem (somatic cell evolution is important); and huge heterogeneity (genetic components lack a shared pattern). The common realizations were the lack of a direct relationship between DNA sequences and biological functional elements, and the unexplainable unknown.

Digesting these comments, one cannot help but wonder whether or not genetics/genomics is facing a paradigm shift to replace the century long gene-centric and genetic determinist framework of genetics, as these big surprises, in fact, can no longer be explained by the current framework of genetic theory, and a new concept of inheritance is needed.

Since molecular genetics/genomics serve as the theoretical basis for current cancer research, it is obvious that the confusion in genetic theory will have a direct impact on cancer research, which is precisely reflected by the data generated from cancer genome sequencing efforts.

Such realizations call for thinking big, thinking different, and thinking out of the (current framework) box. Coincidentally, the former director of US National Cancer Institute (NCI), Dr. Harold Varmus, has also promoted the importance of thinking… for budgetary reasons. Varmus quoted the father of nuclear physics Ernest Rutherford in his recent resignation letter to the NCI staff: "… we've run out of money, it is time to start thinking."

Regardless of circumstance, it is clear that switching from "just do it" to critical thinking is the only way out of the confusion, as the push for methodologies and large data sets has not delivered. Serious, candid debate is the key to success.

To be fair, there *have* been some minimal academic debates. In addition to specific topics in conferences, some journals have opened sessions of "point and counterpoint debating". NCI has offered grant opportunities for answering "provocative questions" in recent years. However, the majority of debates have been held within the accepted framework, which prevented discussions of more fundamental principles. Occasionally, when some serious issues that questioned the theoretical basis and the direction of current cancer research were raised, they were met with ignorance or even heavy resistance to dismiss the efforts. Typical responses included (but were not limited to):

"While there are some surprises, science can correct itself! Just work hard and collect more data…"

"Criticism towards the gene mutation theory is unbalanced and unnecessary and it is attacking a straw man…"

"Who doesn't already know the heterogeneity and complexity of cancer? But as a scientist, I am only interested in the elegant molecular understanding. I am a reductionist, and a very proud one."

"These arguments are true but challenging the foundation of current theory is not good for the research community. We need to work together rather than divide the community..."

"We're waving, not drowning!"

The sentiment of the community majority seems to be that many important paradoxes will be automatically solved by accumulating data without vigorously asking questions! The unfortunate fact is that, like any field of science, the field will never truly advance without the correct framework.

The main goal of this book is to initiate essential debates and call for a new framework for future cancer research. Hundreds of paradoxes and specific questions are addressed including some key myths. The focus of discussion is to define questions, including the rationale of various cancer theories, the alternative interpretations of some popular experiments/concepts, and the limitations of general inference based on exceptional cases in cancer research. By debating many controversial aspects of current cancer research, this book will not only point out the limitations of some popular concepts, but also offer new insight for genome-based somatic cell evolution. In particular, the newly introduced genome theory will be discussed.

To engage readers, we have chosen a more candid style of discussion. Some of these analyses may even be out of the readers' comfort zones, as we still recall the great discomfort we felt when we first realized the shocking disagreement between our previous traditional knowledge and the reality of cancer. We have also used as much of our own data and concepts as possible to tell the story, and because of this, we owe an apology to the many other researchers who have made contributions to cancer evolutionary study, yet due to length restrictions, were not included. We invite all readers to debate with us and examine the genome theory we are promoting. We hope readers will benefit from this debating process and begin searching for their own questions and answers.

Debating matters, especially when the status quo is not working.

Acknowledgements

There are many people on my thank-you list. My research team members Joshua Stevens, Batoul Abdallah, and Steven Horne have joined me to coauthor Chapters 4, 6, and 7 respectively. I would like to thank Steven Horne, Barbara Spyropoulos, Joshua Stevens, Steve Bremer, and Batoul Abdallah for editing the manuscript and improving diagrams, without whom it would have taken much longer to craft this book. I would like to thank other current and former members of my research team, including Guo Liu, Karen Ye, Lesley Lawrenson, Wei Lu, and Christine Ye for their efforts and support.

My sincere thanks goes to many of my mentors/colleagues/friends who have encouraged me over the years. They have patiently listened and read my ideas, helped me better articulate my viewpoints, and provided constructive criticism where needed. They are Bill Brinkley, Mina Bissell, Linda Cannizzaro, David Chen, Don Coffey, Bernard Crespi, Jim Crow, Jim Davie, Peter Duesberg, Jing-Bing Fan, Russell Finley, Rafael Fridman, Y-B Fu, Rafe Furst, Robert Gatenby, Wayt Gibbs, David Gisselsson, Arny Glazier, Alex Gow, Morris Goodman, Root Gorelick, Larry Grossman, Gloria Heppner, Lee S. Hong, Sui Huang, Maik Huttemann, Gregory Kapatos, Steve Krawetz, Markku Kurkinen, Yur Lazebnik, Li Li, Joshua Liao, Rong Li, Leonard Lipovich, Larry Loeb, Carlo Maley, George Miklos, Fred Miller, OJ Miller, Peter Moens, Avraham Raz, Brian Reid, Harry Rubin, Wael Sakr, Sureyya Savaşan, L-J Shi, Jeremy Squire, Gary Stein, Joachim Sturmberg, John Tomkiel, Jeffrey Tseng, Lap-Chee Tsui, Bert Vogelstein, Douglas Wallace, Alan Wang, Derek Wildman, Adam Wilkins, H-Y Wu, G-S Wu, Y-M Xie, Yang Yang, Holly Yu, J-W Yu, Kezhong Zhang, and Ren Zhang.

I owe my gratitude to the essential support I received from my editors at World Scientific: Maranda Ward for her important initiation of this project, and Darilyn Yap for her valuable editorial help.

Utmost love and thanks to my wife, Christine, and children, Julie and Eric, who have heard my "genome story" no less than too often. Their enthusiasm has compelled me to begin writing when ideas come to mind even well before dawn.

Contents

Chapter 1

Introduction: Why Debate Cancer, and Why Now?

Cancer, the eighth leading cause of death in the 1900s in the US, has now surpassed heart disease as the primary cause of death in people under the age of 85 [Twombly, 2005]. In recent years, cancer has become the number one killer in the UK, France, Canada and China, just to name a few. The World Health Organization estimates that cancer accounted for 7.6 million deaths (approximately 13% of all deaths) in 2008, and will likely reach 13.1 million deaths by 2030 [WHO, 2014]. To put these numbers in perspective, about 41% of Americans will be diagnosed with cancer at some point in their lives, and 21% will die from it [Annual report of president's cancer panel, 2008–2009]. These stunning statistics are a true challenge and are very disconcerting for most cancer researchers. Long before President Nixon declared the war on cancer in 1971, extreme efforts were being made by researchers to decipher the mechanisms of how cancer works in the hope of producing a cure. Along the way, the gene mutation theory of cancer was developed and many key cancer genes/pathways were identified. Today entire cancer genomes are sequenced in order to identify the genetic landscape of cancer. With this molecular knowledge and impressive technologies available, we not only can create cancer in petri dishes and in a variety of animal models, but can also cure them effectively in the laboratory setting. The decades' long promise is that these basic research efforts would soon translate to clinical reality. "Win the war on cancer", and "eliminate cancer in our generation" have become battle cries, not just in the cancer research community, but also by the nation at large.

Over 300 billion dollars have been spent (since 1971), scores of research papers have been published, countless "breakthroughs" have been announced in newspapers, TV networks and university press releases. Unfortunately we have only seen real success in very few types of cancer, which account for a very small portion of all cancer patients. The ultimate goal of achieving beneficial translational strategies at the clinical level is still largely unfulfilled. This raises the question "Why?"

1.1 The Progress

Cancer has been observed since antiquity with recorded cases as far back circa 2500 BC in ancient Egypt. Impressive scientific observations led to the linkage of environmental causes since the mid-1700s, including the association between tobacco snuff and nose cancer, and between soot and scrotal cancer in chimney sweeps. The transition between the 19th and 20th centuries saw rapid growth in research driven by newly established biological concepts and technologies such as re-discovered laws of genetics, chromosome theory, tumor transplantation and cell culture methods, the isolation of tumor viruses, and by applying earlier genetics concepts and developmental biology to the study of cancer [Bernard *et al.*, 2008]. Among these seminal works, Theodor Boveri (1914), the father of chromosome theory, suggested that cancer was due to defects of chromosomes as chromosomes are the carriers of heritable information [Boveri, 1914; 2008].

The history of modern cancer research nicely mirrors the development of biological concepts and technologies. With the blossoming of biochemistry (1920 to 1950) and the subsequent arrival of molecular biology (since 1950), the research landscape drastically changed with the focus shifting to gene-centric theories and analyses (since 1970s). Currently, gene/epigenetic/pathway profiling coupled with various large scale genomics, proteomics and metabolomics approaches (since 2000) have generated yet another wave of excitement and huge amounts of data, again with the promise of understanding the mechanism of cancer once for all.

From a molecular medical point of view, cancer is a disease entity defined by uncontrolled cell proliferation with the ability to invade normal tissue/organs (both adjacent and distal). The cancer research strategy seems straightforward: understand the molecular basis of the key hallmarks of cancer and then specifically target these key features, especially cell overgrowth and invasiveness. It is generally accepted that the cancer process, similar to some other diseases, can be dissected into a series of stepwise stages where the early stages of this process should provide specific targets for molecular diagnosis and treatment. This concept has been the rationale for molecular-based approaches that have dominated cancer research for over a half of a century.

Gene-based cancer research has been an exciting ride with one breakthrough after another [Mukherjee, 2010]. Along the way, a series of landmark molecular studies has promised to finally solve the mystery of cancer. Various types of cancer gene mutations were identified, and increased familial cancers were linked to specific gene alleles with impaired function. The identification and characterization of cancer genes and application of these molecular targets in disease diagnosis and treatment have dominated current cancer research. With various -omics technologies and increased computational power, we should now be entering the victory lap.

1.2 The Challenges

The development of various powerful molecular methodologies has further quickened the pace of discovery, bringing us ever closer to the goal of finding the entire set of key cancer genes, which has previously predicted to not exceed a handful. However the picture becomes more complex each year as we find more cancer genes than anticipated. Cancer genes now include oncogenes, tumor suppressor genes, cell death genes, expression regulating genes, DNA repair genes, cell cycle genes, caretaker genes, metabolic genes, chromosomal machinery genes, stress response genes, developmental genes, epigenetic regulation genes, tissue architecture genes, blood supply genes, mitochondria genes, and immunological genes just to name a few [Heng *et al.*, 2009]. Many of them have multiple functions depending on the conditions [Horne *et al.*, 2015a]. Furthermore, the same gene within different karyotypes can display different or even conflicting functions. As pointed out by H. Harris, cancer research has become like the fashion world where there is a high level of excitement whenever a new molecular mechanism (such as a cluster of genes) is discovered, only to be replaced when the next gene cluster is uncovered [Harris, 2005]. This trend will for sure continue, as new fashions will certainly bring more gene clusters to the table.

Meanwhile, clinical implications of the identified gene mutations have been limited despite the very impressive lab data, excellent results in some animal models and noted success in familial cancers. For most solid tumors, early surgical removal results in the best patient outlook, but our large amount of accumulated molecular knowledge is of little help once metastasis occurs. Many chemo- and target-specific therapies initially reduce tumor size. However, this often does not translate into long term benefits, as many of these sky high expensive therapies prolong patients' lives by only weeks or months at best and do not significantly improve the quality of life [Fojo and Grady, 2009; Mittral, 2007]. Therapeutic dosages of various drugs are typically designed to be just under the maximum dose tolerated by the general population, which in some cases has the disastrous effect of pushing these compromised patients over the edge.

This gap between molecular understanding (often based on exceptional models) and clinical implication is enormous and hard to cross. The majority of game-changing results from basic research have failed in the clinic, and the translational success is extremely low, giving people the impression that bridging the gap is like crossing the valley of death [Butler, 2008]. This serious situation deserves more attention and needs to be discussed within the cancer research community. Unfortunately, most researchers tend to play up the positive aspects of experimental progress and disqualify any discussion regarding the lack of translational relevance to clinical cases as "negative thinking". In many researchers' minds, cancer should be looked at as an issue of "half empty or half full". As more genes are identified, it is simply a question of when the cup will be full.

Sadly, we have yet to reach the "half empty or half full" stage when it comes to the understanding of cancer. Current efforts have produced a great deal of accumulated molecular data and associated inferences but in clinical terms, we have very little to show for these efforts. This strange situation undoubtedly questions the validity of the theoretical basis. Interestingly, the executive summary of the 2010–2011 annual report of the US President's panel, titled "The future of cancer research: accelerating scientific innovation", contained the following rare statement [Annual report of the president's panel (2010–2011)]:

> *"America's investment in cancer research has vastly expanded and deepened our understanding of the many diseases called cancer. Some of the genetic and environmental factors and biologic mechanisms that cause or contribute to cancer development, progression, and spread have been elucidated. This knowledge has led to the development of diverse interventions to reduce risk of cancer and more effectively treat some cancers, enabling many individuals to survive diseases that previously were almost universally fatal.* ***Although notable, these achievements do not obscure the fact that cancer prevention and cure remain largely elusive****..."*

Despite the high praise for scientific achievements, this report undoubtedly underscores the gap between the impressive progress of molecular-based knowledge and clear lack of working achievements that effectively benefit most cancer patients. Beyond achieving a theoretical understanding of cancer, the ultimate goal is to benefit patients. Without real clinical success, no one can confidently claim that we are winning the war on cancer simply by accumulating molecular knowledge such as identifying and successfully targeting cancer genes or cancer specific pathways in the laboratory.

1.3 Hard Questions

Not all share this honest evaluation of our unexpectedly poor progress in the war on cancer, and few researchers have seriously considered this gap as a fundamental problem. The basic tenet of our scientific training leads us to respond to this lack of clinical progress with statements such as "*As long as the research data are promising, the accumulated knowledge will someday cure cancer*" and "*Science takes time to deliver in the clinic; the only way out is to keep digging*." Rather than question the theoretical concepts of current cancer research, many prefer to focus on the details surrounding specific techniques, particular experimental models or newer molecular targets. The belief is still the same: researchers have to find the key cause by collecting more data and applying more sophisticated analyses. If it is not a few key gene mutations, it must be a large number of drivers; if it is not caused by gene mutation, it must be epigenetic. If it is not DNA, it must be RNA. If it is not in the coding regions, then it must be in non-coding regions, and so on. Many focus on adding to the already overabundance of data leaving little time for researchers to critically read recent publications, synthesize one's own data, or think outside of the (gene) box. The holy grail of cancer gene mutations might lie in the gene we have just discovered or perhaps one still in the freezer waiting to be isolated. And so it goes.

This situation has serious consequences for current cancer research. On the surface, "just do it" seems to reflect the positive attitude of the researchers. But when science runs out of new concepts, doing the same thing over and over again and expecting different results will only damage science. According to Einstein, it is insanity. Or at least, it is intellectual dishonesty.

The mindset that the long heralded gene-based answer was hiding in the genetic noise of cancer samples was certainly behind the cancer genome sequencing project officially launched in 2005 as the Cancer Genome Atlas (TCGA) project [Collins and Barker, 2007; Garber, 2005; Heng, 2007a]. If common gene mutations are the key for cancer, why not sequence them all directly from large numbers of cancer patients with different types of tumors? This unbiased and "big science" type of research should have ended a decade's long effort to identify individual gene mutations once and for all and solve the issue of heterogeneity since large numbers of clinical samples would reveal the universally significant mutations and wash out less important "bystanders". If no stone was left unturned, surely the Achilles' heel of cancer would be identified. One powerful argument was that this approach would light up the entire genome (a metaphor similar to lighting up every single lamp post to find a lost key). The question that no one dared to ask however was, "What if common gene mutations are not the key to cancer?"

Ten years later, after tens of thousands of samples have been sequenced, what has been found is not common cancer gene mutation patterns but rather an overwhelming level of genetic heterogeneity that occurs as both inter- and intra- tumor heterogeneity. This blows away the decades' long belief that common cancer genes will be identified by comparing DNA sequencing between normal and cancer cells [Hayden, 2010a].

Like any political issue, results can be spun in many ways. The supporters of the cancer genome sequencing project tell people that the genome sequencing technologies have revealed the complexity of cancer. Major national newspapers have reported that genetic patterns have been found in the now decoded cancer genome. Instead of reporting the disappointing clinical findings, they insist that continued sequencing of the highly diverse gene mutations will result in personalized cancer diagnosis and treatment. Welcome to the era of sequencing everything!

Dissection of data from each genome sequencing publication is difficult due to the sheer volume of data. It requires persistence to critically analyze these reports and understand the real picture the reports paint. Many have only time and interest to read the often glowing title and abstract, making it hard to judge these more advanced technologies and the huge diversity of data they produce. Most papers do not provide a simple list of all the significant gene mutations detected and their clinical relevancy, but rather focus on the author's choice of few mutations/pathways that make sense for storytelling.

A few have publically voiced their concerns of continuing to sequence yet more samples [Fox *et al.*, 2009; Heng *et al.*, 2011a; Nicholson, 2013; Strauss, 2013]. One such powerful voice is from a cancer guru, Bert Vogelstein, who proposed the model of stepwise progression of cancer based on genetic alterations, and led the first large scale cancer genome sequencing in a number of major cancer types. Vogelstein and others further stated that additional large-scale sequencing would not promote our understanding of the cancer landscape as hundreds of samples have already illustrated the overall complexity and, moreover, any such data would not change the big picture [Vogelstein *et al.*, 2013].

Understandably, his view is at odds with sequence factory operators. Apparently, few — including the NCI (US National Cancer Institute) leadership — are really listening to him on this issue as the strategy of continuing to sequence is gaining popularity across the globe [Kaiser, 2010; Ledford, 2010]. His advice to stop sequencing more samples was not an easy one and should be very influential. Yet researchers who have followed his lead for decades now choose to ignore this valuable advice. Is it because people only believe what they want to believe? Why is it so difficult to critically evaluate the cancer genome sequencing project and its future? What is the potential harm of focusing on sequencing more samples? And is it the right time now to offer whole genome sequencing to less informed patients without understanding the true meaning of the large number of gene mutations and genome alterations?

1.4 New Paradigm

In the past several decades, numerous promising strategies have routinely come and gone in the cycle of hope/promise/disappointment that seems to be a key feature of cancer research. The prediction that cancer will soon be cured has been claimed so many times. For example, the targeted anti-angiogenesis therapy was hailed as the beginning of a new era in cancer therapy and led to a firm prediction by some leading molecular biologists that the NCI would soon close its doors. Unfortunately, angiogenesis inhibition was not the silver bullet and, like so many other promising ideas in cancer research, it did not improve patient outcomes [Hayden, 2009]. Yet not long ago, based on the promise of the human genome project, the leadership of the NCI declared that cancer would be cured by 2015. This obviously unrealistic goal was enthusiastically supported by many scholars and organizations including the American Association of Cancer Research. The 2015 deadline is now here with no cure in sight.

It seems that history has a pesky habit of repeating itself. It is easy to point out the limitations of some historical approaches and predictions, but difficult to use the lens of history to critically analyze our current strategies and promises, even though what we are doing today will be tomorrow's history.

Not surprisingly, the cancer genome sequencing project again offers the biggest promise to date. The moon shots project was launched in 2013 — based on the claimed and projected success of the cancer genome sequencing project to cure individual cancer patients using their genetic profiles — scheduled to be achieved within 10 years (now only 8 years left). But in reality, exactly because of the mind-boggling heterogeneity revealed by the cancer genome sequencing project, the goal of a cure has become more than ever elusive, which has surprisingly led some leading researchers to admit that the current approaches are not working and we do not even know how to interpret the massive amounts of data (Weinberg, 2014)!

There are many reasons why cancer research frequently repeats cycles of big-promise/huge-disappointment/confusion. In addition to the influence of various exciting technologies and impressive data generation capabilities, the increased pressure from patients and colleagues/funding agencies, and even personal gains of researchers all contribute. Those who make bold promises benefit from increased fame and funding but unfortunately are not held accountable when their key predictions do not meet their original promises. Furthermore, perhaps one most critical reason is the lack of critical thinking in the field and no vigorous debates directed towards the dominating cancer gene mutation theory, especially when it fails to explain many paradoxes as well as clinical reality.

Without the much needed debate, this serious scientific issue comes down to a subject of "he said, she said". Whom do we believe? Of course, following the popular viewpoints becomes the safest strategy (but often not the most useful). The tradition of scientific debating is gradually lost, and different ideas are effectively suppressed, especially those that challenge the status quo.

To break up the tragic cycle and search for a new direction, we must discard false paradigms before committing any more resources and human effort to any other specific premise. We must critically analyze the rationale and the conceptual basis of current dominant beliefs in light of the –omics data and the gap between basic research and the clinic. Scientific logic and honesty must trump politics, personal gains, traditions and even arguing for the sake of arguing. We need to readdress the concepts on which these popular models are based, especially when the model has failed multiple times. The limitations of models in general should not be used as an excuse, as the main goal for cancer research is its clinical implications. In particular, we must realize the limitations of scientific inference, downplay the power of wishful thinking and hand waving, and pay more attention to some key paradoxes in the field.

Fortunately, many paradoxes in current cancer research can be solved by embracing a new evolutionary thinking that considers cancer as a genome alteration-mediated disease defined by a somatic cell evolutionary process, where system heterogeneity is the key. Specifically, addressing the following five types of questions have initiated the new conceptual framework of genome-based cancer evolutionary theory.

(1) The question of cancer heterogeneity: If cancer represents an evolutionary process, and the key factor for evolution selection is variation, what types of genetic and non-genetic heterogeneity matter the most for cancer evolution? Is it gene mutation? Is it epigene alteration or genome variation? Why is there so much heterogeneity at multiple genetic levels in the first place? What is the key contribution of this heterogeneity to normal cellular function and to cancer?

(2) The question of inheritance in cancer: Which genetic organization level defines the genetics of cancer cell: genes, epigenes or genomes? Why does gene-defined "parts inheritance" differ from genome-defined "system inheritance" or the blueprint? Is genetic inheritance precise or fuzzy?

(3) The question of the evolutionary mechanism of cancer: What is the dominant pattern of cancer evolution? Is it gene mutation mediated stepwise accumulation or punctuated genome re-organization? What is the genetic basis for different phases of somatic cell evolution? Is the mechanism of long-term population adaptation different from the population survival mechanism under high stress? Does cancer evolution follow Darwinian principles?

(4) The question of system constraint for cancer evolution: If the genome represents a level of constraint for genes, how can other cellular ecological constraints impact the genome? What are these factors that can contribute to the breakdown of system constraint and favor cancer

evolution? And what is the general mechanism to unify these diverse molecular mechanisms?

(5) The question of the knowledge gap between basic research and the clinic: Why is the knowledge gap between cancer gene mutation theory and clinical utility so wide? How do we apply the new genome theory to explain the mechanisms of rapid cancer evolution including drug resistance? What is the evolutionary role of outliers? Does this unique cancer evolutionary insight offer clinical implications for diagnosis and treatment? Why does fuzzy inheritance challenge the concept of precision medicine in cancer?

These issues will be debated throughout the book, which ultimately will lead to a sensitive yet profound question: Is the conceptual framework of the gene mutation theory still relevant in terms of guiding future cancer research? If the answer is "yes", then we must define new technical strategies as clearly current gene mutation-based cancer research is heading in the wrong direction. (Increased gene-based knowledge can hardly apply to the clinic!) If, however, the answer is "no", what new framework should be used and why (epigenetic-based, or genome-based, or tissue-based theories)? Since cancer genome sequencing has generated sufficient data, and more importantly, a high degree of confusion, the time is now to debate this issue. Such debate will impact the future of this field.

Obviously, addressing such questions is neither easy nor popular. According to Thomas Kuhn, when there are enough significant anomalies against a current paradigm, the scientific discipline will enter a state of crisis. Typical features of such a crisis include scientists on either side of the paradigmatic divide lacking the ability to understand the other's viewpoint or reasoning; the intensification of an intellectual "battle" between new ideas and the old paradigm; and a tremendous disinterest within the research community to provide a comprehensive scientific analysis to explain the inconsistencies despite the fact that the old paradigm is full of mistaken assumptions and failed predictions. Debate within current cancer research is needed to reveal these anomalies, an exercise that will undoubtedly result in the abandonment of the current paradigm and the search for a new one.

1.5 Outline

In the following chapters, we will try to debate issues in cancer research ranging from basic concepts to experimental designs/explanations. Each chapter will focus on the numerous paradoxes, common misconceptions and challenging issues related to a major topic. This discussion will be followed by hard and sometimes uncomfortable pertinent questions and critical analysis. We will debate key facts, key concepts and key interpretations of some well-known data or phenomena to examine seemingly important concepts based on artificially created constructs in order to expose any shaky logic and at the same time, explore highly significant developments that arise from what initially appear to be non-significant findings.

Our main goal is to introduce a new paradigm of genome theory as without it, the old paradigm will never be replaced no matter how outdated. During this debate, the key limitations of the current gene mutation theory of cancer will be critically reviewed (Chapter 2) followed by the brief description and commentary on over a dozen noted alternative theories of cancer (Chapter 3). To deliver this convincing message, the current status of the cancer genome sequencing project will be evaluated, which calls for a new framework (Chapter 4). Chapter 5 introduces the genome theory of cancer based on the discovery of two phases of cancer evolution, which are genome alteration-mediated macrocellular evolution and gene/epigene-mediated microcellular evolution. By redefining system inheritance, stochastic genome variation represents the driving force for cancer evolution that can unify diverse molecular pathways. Cancer population behavior will be further discussed in Chapter 6, leading to discussion regarding the recent discovery of a new form of inheritance termed fuzzy inheritance. The realization that genetic information is not as precise as we previously thought especially at the somatic cell level is highly significant; it not only provides the basis for understanding cancer heterogeneity, but also offers explanation for the missing heritability of common diseases, why most cancer gene mutations display low frequencies within the patient population and why it is challenging to target them. Chapter 7 will engage whether or not cancer should be considered as a new species as well as the potential contribution of cancer evolutionary study to evolutionary theory in general. In Chapter 8, a number of important paradoxes will be discussed with application of the genome theory. Finally in the epilogue, some key messages of this book as well as future perspectives are briefly summarized.

To illustrate that a change in paradigm is long overdue, Kuhnian standards will be used to evaluate or grade current cancer research [Kuhn, 1962; Strohman, 1999]. This includes the incommensurability between believers and non-believers of the current paradigm, the lost confidence by some scientists within the field, encountering the increased opposition from the establishment also based on non-scientific reasons, and finally, the tremendous disinterest in trying to explain inconsistencies and provide a comprehensive scientific analysis seeking to answer these questions that might actually give direction to future research and technology.

In addition to re-evaluating some generally accepted concepts and famous case studies, we will also address the many paradoxes and untold stories associated with these individual success stories. New analyses and re-syntheses will provide a fresh and holistic picture of the current challenges in cancer research. We hope this new perspective will make you agree that there has been enough "business as usual" and it is now time for a real change.

It should be noted that the main goal of this book is not to provide a comprehensive coverage of current molecular cancer research or to systematically list many molecular details but rather to initiate much needed debate that hopefully leads to acceptance of a more realistic paradigm. To achieve this goal, we have raised hundreds of important questions that will effectively illustrate our points. With this open debate, we believe readers will appreciate this new perspective and will join in our search for a new direction, finally reaching the point of no return with regards to the gene-based approach. Please join us for this exciting, sometimes confusing and even painful, but ultimately an eye opening and game changing journey.

Chapter 2

The Gene Mutation Theory of Cancer

2.1 Summary

Due to rapid technological development, current cancer research is driven by the massive generation of molecular data. With so much data and so little time, only selective information is analyzed and published. For example, individual reports focus on only tiny portions of the altered genes or pathways in global gene expression studies according to the investigator's hypotheses or scope of analysis; despite the hundreds to thousands of gene mutations that can be detected from a given case by whole genome sequencing, only a few will be chosen as important ones. The vast majority of dynamics is dismissed as "noise" and is simply ignored. The even bigger problem in the face of so much data is that little attention is paid to the theoretical basis behind data generation and synthesis, which leads to increasing confusion in the field: the explosion of molecular discoveries and knowledge has very limited relevance in the clinic. We work hard, but are we heading in the right direction? In this chapter, we briefly review and re-evaluate the history of the gene mutation theory of cancer — its key frameworks and supporting experiments, its initial promise, and the increasing challenges it brings. Our conclusion is that it is the time to search for a new framework and drastically change this course of current cancer research.

2.2 Brief History: Promises

The gene mutation theory of cancer, also known as the somatic mutation theory of cancer, is the theoretical framework of current molecular-based cancer research. According to this theory, "cancer begins with a genetic change in a single cell that passes it on to its progeny, thereby generating a clone of malignant cells" [Vaux, 2011]. Since chromosomal changes have traditionally been considered either incidental or as late events resulting from gene mutations, the somatic mutation theory has focused on gene mutations within the gene-centric framework. The reductionist's attractive and convincing hope was that this theory would lead to the mechanistic understanding of cancer by identifying a series of key gene mutations, as cancer is the "disease of gene mutations". The complexity of cancer would be dissected into simplified molecular principles, and a small number of molecular events controlled by a handful of cancer gene mutations would explain the basis of cancer [Hanahan and Weinberg, 2000; Heng, 2007a; Heng *et al.*, 2010a; Horne *et al.*, 2015a; Weinberg, 2014].

The somatic cell gene mutation theory of cancer has deep roots. Ernest Tyzzer first coined the term "somatic mutation" in 1916 to explain the permanent modification of somatic tissue observed in tumors. Although Theodor Boveri never mentioned gene mutation, his theory inspired the linking of chromosomal aberrations to cancer. Thomas Morgan and Calvin Bridges speculated on the theoretical possibility of cancerogenesis via the recurrent somatic mutation of a specific gene [Wundwelich, 2006]. K.H. Bauer later developed the theory that neoplasia occurs as a consequence of altered genes. Bauer's theory was developed in the 1920s [Bauer, 1928], long before the establishment of the molecular definition of genes and was supported by a number of lines of indirect evidence in the 1940s. It was already known that X-rays, ultraviolet radiation, as well as some carcinogens could induce gene mutations. Early work showed that carcinogenesis could be considered to occur in two stages: initiation and promotion. Experiments correlating mutagenic and carcinogenic potencies were performed in various species including tumor induction through transplantation. However, evidence was often inconsistent and contradictory. For example, even within the same publication, there was often strong evidence both supporting and refuting the gene mutation theory [Berenblum and Shubik, 1949]. Nevertheless, it was suggested that several successive mutations in the same cell, perhaps seven for human cancer, would be necessary [Muller, 1951]. The rationale for considering that cancer could arise from more than one gene mutation was based on the observation that cancer is often an age-related disease. If one mutation caused cancer, cancer incidence should be equally distributed among different age groups exposed to cancer causing agents following some latency period. This latency varied for each cancer causing agent, for example, X-ray associated cancer was estimated as 9 years [Bauer, 1949].

Based on the observation that the frequency of cancer seemed to increase according to the sixth power of age, Fisher and Hollomon suggested that approximately six or seven cells had to be transformed independently in order to form cancer [Fisher and Hollomon, 1951]. In contrast, Nordling proposed the multi-mutation theory of cancer to explain the same relationship between frequency of cancer and age of the population. He explained this correlation by assuming that cancer formation requires the accumulation of six consecutive mutations [Nordling, 1953]. Soon after, Armitage and Doll published their landmark paper on the multistage model, which rejected Fisher and Hallomon's theory and support Nordling's concept of successive mutations, based on considering the pattern of cancer incidence in response to carcinogens [Armitage and Doll, 1954]. A few years later, they published a two-stage theory of carcinogenesis based on age distributions of cancer patients [Armitage and Doll, 1957].

Fisher proposed the growth model to support Armitage and Doll's work. Interestingly, he summarized the rationale of the multiple-mutation concept:

> "*Based upon these multiple-mutation ideas, the most realistic theory now would seem to require the following sequence of events: (1) Mutation; (2) clone growth from the mutated cell; (3) another mutation somewhere in the clone; (4) growth of a new clone from the cell with the second mutation; (5) a third mutation in the second clone and so on until a cancer clone is the result.*" [Fisher, 1958].

A similar analyses was also used by Knudson to conceptualize the "two hit" theory of cancer [Knudson, 1971]. By performing a statistical analysis of 48 retinoblastoma (RB) cases, Knudson observed that the inherited and sporadic forms of the disease display different spectra. Inherited retinoblastoma occurred at a younger age than the sporadic type, and children with inherited retinoblastoma often developed the tumors in both eyes in contrast to the sporadic form. Knudsen postulated that two "hits" to a gene were therefore necessary to cause cancer. In children with inherited retinoblastoma, the first hit was inherited in the DNA (germline mutation), and any second hit (somatic cell mutation) would rapidly lead to cancer. In non-inherited sporadic retinoblastoma, two somatic mutations had to occur prior to tumor formation, explaining the age difference. This "two hits" concept later had a huge impact on molecular cancer research upon the cloning of RB genes.

Later in the 70s, Bruce Ames developed a bacterial test system for measuring the mutagenic potency of various compounds. This system was used to analyze a number of compounds known to cause cancer in laboratory rodents, which allowed him to report a correlation over 6 orders of magnitude between the mutagenic and carcinogenic potencies of these compounds [McCann *et al.*, 1975]. The idea of multiple mutations causing cancer was also explained by Nowell, who applied the concept of somatic cell evolution, originating the idea that clonal evolution is a key to cancer and postulating that a series of genetic alterations are required for cancer evolution to be successful [Nowell, 1976].

By summarizing the above mentioned reports that span over half a century of research, the following conclusions are clear: (1) age-related cancer can be linked to the accumulation of gene mutations; (2) carcinogens cause cancer by damaging genes of target cells; (3) mutation of normal cellular genes yield cancer, similar to the result of an oncogene in Rous sarcoma; and (4) stepwise somatic cell evolution is involved in cancer.

Thus it seems that different researchers have pointed out the same thing: a clear linkage between multiple gene hits and cancer. However, these statements all depend on the accuracy of the assumptions that the individual gene defines cellular inheritance and that cancer is a stepwise evolutionary process. As we will discuss more in later sections, these key assumptions are not necessarily true, as there are many other types of genomic alterations beyond gene mutations that can be used to better explain the same phenomenon.

On the molecular research front, solid examples were needed to directly link specific gene mutations or allelic variants to particular types of cancer. Systematic analyses were also required to directly validate the gene mutation theory beyond the assumptions from which it was derived. Over the past four decades, a massive research effort has been driven by enormous research funding suddenly available through the War on Cancer. The initial enemy #1 was the reverse transcriptase enzyme and DNA tumor viruses. This was replaced by the "cancer gene" (oncogenes and tumor suppressor genes). The latest targets are the entire genomic landscape of cancer and cancer-specific gene modules or networks. Throughout the past five decades, there has been one wave after another of seemingly gene-based advances, each armed with new terminologies and convincing predictions of success. Most of this research can be classified into four categories that collectively developed and validated the gene mutation theory:

2.2.1 *Category 1: The identification and characterization of cancer genes and the demonstration that gene mutations are the basis of carcinogenesis*

The first oncogene (Src) was identified in the Rous Sarcoma Virus, a chicken retrovirus by Peter Duesberg and Peter Vogt in 1970 [Duesberg and Vogt, 1970]. In 1976 Dominique Stehelin, Michael Bishop and Harold Varmus demonstrated that activated oncogenes were proto-oncogenes present in the genomes of many species including humans. In 1982, the Ras gene was identified and subsequently found to be mutated in a variety of tumor samples. Soon afterward, the Bcr-Abl fusion gene was identified in chronic myeloid leukemia (CML) and served as the new paradigm for cancer research, explaining the molecular basis of the Ph-chromosome. Within a few short years, various forms of familial cancer types were linked to specific gene mutations including p53 in Li-Fraumeni syndrome, and BRCA1 in hereditary breast-ovarian cancer syndrome. These exciting discoveries collectively supported two lines of thinking in the conceptual framework of the gene mutational theory of cancer. The first concept was the oncogene or "jammed accelerator" theory. Proto-oncogene coded proteins help to regulate cell growth and differentiation, which gain oncogene function upon their mutation or overexpression. The link between proto-oncogenes and activated oncogenes provided a genetic explanation of how mutated normal genes could be responsible for unregulated growth, the key characteristic of cancer. A second concept was the "missing breaker" or tumor suppressor idea. The concept of tumor suppressors, particularly the case of the Rb *tumor suppressor,* supported the "two hit" theory or Knudson hypothesis. Again, the key to this concept is the regulation of cell growth (to prevent out-of-control growth). This Yin/Yang explanation seemed to be a perfect explanation of cancer. With these successes, the identification of key universal mutations leading to cancer became the obvious priority for cancer research. From 1983 to 1993, many important oncogenes were identified including the myc, fos, and akt genes, supporting the concept that oncogenes are normal genes activated by mutation or over-expression. Similarly many tumor suppressor genes were isolated including the famous TP53 and APC genes, supporting the importance of tumor suppression. Following their identification, experiments using combinations of cancer genes to transform normal human cells *in vitro* and in the OncoMouse system seem to have proven that cancer genes can cause cancer in both *in vitro* and animal models [Mukherjee, 2010].

2.2.2 *Category 2: The identification of the sequential order of gene mutations further validated the gene mutation theory*

It was well established that cancer formation could be divided into two distinct phases. According to Friedewald and Rous' viral model that was supported by chemical carcinogen experiments, [Friedewald and Rous, 1944] carcinogenesis was composed of an "Initiating Process," responsible for the conversion of some normal cells into latent tumors cells, and a "Promoting Process," where these latent tumor cells developed into full blown tumors. For example, a single topical dose of a polycyclic aromatic hydrocarbon functions as cancer initiator, and if followed by repeated topical doses of croton oil (promotion), carcinogenesis occurs on mouse skin [Berenblum and Shubik, 1947]. This model has been expanded to many rodent tissue types, including colon, esophagus, liver, lung, mammary gland, and stomach.

The established link between gene mutation and cellular transformation promoted the search for the mutations that would explain the molecular detail of the two phases of cancer formation. Determination of the order of gene mutations for each cancer type would provide specific reliable biomarkers for early diagnosis and should allow early stage tumors to be specifically targeted. Sequential genetic alterations have been observed in many blood cancers, the archetype being the early occurrence of the BCR-ABL fusion gene in CML (even though such sequential events are more fully illustrated by chromosomal alterations).

Such identified patterns were lacking in solid tumors, so when Vogelstein proposed a sequential genetic model of colorectal cancer (sequential mutation of APC, RAS and p53) [Fearon and Vogelstein, 1990], it was quickly accepted by the research community. A great deal of research data has emerged to support this model, including mathematical modeling and additional gene mutation identification for different types of cancer. Interestingly, in addition to their seminal gene mutation-based work, Vogelstein's group also actively promotes the evolutionary concept of cancer, and has nicely incorporated the notion of genomic instability into their explanation of how a series of gene mutations contributes to cancer. In addition, his group's combination of basic and clinical research has provided direct experimental support and some clinical validation. The stepwise model of colon cancer tumorigenesis gave hope to the possibility of finally identifying initial and sequential genetic events which would vastly improve early diagnosis and treatment. Similar to the CML model of blood cancer [Rowley, 1998; Zhang and Rowley, 2011], this landmark model unified the field of solid tumor research in search of the stepwise genetic alterations that lead to cancer.

2.2.3 *Category 3: The hallmarks of cancer rationalized gene based cancer research*

Vogelstein's colon cancer model energized the field in its search for similar sequential gene mutations in other types of cancer. Paradoxically, a large number of gene mutations were identified, but many of them were neither typical oncogenes nor tumor suppressor genes. The order of gene mutation was also found to be highly complex with more exceptions emerging than would be expected if some stepwise accumulation of gene mutations with a predetermined order was a general rule. To solve this confusion and integrate all the newly identified cancer genes within some known molecular principles, the "Hallmarks of Cancer" concept was proposed in 2000 by Douglas Hanahan and Robert Weinberg in a highly cited review article published in the journal *Cell* [Hanahan and Weinberg, 2000].

The conceptual basis for this idea is that the complexity of cancer can be reduced to a small number of underlying phenotypes/molecular principles where all cancers share these same basic principles or common traits called "hallmarks". Initially, the list included six hallmarks: (1) Cancer cells stimulate their own growth; (2) cancer cells resist inhibitory signals that stop their growth; (3) cancer cells resist their own apoptosis; (4) cancer cells stimulate angiogenesis; (5) cancer cells can multiply forever; and (6) cancer cells are invasive (metastasis). Ten years later four more hallmarks were added, including (1) abnormal metabolic pathways; (2) evasion of the immune system; (3) chromosome abnormalities and unstable DNA; and (4) inflammation [Hanahan and Weinberg, 2011]. The hallmarks of cancer publication has unified researchers studying diverse gene mutations, as it allowed the connection (directly or indirectly) of these mutations to the hallmarks. In addition, this work represents a departure from the prediction of original gene mutation theory of cancer that only a handful of cancer genes would be identified. This concept perhaps is the most widely used concept of cancer in cancer seminars, publications, and grant applications. The hallmarks instilled confidence in molecular cancer biologists. Their work was legitimized if it could be linked to one of the hallmarks of cancer.

2.2.4 *Category 4: The clinical validation of the theory*

Much needed validation of the gene mutation theory was obtained by detecting most key cancer gene mutations in patients. In particular, the strongest validation comes from the clinical success of targeting a few specific gene mutations or profiles in certain cancer types [Vaux, 2011].

The most highlighted example is imatinib treatment in the chronic phase of chronic myeloid leukemia (CML-CP) which directly blocks the activity of the BCR-ABL fusion gene [Druker *et al.*, 1996]. Another of several success stories of target specific treatment generating good responses in patients with specific gene mutations includes treating PML-RARA-positive acute promyelocytic leukemia (APL) patients with a combination of arsenic trioxide and all-trans retinoic acid [de The and Chen, 2010; Horne *et al.*, 2013a].

Together, these four concepts coupled with the large amount of experimental data in many linear models and their interpretations have seemingly "validated" the gene mutation theory. More fundamentally, the cancer gene mutation theory fits well with the gene-centric assumption that genes provide the genetic blueprints of biological systems and that a number of defective genes must be the basis for cancer. Furthermore, these concepts also fit perfectly within the neo-Darwinian concept that an accumulation of small changes (gene mutations) leads to big changes (the transition from normal cell to cancer). In the eyes of many cancer biologists, tying together these concepts left only one job: find these abnormal genes and fix them.

2.3 Problems

Despite general acceptance and domination of the gene mutation theory in cancer research for the past four to five decades, there have been many important yet unaddressed problems. First and foremost is that all of these elegant theories are diminished by their inability to be transferred to the clinical setting. Rather than recognizing these limitations and returning to the drawing board to develop new frameworks, few are willing to question the very basis of the working model. A common strategy is to switch assumed key target genes and try one's luck with another gene. If oncogenes are not enough, then include tumor suppressor genes. If gene mutations are not enough, try epigenetics. If it is not coding sequences, then investigate non-coding sequences. The common factor in all of these concepts is the attempt to reduce cancer to a small number of recurrent, specific molecular changes/principles. Researchers hope to hit that home run and broadcast the message that, "We cannot give up!" and, "We just need more funding and to work harder to collect more data." Indeed, as judged by news articles that continually appear in both the popular and scientific press, exciting breakthroughs in basic cancer research suggest that the cure is just around the corner. Unfortunately there are many elephants in the room when it comes to the current concept of cancer. Why does more research based on the gene mutation concept lead to more confusion? Why have all of the game-changing basic research discoveries fallen so short in the clinic? What if the modern conceptual framework of cancer — the gene mutation theory — is wrong?

For years, we have been asking these questions and searching for answers. Yes, cancer is very complicated. We have to find a way to understand and combat it. But we must be willing to identify the correct level of study for data collection and allow ourselves to recognize and abandon failing paradigms. In the case of cancer, establishing a correct framework first gives direction and then hard work gets results. The real world goal of improving cancer treatment is much more important than simply exploring theoretical concepts. Concepts — no matter how elegant — must be changed or modified when they fail to explain reality. The following discussion highlights some of the shortcomings of current cancer research, which we will further discuss in following chapters.

2.3.1 *New conceptual frameworks are needed to solve the many paradoxes in cancer research*

The key prediction of the somatic mutation theory of cancer has been the small number of gene mutations hypothesized to drive cancer. But in reality, the number of cancer genes continues to expand rapidly, exceeding far more than the initial estimation that there would be less than a handful of gene mutations [Weinberg, 1982].

The prediction of a small number of common gene mutations was highly significant for the legitimacy of the gene mutation theory. It was not only predicted by all the landmark models we previously mentioned, but it also provided genetic certainty to cancer development and offered therapeutic targets. Historically speaking, the capability of identifying a defined genetic pattern of cancer (such as limited common gene mutations) was a key advantage of the cancer gene mutation theory over the chromosome aberration-based cancer theory. In fact, the lack of a specific pattern of chromosome aberrations among the majority of cancers was the main reason to move to the gene mutation theory of cancer. With the identification of thousands of gene mutations in various cancers, the question becomes which ones are the drivers and which one are the passengers (see Chapter 4). Now some researchers even suggest that it is not the diverse gene mutations themselves that matter, but rather a relatively limited number of pathways. Others think the key is to identify "master regulators" within the genetic network that may not be mutated but play significant roles in cancer pathways. It seems that everyone still wants to identify key magic targets. Despite decades of research and billions of dollars spent, much to the surprise of many in the field the crucial question remains unanswered: are gene mutations the real cause of the majority of human cancers?

There are many other key predictions from the gene mutation theory that are not correct as well. The causation of cancer by gene mutations is no longer a certainty. For example, the stepwise accumulation of gene mutations driving cancer evolution concept is now questionable; the identification of genetic markers for earlier diagnosis has not been very successful; and the promise of targeting gene mutations to cure cancer does not work in most cancer types. In physics, when key theoretical predictions do not stand the test of reality, the theory is replaced with a new framework. Cancer researchers need to emulate physicists in this case; otherwise we will continue to be stuck with many of these incorrect ideas.

New frameworks must recognize the complexity of cancer and realize the limitation of current research tools. Physics deals with systems with less heterogeneity than biological systems. Solving the three-body problem in such a system is not on the horizon. Biologists face a much more daunting task in solving the cancer problem, as it requires understanding of at least dozens of independent agents within a much more heterogeneous and dynamic system. This point has recently been accepted even by some key advocates of the cancer gene mutation theory [Weinberg, 2014].

2.3.2 *The complex relationship between genotypes and phenotypes of cancer challenges the theoretical basis of the cancer gene mutation theory*

Clearly there are many genetic/epigenetic and environmental layers/factors that play a role in each cancer case. These factors as a whole reduce the direct contribution of individual gene mutations to the cancer phenotype, resulting in the lack of one-to-one correlations between gene mutation and cancer phenotype. In many experimental models, specific gene mutations lead to cancer-related phenotypes in a more or less linear fashion. However, the complexity observed in the clinical setting overshadows this linear relationship, squelching the predictability of the experimentally derived cancer mutation. The following facts support this conclusion: (a) There is a large amount of diverse gene mutations and genome alterations involved in a given cancer type and a specific tumor. (b) The genetic landscape of a given tumor continually changes during cancer evolution. (c) Patients with the same key gene mutation profiles often have very different outcomes. Even for monozygotic twins with the same gene mutation profiles, the disease phenotype is not surely predictable despite a stronger genotype-phenotype linkage between them relative to the general population. (d) Personal genome sequencing surprisingly revealed that many normal individuals also host some typical cancer gene mutations, suggesting gene mutations alone are not enough to cause or predict cancer.

2.3.3 *The rationale of focusing on specific gene mutation(s) of interest has led to unintentional cherry picking*

Since many cancer cell populations are a collection of cells with different gene mutations, investigators can often find their favored gene mutation in these cells in a particular experimental setting. In fact, the heterogeneity of cancer can mislead researchers, reflected by the many conflicting reports of the same cancer type. More seriously, the influence of the gene centric concept often masks the bigger picture of genome level alterations, as concentration is on the search for molecular mechanisms (See Chapter 5). Even within the cancer biology field, many believe that the genomes of cancers are not different from those of normal humans except for mutations/epigenetic alterations in a few key genes. However, the karyotypes (a window into the organization of the genome) of cancers are nearly universally altered. This is especially important because the genome is the platform of emergent genetic properties, and it is the karyotype rather than individual genes that define the system inheritance of a bio-system [Heng, 2009; Heng *et al.*, 2011a; Chapters 5–7].

2.3.4 *Many elegant experiments supporting the cancer mutation theory are performed in well-designed "noiseless" and linear systems*

These experiments often lead to the conclusion that specific gene mutations and pathways directly cause cancer and that blocking these mutations and pathways are probable means of curing cancer. Unfortunately there is a huge gap between these model systems and cancer in humans (see Chapters 5–6 and 8). There is an almost constant news cycle of exciting research breakthroughs and a much quieter response of disappointment years later when the breakthroughs fail to manifest themselves in the clinic. Nevertheless, the researchers who make these basic discoveries often still enjoy the exultation of the field for their success.

2.3.5 *All too often researchers tend to generalize their findings and fill conceptual gaps with inferences*

Many exceptions were laced together to support the central idea to synthesize the gene mutation theory of cancer. These exceptions look elegant and convincing on their own, but whether or not they can be applied to general cases was initially unknown. The knowledge derived from some blood cancer types is applied to solid tumors; the contribution of some key gene mutations to familial cancer types become the basis for most sporadic cancers; and short term cell killing effects is used to predict long term cancer response... Now, it is clear that many of these attractive inferences are not working! For example, the genetic landscape of familial cancers and sporadic cancers are quite different. Even for the familial cancer cases, unpredictability is high. As for diagnosis and treatment, the benefits have been rather limited for most sporadic cancers. The logical conclusion is, if the theory does not fit the clinical reality, something essential must be wrong.

2.4 Re-examining the Keystones

To illustrate the fundamental problems of the current cancer gene mutation theory, we must re-examine some of the key concepts and the interpretations of those milestone experiments on which the theory is based. Not surprisingly, the new –omics data are at odds with many of the original data interpretations. In addition, there are also many obvious new avenues for re-interpretation of old experiments which only become evident with a new mindset: changing the framework from the gene-centric concept to genome-mediated cancer evolution. In addition, we must focus not on simplified linear experimental models but on the clinical realities of cancer. The following are brief critical discussions of the four conceptual bases of the gene mutation theory of cancer. More detailed discussion can be found throughout the rest of the book.

2.4.1 *The certainty of oncogenes or the probability of genome alteration? The tale of cell transformation by oncogenes*

Experiments demonstrating that oncogenes lead to transformation of normal cells represent a key basis for current cancer theory. However, these experiments can be interpreted very differently by including genome alterations in the equation. Oncogenic transformation was initially demonstrated in a series of experiments in the late 1970s and into the 1980s with keystone work by Robert Weinberg's laboratory. This milestone discovery served as the turning point of the reductionist movement in cancer research. In this work, Weinberg first demonstrated that transformation was caused by transfection of genomic DNA isolated from chemically transformed mouse cells into NIH 3T3 mouse fibroblasts. After the confirmation that cancer DNA was indeed introduced into transformed NIH 3T3 cells with transformation frequencies higher than controls, the general conclusion reached through their original data combined with inferences derived from other publications was that the transfer of a single gene from cancer cells into normal cells is sufficient for transformation. Weinberg's group went on to identify Ras mutations in bladder tumors. Subsequently, they used combinations of transfected oncogenes to transform mouse, rat and finally human cells. Their assumption from this work was that three gene mutations/alterations were needed to transform human cells. However, their important conclusion that oncogenes drive cancer was not based on solid and fully understood evidence but rather on an incomplete picture based on many inferences.

First, NIH 3T3 cells are not normal cells. These mouse embryonic fibroblasts are hypertriploid with a modal chromosome number of 68, which occurs in 30% of cells. These cells are known to spontaneously transform in continuous culture. For example, Harry Rubin from UC Berkeley found that NIH 3T3 cells from all different sources transformed spontaneously. The propensity for these cells to transform explains why this cell line was so widely used in oncogenic transformation studies as they are already in a preneoplastic state (very close to transformation) and very much subject to progressive selection. Unfortunately, as a tradition of molecular biology, very few have paid attention to the altered genome of these cells. The chromosomal change is the elephant in the room.

Second, transfection with DNA from chemically transformed cells only increased the probability of transformation. As recalled by Dr. Weinberg, they were working with an experimental system that had a weak signal and strong noise. It was only after they carried out blinded experiments which provided evidence that the DNA prepared from methylcholanthrene-transformed mouse cells induced far more transformed foci than did control normal mouse DNA [Shih *et al.*, 1979]. However, Harry Rubin also achieved cellular transformation by simply increasing the culture time without transfection of any tumor DNA! Clearly, the conclusion was a leap of faith that cancer cells induced by chemicals indeed carried mutants that could transform genetic information. Precisely speaking, any oncogene transfer only accelerated transformation in this experimental system. A more direct factor must exist if transformation occurred even without the addition of an oncogene [Heng *et al.*, 2006a–c; Heng, 2007a]. Decades later it was illustrated that aneuploidy affects spontaneous immortalization in primary mouse cells [Williams *et al.*, 2008].

Third, many key facts and experimental artifacts found in the experimental systems that differ from tumors in humans were overlooked. Experimental techniques were capable of introducing many copies of a given oncogene resulting in expression much higher than the expression of the same oncogenes in primary human tumors. In practice, introducing multiple copies of the same gene differs significantly from a single copy gene mutation within a specific genome location that causes transformation. In addition, not all cells with the introduced oncogenes were transformed. In fact, the transformation rate of those with the oncogenes was very low. Furthermore, the heterogeneity of transformation was ignored by statistical averaging, giving the impression of a one-to-one causation between introduced oncogene and transformation.

Fourth, the synthesis of the oncogene mutation theory came about through different lines of evidence surrounding Ras mutations that were over-generalizations based on a variety of exceptions. For example, the discovery that the point-mutated Ras oncogene detected in human bladder carcinoma played an important role in the acceptance that oncogenes cause cancer as it validated the function of ras suggested by cellular transformation experiments. However, according to the COSMIC (Catalogue Of Somatic Mutations In Cancer) database, Ras is not even in the top 20 most frequently mutated genes in bladder carcinoma. However, the overwhelming majority of these tumors do display chromosomal aberrations and random gene mutations/epigenetic variations.

These issues clearly show the limitation of the gene mutation theory of cancer, especially when applied to clinical settings. As a result, the explanations of how an oncogene causes cancer became increasingly complex even in experimental systems. The number of oncogenes required for transformation varies in different systems from one oncogene transforming an unstable mouse cell line (NIH 3T3), two oncogenes transforming rat embryo fibroblasts (ras plus myc), and three key genes transforming human cells (the ectopic expression of the telomerase catalytic subunit, hTERT, plus two oncogenes, the SV40 large-T oncoprotein and H-ras). In addition to results that vary with the cell systems in which transformation is induced, some of these milestone experiments have been replicated by other researchers with great difficulty, if at all, especially when different cell systems with unstable genomes are used.

For example, when investigators used different cell lines to compare the capability of Weinberg's famous three gene combination to transform normal human diploid fibroblasts [Hahn *et al.*, 1999], these defined genetic alterations were not always sufficient for transformation, suggesting the existence of still unidentified factors that determine the transformability of normal human fibroblasts. Interestingly, transformation capability is associated with ploidy. Changes in ploidy are indicative of loss of genome/chromosomal stability by various changes in cell lines examined, and such chromosome abnormalities could be critically involved in the transformation of these normal human fibroblasts [Akagi *et al.*, 2003; Heng, 2007a]. Similar observations that karyotype alterations are commonly associated with oncogenic transformation have been made from many cell lines including primary human mammary epithelial cells (HMEC) [Elenbaas *et al.*, 2001]. When the "less likely possibility" was examined that other requisite genetic changes had occurred in these cells prior to their manipulation or during their *in vitro* passaging prior to injection into mice, the answer was an absolute yes. Not only the karyotype of a near-diploid nontumorigenic HMEC cell line prior to oncogene introduction was not normal, the cells with these oncogenes displayed further genome level alterations. Unfortunately, when reading these reports, most readers have focused on the specific oncogenes and ignored the "other possibility" such as genome changes, which in fact is the key.

Not surprisingly, more oncogenes are needed for human cell transformation in some systems, and Weinberg recently admitted that this classic three gene combination is not sufficient for direct tumorigenic conversion of normal human epithelial and fibroblast cells. In fact, five or more introduced gene mutations are needed [Weinberg, 2014]. More importantly, cellular transformation is coupled with karyotype changes, and karyotype change usually is more powerful than individual gene mutations. As we discuss in Chapters 4 and 5, most genes mutations and alterations can be linked to genome instability-mediated cancer evolution.

2.4.2 *The sequential order of gene accumulation: A beautiful model demolished by ugly facts*

In the early 90s, Bert Vogelstein's group at Johns Hopkins proposed the famous stepwise model of colorectal cancer where APC, TP53, and Ras accumulated during the progression towards cancer. This hypothesis was born out of genetic analyses of various stages of colorectal polyps and tumors [Fearon and Vogelstein, 1990]. This highly influential concept, also known as the Vogelgram, was based on one of the best characterized cancer progression models for which the key molecular steps have been identified. Because of this work, Vogelstein claimed "Cancer is caused by sequential mutations of specific oncogenes and tumor suppressor genes". This demonstration that a specific set of gene mutations are associated with distinct histopathological stages of the colorectal cancer process was widely embraced by the field.

Interestingly, their initial publication did not focus on the sequential order of genetic change. In fact, they concluded that it was the accumulation of mutations and not their order that was most important. Even though the APC mutation was the first step in their model followed by other gene mutations like K-ras and p53, not all tumors followed the same pattern. Regardless, the sequential gene mutation accumulation concept became very popular. Perhaps the overwhelmingly enthusiastic reception by other investigators influenced the Vogelstein team to favor the sequential model of gene accumulation in cancer. Does this situation represent an example of strong feedback causing the authors to extend their conclusions beyond what their data support? Or was it perhaps due to others misreading their conclusions [Weinberg, 2014]? Either way, the Vogelgram has become a dominant concept.

The Vogelgram supports the stepwise evolutionary model. Accumulation of gene mutations are proposed to occur through waves of clonal expansion where a new driver mutation leads to a selective advantage and clonal proliferation, which can provide the opportunity for the next gene mutation to accumulate. This idea presents a logical way to achieve accumulation of the 6–8 mutations required for cancer in one cell — the core concept of the gene mutation theory of cancer. Unfortunately, the key problem, however, is that this model does not fit the clinical realities of most cancer types. Quite the opposite: in the real clinical situation, stochastic genetic changes dominate, and there is no obvious gene mutation accumulation within the patient population for the majority of cancer types.

We have presented the following thought experiment to illustrate this dilemma.

As illustrated in the upper panel of **Fig. 2-1**, if the fixed stepwise accumulation (like the Vogelgram of various genetic changes — from mutation "A" to "E") is true for cancer progression, increased frequencies of "E" to "A" should be observed within cancer patient population. According to this thought experiment, both the earlier genetic changes and the later changes are linked by the stepwise clonal expansion. Each individual tumor of a given type is traditionally thought to progress from the same common gene mutations, accumulating additional mutations that occur during progression. Thus earlier changes (the initial gene mutations) will occur at the highest frequencies when a population of individual patients is profiled.

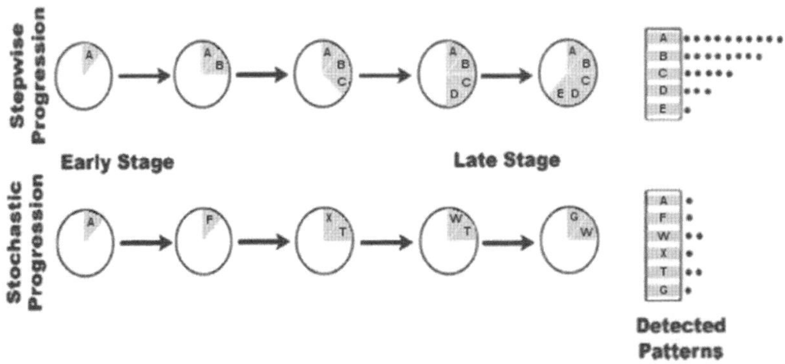

Fig. 2-1. The distribution pattern of gene mutations in the patient population reflects the pattern of the genetic landscape during cancer progression.

Letters A–G, T, W and X represent different gene mutations, different circles represent various stages of cancer evolution for each case. The upper panel shows the accumulation of gene mutations from A to E; the lower panel displays stochastic gene mutation during progression. The detected patterns represent the gene distribution profile for the patient population. This figure was modified from Heng 2007a with permission.

However, if the clinical mutation distribution pattern does not follow the upper panel but follows the lower panel, the explanation should be that there is no stepwise gene mutation accumulation, and the genetic landscape of an individual case is highly stochastic.

When comparing different types of cancer, CML fits the upper panel. For most solid tumors, however, the distribution pattern only fits the lower panel, suggesting that the well-known model of stepwise gene accumulation is not working. For those pioneers of the multiple-mutation theory of cancer, they understood the importance of using the reality to judge the theory. For example, J.C. Fisher has pointed that, *"multiple-mutation theories of carcinogenesis will have to stand or fall through the degree of their agreement with observations of the process of the diseases in afflicted organisms, and with the increases in the cancer-rate produced by age, carcinogenic materials, and radiation."* [Fisher,

1958] Clearly, a medical theory has to fall when it has so little agreement to clinical facts.

The vast accumulation of Cancer Genome Sequencing (CGS) data further supports our reasoning and questions the gene mutation theory. The earlier changes may not predictably lead to the later changes. Often, the earlier changes are lost or replaced during cancer progression and are not co-detected with the later changes. Examination of the genetic profiles of the end products of cancer evolution reveals a high level of heterogeneity of genetic changes among patients and it is difficult to identify any dominant common patterns of mutation.

The following observations undermine the notion of fixed stepwise progression at either the genome or gene levels.

(1) There is a high level of genomic heterogeneity in most human tumors where the majority of oncogenes or tumor suppressor genes are mutated with low penetration and there is limited correlation between gene mutation signatures and specific cancer type or stage.

(2) Even in more linear, specific mouse models (knockout, overexpression, or mutation of a specific cancer gene), different pathways are clearly involved in various tumors despite the fact that these tumors can be linked to the same gene alteration.

(3) Non-genetic environmental factors account for at least two-thirds of all cancer cases, and most sporadic cancers have distinct genomic landscapes (karyotypes and diverse gene mutations).

(4) Hematological malignancies and solid tumors are different in terms of system characteristics and the patterns of their evolution. For example, despite the major efforts to identify the sequential order of gene mutations and chromosomal aberrations during tumor progression, unlike CML, most solid tumors contain heterogeneous populations with stochastic genome alterations and mutations. The penetration of a fusion gene is also much lower in solid tumors compared with many hematological cancers [Horne *et al.,* 2013a]. One very good example is the fusion gene TMPRSS2-ERG discovered in prostate cancer. Despite the initial excitement, we were not convinced about its high penetration in prostate cancer samples based on the high karyotypic heterogeneity detected from prostate cancer which drastically differs from the high-penetration fusion gene of BCR-ABL in CML chronic phase. The recent

sequencing efforts have revealed much lower frequencies in prostate cancer samples.

As we will discuss more fully in Chapters 4 and 5, it has now become obvious that the simple and linear Vogelgram no longer seems to hold water due to the highly diverse pathways observed during stochastic cancer evolution [Heng, 2007a; Weinberg, 2014], which was also acknowledged by Dr. Vogelstein a few years ago following their own sequencing effort [Wood *et al.,* 2007]. Yet despite these data and calls from both Weinberg and Vogelstein to change the field's view from the stepwise accumulation of gene mutations, it is much harder to change the mindset of the majority of researchers.

Prior to the cancer genome sequencing era, we asked many cancer researchers to predict how many cancer types would display common and dominant genetic changes. Nearly all envisioned that most would and that even though some cancer types display complicated genetic changes, new technologies surely would identify common patterns. They often cited the examples of BCR-ABL fusion gene created in the Philadelphia chromosome in CML and especially the APC-H-ras-TP53 pattern of mutations in colon cancer as proof that such a pattern must exist in all other tumor types. Now that large scale cancer genome sequencing data disprove these notions (See Chapter 4), we asked the same questions again. Typical responses were that while there may not be a common gene mutation pattern, there should be common patterns of pathways or that there must be a mutation pattern in non-coding RNA or epigenetic changes. Surprisingly, a large number of cancer researchers continue to cite decades old literature supporting their old mindset as if the large-scale data from various –omics project never existed, completely ignoring the new data that clearly show there is no specific pattern of mutation in most cancers. Even today, at seminars or grand rounds throughout the world, many speakers will still use the Vogelgram in their introduction slides or as the main conceptual basis for their research project.

Knowing the facts from both the massive diverse sequencing data and that both Vogelstein and Weinberg, perhaps two of the most influential leaders in gene mutation theory of cancer, have publicly acknowledged the key limitations of their own theories, we cannot help but feel we are stepping into a time machine while listening to these seminars or reading these papers.

2.4.3 *Hallmarks of cancer: The illusion of using basic molecular principles to fight cancer*

If one set of gene mutation combinations cannot explain cancer, how about a common set of key features? Since diverse gene mutations and molecular pathways are involved in cancer, can a few molecular principles be used to understand cancer in a clinically relevant manner? In 2000, Weinberg and Hanahan published their widely revered *Cell* paper discussed above that proposed six hallmarks of cancer. But despite over a decade's worth of fanfare for this article, do these hallmarks represent an advance?

The hallmarks of cancer have rationalized the identification of genes and pathways in cancer by offering a connection to these hallmarks. Not surprisingly, many gene mutations have been classified with these hallmarks, and the Weinberg and Hanahan paper has become one of the highest cited publications. However, this concept has failed to reveal a general mechanism of cancer, and it has failed to deliver clinical implications. While it describes the common phenotypes of cancer, it does not provide the new framework the field desperately needed. Recently, as acknowledged by Weinberg himself, the hallmark concept is not a much needed conceptual paradigm. We fully agree with Dr. Weinberg's estimation. The following is the reasoning for this (more detailed discussions can be found in Horne *et al.,* 2015a).

First and foremost, the hallmarks of cancer are not a theory but rather a summation of cancer phenotypes that can be linked to the diverse molecular mechanisms. These phenotypes do not provide a theoretical understanding of how and why we get cancer. Cancer is a complex disease not simply explained by a collection of separate cellular hallmarks. Understanding each individual hallmark will not necessarily lead to understanding the dynamics of the disease.

Second, while the hallmarks have energized the gene mutation field by linking diverse cancer genes to a variety of hallmarks, the concept of hallmarks actually failed to act as a guide for the phenotype of cancer. In contrast, in the name of studying the hallmarks of cancer, much effort has focused on molecular characterization of gene mutations which proved to be limited at the clinic level due to the low penetration of most gene mutations within patient populations. Instead of categorizing rarely occurring mutations, more attention should be placed on how the system evolves within the evolutionary context that is coupled to the different phenotypes.

Third, there is a high degree of overlap in these hallmarks and they are highly dynamic. A hallmark may require additional hallmarks (e.g. metastasis involves cell death, proliferation, inflammation and avoidance of destruction by the immune system), large numbers of different genes can be classified by the same hallmark while the same gene can be involved in different hallmarks. Furthermore, both the number of hallmarks and identified gene mutations are expanding, making the hallmark concept less clinically relevant. The number of hallmarks has increased from six to ten, yet altogether, they can explain only a small portion of gene mutations identified by CGS. Likely the list of hallmarks will further increase. For example, the stress phenotypes of tumorigenesis have already been suggested [Luo *et al.*, 2009]. Other specific mechanisms linked to cancer such as viral infection and activation of transposable elements (TEs) would also qualify as additional hallmarks [Lee *et al.*, 2012], as would human behavior and environmental setting as they are closely associated with the survival rate of cancer patients and serve as higher level system constraints. Furthermore, the high level of fuzzy inheritance is perhaps the most important hallmark of cancer as it serves as the basis for multiple levels of genetic and non-genetic heterogeneity (Chapter 5–6). Too many hallmarks will dilute the meaning of each hallmark, making the hallmark concept less clinically relevant.

Fourth, the reductionist idea of dissecting the complexity of cancer into simple individual molecular principles is not working. In a given linear experimental system, such a dissection might work as typically only highly selected factors (in a linear relationship) are examined. Often such molecular analyses are based on the average profile of a diverse population within a fixed time window where no real process is studied (Chapter 5 and 6). This practice can create an artificial one-to-one relationship between parts and the overall phenotype. In reality, emergent properties of cancer are not simply based on a one-to-one causative relationship between the parts of the system, especially when evolutionary selection is often based on the unpredictable outliers of heterogeneous cellular populations. The emergent properties of cancer are even more complicated as there are multiple levels of these properties. Self-organization is based on the genome at the individual cell level, and individual genes often have diverse functions depending on the framework of the genome context under which they function. This means that system inheritance is defined by the genomic topology-mediated genome context (Chapter 5). At the cell population level, there is no simple average property in heterogeneous cellular populations [Abdallah *et al.*, 2013]. We should consider a new type of inheritance where heterogeneity is inherited at the population level (for more information, see fuzzy inheritance in Chapter 6). Furthermore, at a micro-environmental interaction level, large numbers of cell populations of distinct cell types interact dynamically. Clearly, self-organization and evolution are two very powerful interactive forces that shape the bio-system and evolution plays an important role in the selection and maintenance of potential emergent properties [Heng, 2009].

Examples of emergent properties that conflict with and/or that can overpower the function of the parts can be observed at many levels of the cancer biosystem. The same gene can have different or even opposite functions. For example, many oncogenes can function as tumor suppressors; cellular mechanisms such as cell death and the immune system can function to eliminate tumors based on a "parts centric view". However, emergent features of these processes during cancer evolution could also be beneficial to cancer! The same challenge is also reflected in many therapeutics as treatment can also promote the evolution of the cancer (Chapter 8).

Fifth, perhaps, also due to the above mentioned reasons, the concept of hallmarks has failed to deliver clinical benefits. On the surface, it seems essential to identify hallmarks of cancer so that they can be differentially targeted. However, due to the evolutionary dynamics of cancers, tracing a key feature is challenging as specific features may come and go during evolution and the initial key factors may or may not play a role in the future. Even worse, the knowledge of individual hallmarks obtained from experimental systems can vary in human tumors and targeting individual hallmarks can cause harm to patients. For example, targeting the tumor blood supply should starve and eliminate tumors. Anti-angiogenesis-based therapies showed huge promise in both *in vitro* and animal models but failed miserably in actual clinical application due to the rapid onset of resistance. Paradoxically, targeting angiogenesis in the tumor accelerated tumor growth [Reynolds *et al.*, 2009]. In another example, inflammation is described as a hallmark of cancer but inflammation is also associated with a reduced risk of prostate cancer progression [Kryvenko *et al.*, 2012].

One can often explain individual cases based on the specific genotype or even link various gene mutations to different hallmarks (the cherry picking strategy), but we might not be able to predict the dynamic contribution of that gene to the hallmark in general. To solve the disappointing failure of targeting individual hallmarks in the clinic, the new strategy is to target a combination of hallmarks either simultaneously or sequentially. But the same challenge remains: cancer is a moving and constantly evolving target and targeting a fixed gene, pathway, or hallmark will not work as the tumor will inevitably evolve away from this targeting. The multiple level landscape model illustrates this principle well (Chapter 5; Chapter 8). This model describes how so many pathways can be associated with achieving fitness and why it is therefore challenging to predict specific pathways or even phenotypes to target.

2.4.4 *Extremely limited clinical success questions the basis of current cancer research*

While it is well known that there is a gap between basic research and its clinical implications, it remains both very surprising and puzzling to cancer researchers that most of the concepts generated from basic research are not applicable to the clinic. Initial clinical discoveries seemed to strongly support the gene mutation theory of cancer, including the identification of specific gene mutations in tumor samples, the identification of commonly shared dominant genetic changes within patient populations, and the impressive success of using imatinib to treat the chronic phase of CML. The problem is that researchers can always find examples to support basic research conclusions, but the vast majority of conflicting clinical cases suggests that these supportive findings are exceptions rather than rules.

Interestingly, basic researchers and clinicians often have different attitudes towards the inconsistency between theory and reality. Basic researchers pay more attention to a simple hypothesis itself rather than a clinical feature of cancer with much higher complexity. In the search for understanding, they test various models to establish an easy-to-manipulate and interpret system with artificially increased certainty, often leading away from the clinical picture. Such efforts generate many elegant experimental systems and discoveries but unfortunately, they offer limited relevance to understanding cancer in the real world (see Chapters 5–8). For example, studying hallmarks and individual potential pathways/gene mutations can generate molecular knowledge from these simple systems, but the knowledge is not about the cancer itself as a complex evolving system but rather the specific isolated feature as it behaves under the defined conditions. When a clinical fact does not meet theoretical expectations, basic researchers often favor the theory and question or blame the data. For example, limited cancer types (such as CML) display a linear progression in their earlier stages. These exceptions have become the backbone of gene mutation theory and their implications in therapy have made them an excellent example to follow despite the fact that most cancer types do not follow such linear models. One common and strong argument is that basic research is different from clinical research. It takes time to uncover true application in real life and basic research should be pushed further regardless of its lack of short-term clinical application. Another common viewpoint is that we need to test more samples based on the premise that testing more samples will wash away the "noise" and finally validate the theory.

Clinical cancer researchers and physicians in particular are familiar with diverse cases, have more appreciation for heterogeneity and exceptions, and often have less confidence in theories that do not fit clinical reality. However, as most of them are not directly involved in basic research, they tend not to be critical of the theories or they lack a platform from which to be critical. In addition, peer pressure is high, and very few want to stand out to against these powerful and sophisticated theories. Many perceived "breakthroughs" are in the literature, thus it is hard for a clinician to remain current and support these high numbers of theories. For many physicians, standard care and their busy schedules also prevent them from challenging current theories. The separation between basic research and clinical reality allows for many basic researchers to continue their efforts even if they lack clinical relevance.

Nevertheless, following the human genome sequencing project, basic researchers and physicians all shared the viewpoint that technology had advanced to the stage where large numbers of clinical samples could and therefore should be systematically analyzed. The hope was that the gene mutation theory of cancer would be finally validated with sophisticated technology and sufficient clinical sample size. Unexpectedly, recent large scale –omics analyses including the cancer genome sequencing project have reached a stunning conclusion: the massive volume of clinical data does not support the key concepts of the gene mutation theory of cancer (see Chapters 3–5). Specifically, these data question the importance of key gene mutations in cancer and challenge the sequential stepwise model of cancer evolution as well as the traditional approaches of focusing on gene/parts characterization. In contrast to the viewpoint that more samples were needed to once and for all identify the genetic landscape of cancer, the analysis of more samples resulted in observation of more heterogeneity.

2.5 Time to Change

Prior to the cancer genome sequencing era, the hope was that if we could profile the genetic landscape in great detail, we could then target key gene mutations or pathways to eliminate cancer. Now, as we discuss in Chapter 5, we have found the landscape of cancer to be highly dynamic and so targets vary among patients. Most importantly, these targets continually change during cancer evolution even within the same patient, especially in the face of aggressive non-surgical treatment.

Some investigators still blame the problem on sample selection and preparation as if the key limitations are just appropriate sample picking and a lack of standardized preparation techniques. Surprisingly, the overwhelming failure of many clinical trials has not prompted cancer biologists to evaluate the theoretical basis of what they are doing. Rather they are convinced that failures are attributable to mere technical errors and continued hard work will uncover the next magic bullet and solve the mystery of cancer once for all.

Such an attitude favors new buzzwords like "large scale data collection", "system analysis" and "computational modeling" based on the old genetic framework of cancer where the application of computational tools to decipher massive amounts of data will reveal the answers and improve therapeutic strategies and outcomes in cancer patients. The truth, they say, must be hiding within the diverse data and as soon as we appropriately interpret them, we will be able to cure cancer. Ironically, some biologists now rely more on reductionist approaches such as using more mathematical or physical models to understand biology, but under the name of studying the biosystem.

We and a few other groups, in contrast, like to think differently about this conundrum. In particular, we promote the development of a new framework by constantly asking the following questions: Why is most of our current basic knowledge of cancer not effectively translating to the clinic? Can this problem be solved through more data collection and modeling? What if the current theory of cancer — the foundation of our knowledge to fight cancer — is flawed? Asking the right question is key. According to Albert Einstein: "We can't solve problems by using the same kind of thinking we used when we created them". Our cancer problem is not an issue of more data collection and better modeling as some suggest, but rather the problem is the current paradigm. In other words, we need to promote paradox-driven cancer research to search for the new paradigm to depart from the one that is not working [Baker, 2012/2013; Heng, 2007a]. Only when we honestly ask these hard questions can a new conceptual framework be created. Without a new paradigm, the accumulation of anomalies for the old paradigm will be continually ignored.

The idea that the current genetic paradigm of cancer is broken is not new. In fact, even prior to the cancer genome sequencing era, a handful of scholars from different backgrounds were critical of the cancer gene mutation theory. Despite having different rationales and experimental systems, they realized that a new paradigm was and remains urgently needed as the dominant paradigm no longer works. Importantly, an ever-increasing number of voices is calling for new paradigms to replace the gene mutation-dominated current paradigm. The next chapter will briefly introduce some examples.

Chapter 3

Alternative Theories to Explain Cancer

3.1 Summary

The conclusion that the gene mutation theory of cancer is problematic has been reached by different scholars with diverse backgrounds. Limitations of the current conceptual framework of cancer have promoted a search for better theories. To give readers an overview of the current landscape of cancer research, the main competing theories will be briefly outlined in this chapter. This will illustrate that despite the popularity of the current mainstream theory, different thoughts and approaches have been proposed and pursued within the cancer research community. Many of these theories have been around for decades and clearly deserve serious evaluation, especially in light of many of their predictions being confirmed by current cancer research. In addition to the theories themselves, discussion will focus on the various scientists that played key roles in the development of these theories, the rationales behind the theories, the main conclusions, and potential challenges each theory faces.

3.2 The Rationale of Searching for New Theories

There are many alternative theories of cancer. Most of them aim to solve some of the key limitations of the gene mutation theory, but also offer different frameworks to describe the mechanism of cancer and its clinical implications.

Many of these theories can be grouped into different categories based on the following questions. There are certain degrees that overlap among these categories, and some of them have already been adopted by current cancer gene mutation theory:

(1) What is the main common driver of cancer? If it is not a common gene mutation, is it aneuploidy, non-genetic (including epigenetic regulation), a specific network interaction, a metabolic type or the micro-environment? Of the many theories of cancer put forth throughout centuries, many of them can be classified into epigenetic and genetic concepts of inheritance, with the latter further divided into gene-based and genome-based inheritance. On the surface, the significance of arguing which genomic level of information to focus upon (e.g. genome, gene, epigene level) seems unimportant.

Why not study all the levels together? There are many answers to this question. The limited resources of cancer research demands prioritization. Gene level focus and emphasis resulted from gene accessibility. Mendel's description of the gene followed by the molecular era where gene functions were identified made the gene an obvious target and a priority. Recent knowledge regarding the limitations of information derived from the gene mutation level (the entire focus so far) have brought about the idea that the genome is the correct level for study and is capable of providing insight to the questions being asked. For example, studies at the genome level have made understanding macroevolution possible by following karyotypic evolution rather than gene mutation or methylation patterns. As we discuss in this book (see Chapters 5 and 6), these different levels of inheritance define the evolutionary pattern of cancer and provide the rationale for future research with drastically different strategies.

(2) Are specific cell types or structures more important than others for cancer formation? For example, should we focus on cancer stem cells or stromal cells? In addition to understanding the mechanism of cancer, answering this question might provide preferable targets.

(3) Which organizational level should be the focus of cancer research? Should it be the cellular level, tissue level or some other higher or lower level? Cancer is a disease, and the management of a disease should not be limited to gene mutations or pathways or molecular hallmarks. There

are so many influential aspects including social interaction, emotional support, as well as an individual's overall health status. All these levels above the genome can provide constraint for cancer evolution.

(4) Are genetic factors or environmental factors more important for cancer formation? What relative contributions are made by viral or bacterial infections, carcinogens, or even lifestyle, when compared with gene mutations? Knowing the limitation of genetics in cancer evolution and that there is no magic bullet for most cancers makes prevention the key to reducing cancer burden.

(5) Does cancer represent a new species or reflect the execution of an ancient program pre-loaded into the genomes of all multicellular organisms (Chapter 7)? Is cancer a byproduct (trade-off) of somatic cell evolution, which is important for cellular adaptation? Such analyses will put cancer into the evolutionary perspective, as cancer may be ultimately linked to many needed functions including bio-protection and repair.

Since it is not our main purpose to describe each of the alternative theories in detail, we would like to briefly discuss some theories by focusing on their novel aspects and differences in contrast to the current gene mutation theory. To help readers receive a clear message, we also offer further explanations as well as some critical analyses, especially where our own research experience allows.

3.3 Aneuploidy Theory: Using a Chain Reaction of Aneuploidization to Explain the Multistep Process of Carcinogenesis

Aneuploidy, the abnormal number of chromosomes within a cell, is consistently observed in the vast majority of cancers. Despite its universal existence and that its link to cancer was first proposed by Boveri over 100 years ago, the idea of aneuploidy driving tumor progression has been ignored until recently. According to the gene mutation theory, chromosomal changes are secondary to gene mutations, and chromosomal-based research has lower molecular resolution than gene-based molecular studies. Furthermore, the majority of aneuploidy detected in cancer seems to be non-clonal, which does not fit the genetic determinist expectation of a stepwise progression or fixed common causes of cancer.

Peter Duesberg's group at UC Berkeley has pushed the importance of aneuploidy in cancer since the late 1990s [Duesberg *et al.*, 1998]. Through a series of studies linking aneuploidy to genetic instability, cellular transformation, and drug resistance, they later proposed that aneuploidy is the primary cause of genomic instability in neoplastic and preneoplastic cells. Their hypothesis is that independent of mutation, aneuploidy destabilizes the karyotype by corrupting highly conserved teams of proteins that segregate, synthesize and repair chromosomes. They stated that the impact of aneuploidy is several 1000-fold larger than that of gene mutation, because entire chromosomes are involved. They conclude that the transition of stable diploid to unstable aneuploidy cell species is the primary cause of preneoplastic and neoplastic genomic instability and of cancer, and that mutations are secondary.

In the early 2000s, a few mainstream journals published reviews, opinions, and debates on the importance of CIN in solid tumors with theoretical and experimental evidence on the role of chromosomal changes in the initiation of cancer [Cahill *et al.*, 1999; Gisselsson, 2003; Marx, 2002; Sieber *et al.*, 2003]. An influential piece written by Wayt Gibbs in the Scientific American garnered needed attention by pointing out the promise of the chromosome-based cancer theory and the limitations of the popular gene mutation theory [Gibbs, 2003].

One of the initial difficulties (which still remains) of promoting the aneuploidy theory in the field of cancer research was outlining a new theoretical framework to replace the dominant gene-based theory most cancer biologists subscribe to. New ideas were needed to explain how aneuploidy can lead to cancer without the basis of specific known cancer genes. To achieve this, it is necessary to redefine the genes and chromosomes-mediated inheritance, and to analyze the evolutionary role each play in cancer initiation and progression, and finally explain the emergent relationship between gene (parts) and the genome (system) between genome topology and transcriptome, and between individual cell and population [Heng *et al.*, 2011a; Stevens *et al.*, 2013a]. Clearly, a new level of inheritance and pattern of somatic cell evolution is needed to explain new system emergence without focusing on individual genes (or parts of the system). Our group has actively contributed in these aspects (see Chapter 5).

Over the past decade, increased publications have attempted to address the mechanisms whereby aneuploidy could cause or contribute to cancer. It seems that everyone is thinking about aneuploidy but they all have different perspectives on it. While many of these publications are worthy of discussion, we will briefly mention the following groups' work to present different aspects/ideas of aneuploidy and cancer, in addition to the various studies linking specific genes/pathways to aneuploidy [Heng *et al.*, 2013b].

Kirsch's group in NCI analyzed the NCI-60 drug-screening panel using spectral karyotyping. These 60 diverse cancer cell lines were established by the National Cancer Institute for the purpose of anticancer drug discovery. They compared karyotypic complexity and heterogeneity (metaphase-to-metaphase variations) based on three categories including ploidy, numerical changes, and structural rearrangements. A wide variation in all three parameters was evident. The similarities between karyotypic changes in cancer cell lines and primary tumors were also observed. They also proposed the concept to use karyotypic "state" as a potential determinant for anticancer drug discovery [Roschke *et al.*, 2005; Roschke and Kirsch, 2010].

Jeremy Squire's group in Canada completed diverse molecular cytogenetic analyses of various cancer types in order to study the contribution of chromosomal abnormalities by linking alterations affecting specific chromosomal regions to changes in specific molecular pathways. They have proposed the term structural chromosomal instability [(S)-CIN] to distinguish from numerical CIN. They also promoted the value of monitoring both nonclonal and clonal chromosomal aberrations [Bayani *et al.*, 2007].

Thomas Ried's group at the NIH generated artificial trisomies in a karyotypically stable diploid cancer cell line and in immortalized, cytogenetically normal human breast epithelial cells using microcell-mediated chromosome transfer. Chromosomal trisomies resulted in a significant increase in the average transcriptional activity of the trisomic chromosome regardless of chromosome or cell type. Interestingly, this increase affects the expression of numerous genes on other chromosomes as well [Upender *et al.*, 2004]. The message is that there is no simple one-to-one correlation of trisomy to gene expression of genes specific to the additional chromosome, indicating the emergent properties of the chromosome.

Although aneuploidy is linked to cancer progression, Cleveland's group showed a paradoxical role of aneuploidy. They showed aneuploidy could act both in a tumorigenic and tumor suppressing manner. An increased rate of aneuploidy was found to lead to lymphomas and lung tumors in aged mice. However, in cases of chemically- or genetically-induced tumor formation, an elevated rate of aneuploidy was unexpectedly found to inhibit rather than promote tumorigenesis [Weaver *et al.*, 2007]. Similarly, more recent research concluded that low rates of chromosome missegregation could promote tumorigenesis, whereas missegregation of high numbers of chromosomes leads to cell death and tumor suppression [Silk *et al.*, 2013]. As most biological phenomena display multiple and changeable functions depending on context, this finding should not be a surprise. However, it is a source of confusion for investigators of molecular mechanism studies of aneuploidy in cancer, similar to the discovery of many gene mutations that can function both as oncogenes or suppressors. Again, in a more linear animal model, the overall trend or average behavior can be predicted. However, during real evolution in clinical conditions, the winners of evolutionary selection are often the outliers [Abdallah *et al.*, 2013; Chapters 5 and 6], which challenges the way of current aneuploidy studies that focus on individual molecular pathways in linear models.

Interestingly, the popularity of studying aneuploidy has surprisingly grown due to the work of a number of well-established yeast cell biologists. These include but are not limited to David Pellman from Harvard, Angela Amon from MIT and Rong Li from Stowers Institute for Medical Research. These scientists reasoned that to dissect the intrinsic complexity of aneuploidy into individual genetic and molecular mechanisms, a simpler model was needed. They used methods to selectively generate aneuploid strains without causing specific gene mutations in a yeast model system and studied the results of evolution as well as genetic profiles using various cutting-edge genetics and –omics approaches [Mulla *et al.*, 2010]. These studies showed that geometric constraints may have profound effects on genome stability [Storchova *et al.*, 2006]; Aneuploidy directly affects the transcriptome and proteome levels and can generate significant phenotypic variation that could bring about fitness gains under diverse conditions [Pavelka *et al.*, 2010]; Aneuploidy-tolerating mutations can improve fitness and aneuploidy can be linked to specific molecular pathways such as ubiquitin-proteasomal degradation [Torres *et al.*, 2010]; aneuploidy strains exhibited one or more forms of genomic instability [Sheltzer *et al.*, 2011]; and the overall karyotype pattern determines the dynamics of chromosomal instability including chromosomal predictability [Zhu *et al.*, 2012]. All together, these researchers have convincingly showed molecular mechanistic links between aneuploidy, genes/pathways and cancer phenotypes. These results directly addressed the skepticism about the function of aneuploidy in cancer [Williams *et al.*, 2008].

On the clinical research front, there have been promising efforts to use chromosomal aberrations (both aneuploidy and polyploidy) to measure and predict cancer disease progression (see works from Brian Reid, Kornelia Polyak, and Charles Swanton) [Dewhurst *et al.*, 2014; Galipeau *et al.*, 2007; Park *et al.*, 2010; Paulson *et al.*, 2009]. However, the aneuploidy theory still faces the following conceptual and methodological challenges:

(1) Experimental systems are needed to directly examine how aneuploidy contributes to cancer in the human system; to study the relationship between other types of genetic or non-genetic alterations, especially the relationship with chromosomal structural aberrations; and to develop quantitative methods to measure aneuploidy for clinical prediction.

Recently, Valind *et al.* illustrated that, in contrast to the data obtained from yeast, in humans, constitutional aneuploidy itself does not lead to cancer-like chromosomal instability (CIN) [Valind *et al.*, 2013]. This report is significant, as it shows that specific constitutional aneuploidy alone is not sufficient for generating numerical CIN (structural CIN was not examined in this study). Since the majority of cancer cases are not caused by clonal constitutional aneuploidy, it is therefore necessary to further examine the impact of non-recurrent, stochastic aneuploidy for generating all types of CIN [Heng, 2014]. According to the evolutionary concept of cancer, even though aneuploidy often is more effective than individual gene mutations to change the genetic landscape of cancer, there is no one-to-one correlation between aneuploidy to cancer, and the evolutionary process differs from the end products.

(2) There might be fundamental limitations of using simple models to dissect molecular mechanisms of aneuploidy. Current studies simply link aneuploidy to specific gene mutations or pathways (both from cause and consequences point of view). Most experimental models do not address critical issues of cancer evolution including cellular competition/collaboration and emergence of outliers from heterogeneity, which would be difficult to explain using an average pathway or a given gene mutation. In other words, there is a big gap between known pathways and stochastic pathway interactions. That might be the reason why after over a decade of aneuploidy research in cancer, despite the accumulation of large amounts of molecular information, we are still lacking a real conceptual breakthrough.

One promising study deserves more attention [Rancati *et al.*, 2008]. Following the deletion of MYO1, the only copy in yeast that encodes the myosin II protein normally required for cytokinesis, extensive polyploidy and aneuploidy — not simply mutations of a specific gene — were the initial evolutionary changes in yeast selected *in vitro* to overcome the deletion defect. This experiment demonstrated the principle that genome level changes can yield emergent new features without directly fixing the specific deleted gene (parts). If the same phenotype can be achieved by genome level changes that can stochastically involve a large number of parts, then the understanding of each specific part based mechanism becomes less important, especially when there is no one-to-one correlation between emergence and the diverse parts. This idea emphasizes the priority to search for the evolutionary framework of aneuploidy in cancer.

(3) Applying the evolutionary mechanism to understand the impact of aneuploidy on cancer. If the complexity of aneuploidy cannot be satisfactorily explained by the genes whose copy number is directly affected, a general evolutionary mechanism becomes most important, as it unifies a large number of molecular mechanisms. The evolutionary mechanism can clarify the confusion derived from individual functions

of isolated pathways by focusing on a more proper level that has better predictability than individual parts or pathways.

For example, there is a common viewpoint that aneuploidy leads to the deregulation of normal pathways though abnormal expression. Since "abnormal" is bad and often is associated with growth disadvantages, it leads to the "paradox" of aneuploidy. From molecular terms, tumor promotion can be linked to genetic diversity (which increases adaptive potential) and genome instability (which increases the probability of acquiring tumor-promoting genetic alterations), while tumor inhibition can be linked to growth and developmental detects (through proteotoxic stress and p53 activation) [Pfau and Amon, 2012]. Such a linkage list can be quickly expanded to most pathways, with many of them conflicting with each other. In fact, there are many conflicting results regarding the correlation of copy number and expression. Such disagreements will increase when single cell profiling is used.

As soon as we are out of the box of "parts characterization", the paradox is no longer there. From the genome theory of cancer evolution perspective [Heng *et al.*, 2010a; 2013b; Chapter 5], the key for cancer success is to form new genome systems that differ from the host's genome. Therefore, the key to successful cancer evolution is through producing variance, regardless of whether the parts are "good" or "bad". By the way, within the dynamic evolutionary process, good or bad is constantly changing. That is the reason why aneuploidy research needs to adopt this new framework and research strategy.

3.4 Tissue Organization Field Theory (TOFT)

Proposed by Carlos Sonnenschein and Ana Soto in 1999, TOFT represents a major competing theory to somatic mutation theory (including gene mutation and chromosomal aberration basis of cancer) [Sonnenschein *et al.*, 2014; Soto and Sonnenschein, 2011]. The debate involves multiple aspects of cancer research. The TOFT theory has challenged more mainstream approaches to cancer. The following questions are some examples: Should our understanding of cancer be based on gene/cell based reductionist approaches or holistic or "organicist" approaches (the tissue-based "society of cells")? Should cancer be treated as a genetic disease or a developmental disease? Should 3D culture methods replace 2D culture methods?

Specifically, there are two alternative premises of TOFT that contrast with somatic mutation theory. First, the TOFT posits that carcinogenesis takes place at the tissue level of biological organization, as does normal morphogenesis. Cancer causing agents (environment, chemicals, radiation, carcinogens) affect the underlying mesenchyme/stroma, rather than cause mutations in epithelial cells that trigger clonal evolution, leading to the disruption of cell-to-cell and cell-to-extracellular matrix communication. Second, the default state of all cells is proliferation, making carcinogenesis a reversible process. When neoplastic tissues are placed in a normal tissue environment, they may be normalized. It is thus essential to study normal and neoplastic tissue together in their natural architecture [Soto and Sonnenschein, 2011]. During the last 15 years, Sonnenschein and Soto represented a persistent voice that vigorously challenged the somatic theory of cancer by illustrating its conceptual and experimental limitation.

It is important to point out that Mina Bissell's group in Lawrence Berkeley National Laboratory has contributed greatly to the acceptance of the importance of tissue architecture or microenvironment for cancer formation. In 1984, Dolberg and Bissell showed that the Rous sarcoma virus is not able to cause sarcomas in the avian embryo; the injection of the same virus into the wing web of newly hatched chicks causes a rapidly growing sarcomatous tumor, and cultures of fibroblasts derived from chick embryos (CEF) and infected with RSV become rapidly transformed, leading to the conclusion that tumorigenesis cannot simply be explained as solely a consequence of a cancer gene [Dolberg and Bissell, 1984]. Over the past three decades, Bissell's group has studied how the extracellular matrix (ECM) and nuclear environment affect gene expression in normal and malignant tissues [Nelson and Bissell, 2006]. They also popularized 3D cell culture using a matrix gel system by describing a robust and generalized method for the culturing of various human breast cell lines in three dimensions. In one of their milestone publications, they illustrated that the extracellular matrix and its receptors dictate the phenotype of mammary epithelial cells, and the tissue phenotype in this model system is dominant over the cellular genotype. Specifically, in this experiment, when the mutated mammary cells were placed into a normal cellular microenvironment and exposed to inhibitory β1-integrin antibody, these "abnormal" cells behaved normally [Weaver *et al.*, 1997]. Since this discovery, the micro-environment become a hot topic and has been linked to many molecular pathways, gene mutations, micro-RNAs, genomic instability and cancer stem cells.

For example, it is now accepted that the local microenvironment, or niche, of a cancer cell plays important roles in cancer initiation and progression. A major component of the microenvironment is the extracellular matrix, a complex network of macromolecules with distinctive physical, biochemical, and biomechanical properties, which serves as a system constraint. Abnormal ECM affects cancer progression by directly increasing the probability of cancer. ECM anomalies also deregulate function of stromal cells, leading to the emergence of tumor-associated angiogenesis and further variations, and thus leading to the production of environments that favor tumors [Lu *et al.*, 2012].

Similarly, many other hypotheses are also associated with epigenetics and tissue organization. For example, the calcium-based theory of carcinogenesis hypothesizes that increased intracellular calcium levels initiate oncogenesis by breaking gap junctions, disorganizing tissues and activating proto-oncogenes, as most human cancers are initiated by chronic injuries, which involve repeated cell death and raise cell calcium levels [Jaffe, 2005].

Despite the fact that a substantial number of scientists have now acknowledged the important role of the tissue component in carcinogenesis, most of them are tracing this component back to gene mutations or pathways under the gene mutation theory. As pointed by Sonnenschein and Soto, while these scientists have tried to modify the "classical" gene theory to include the importance of tissue organization, most of them still fall within the framework of genetic determinism and fail to follow the framework of the tissue-based theory. In fact, different questions have been asked to address the relationships between tissues and gene mutations [Sonnenschein and Soto, 2008].

Interestingly, Sonnenschein and Soto have rallied against integrating TOFT and somatic mutation theory and concluded that it would not be productive to reconcile the two. This point was reiterated by Baker who said that no compromise between TOFT and somatic mutation theory is good for science [Baker, 2012/2013].

While we share and appreciate many of the TOFT criticisms of the gene mutation theory of cancer and strongly encourage debating the pros and cons of the different frameworks, especially when the current paradigm is not working, we also realize there are many challenges TOFT faces.

First, like many other cancer theories, TOFT fails to pay more attention to the key component of somatic cell evolution. Tissue is made up of populations of various cell types that interact with cancer cells that change over time. Therefore, how does evolution progress at the tissue level and what is the relationship between the tissue and cell populations (both tumor and non-tumor) during somatic evolution? Since inheritance is an important element of evolution, genetic/epigenetic changes cannot be ignored if we are to understand cancer evolution. Genetic changes are not just the products of cancer, but indeed drive the evolutionary process itself (Chapter 5).

Second, cancer evolution differs from a typical developmental process. Even though the disruption of development can lead to cancer, cancer by and large is not a developmental issue. For most developmental processes, the cellular differentiation process does not require genome level alteration, but cancer evolution does. This difference is not appreciated by the TOFT as it considers individual cell status the same in normal tissue and cancer cells. "So far no qualitative difference has been described" [Soto and Sonnenschein, 2011]. They recently further concluded:

"...Moreover, looked at from the cell-level perspective of biological organization, those genomic alterations have not changed the fact that, as with normal 'mother' cells, cancer 'mother' cells also generate two similar 'daughter' cells. Thus, the chromatin complex aberrations (i.e. mutations) that Boveri and his followers have alluded to since 1914 may have either a 'neutral' effect on somatic target cells, or they are just a consequence of the carcinogenic

process rather than its ultimate cause" [Soto and Sonnenschein, 2014].

This statement regarding cancer cells generating similar cells like normal cells is far from accurate. As illustrated by our and others' experiments, most cancer cells do not pass the same or similar genomes! The genomes of two daughter cells often vary widely from each other and that of their mother cell. Therefore, cancer cell populations differ from normal cell populations in terms of how system inheritance is passed [Heng *et al.*, 2006a–c; 2013b; see Chapter 5 and 6 for more details].

The misunderstanding of the genome status of cancer and normal cells also leads to a common ignorance of the importance of the genome. Some experiments have claimed to show the normalization of cancer cells through regulation of tissue architecture regardless of the genome alterations contained within these cancer cells. In fact, this point is one of the strongest arguments for TOFT.

To correctly evaluate this type of tumor reversion at the genome level, a careful analysis is required to define the normal and altered genome. The importance of the genome context to tumor reversion has been noted from the very beginning by original investigators [Illmensee and Mintz, 1976]. Illmensee and Mintz pointed out that a near-normal chromosome complement is sufficient (thus important) for total restoration of orderly gene expression in a normal embryonic environment. Normalization of melanoma genomes has also emphasized the importance of a lack of genetic changes for phenotype reversion, as one of the two ES lines cloned from RAS inducible melanoma nuclei from a total of 57 explanted blastocysts failed to produce teratomas and chimeras, possibly due to the near tetraploid genotype. In addition, R545-1 ES cells failed in later stage development, presumably due to irreversible genetic aberrations in the cancer donor genome [Hochedlinger *et al.*, 2004].

In fact, Jaenisch's group has also showed that the non-reprogrammable genetic modifications within embryonal carcinoma nuclei define the developmental and tumorigenic potential of resulting nuclear transfer ES cells [Blelloch *et al.*, 2004]. They pointed out:

"Although EC (embryonal carcinoma) appeared to be a good candidate for an epigenetic model of tumorigenesis, the failure to alter their developmental and tumorigenic phenotype by NT (nuclear transfer) argues that genetic lesions rather than epigenetic restriction are responsible."

Clearly, more careful examination of many claimed normalization experiments is needed. For example, many "cancer stem cells" are cells with altered karyotypes. Reprogramming will cause further genome changes but will not return altered karyotypes to normal. It is thus important to know how reprogrammed cancer cells behave in patients rather than in more linear animal models, especially with regards to how genome dynamics change during the treatment, and whether or not the converted phenotype of the cancer genome can persist longer durations.

3.5 Various Theories Based on Systems Biology Perspectives

Departing from the individual gene-based deterministic viewpoint, there are collective efforts/ideas to explain cancer from a system perspective by offering new non-genetic frameworks. Such efforts have been boosted in recent years by the large amount of available –omics data and increased computational/modeling capabilities.

Many different concepts share the same considerations such as network structure and dynamics, self-organization principle, state/phenotype switching (gene regulatory network controls gene activities, producing the variety of cellular states), and landscape adaptation. However, they focus on different aspects of the system, and often, different terms are used. A few examples are briefly discussed in the following:

3.5.1 *The cancer attractor concept*

Based on the concept that a complex gene regulatory network controls cell lineage commitment and differentiation, it was proposed that inappropriate regulatory signaling and mutational rewiring of the network could disrupt cellular differentiation and lead to tumorigenesis.

To explain why oncogenesis recapitulates ontogenesis, a formal concept of cancer attractors was used to conceptualize tumorigenesis. By considering cell types as attractors, this theory argues that cancer cells are trapped in abnormal attractor states [Huang *et al.*, 2009]. Huang's group has been pushing the importance of various non-genetic variations in cancer progression, and these non-genetic variations are intimately associated with cell status. Their analyses have generated high interest in the field as such theory could provide understanding to cancer stem cells and epigenetic landscapes, two popular subjects in current cancer research.

There are many studies/ideas illustrating how protein interaction network rewiring can change cell status. One example is to link intrinsically disordered proteins (IDPs) to protein interaction network rewiring, and this would ultimately lead to cancer [Kulkarni *et al.*, 2013; Tompa, 2012].

IDPs are proteins that lack well-defined 3D structures either along their entire length or in specific regions under native, functional conditions (at least under physiological conditions *in vitro*) [Tompa, 2012; Uversky and Dunker, 2010]. It is suggested that IDPs drive the tumorigenic process [Kulkarni *et al.*, 2013; Sonnenschin *et al.*, 2014]. This is based on the following observations/ideas: many hub proteins that interact with multiple partners in the network, are IDPs, which are more significant than end proteins that interact with far fewer partners [Patil *et al.*, 2010]; IDPs can perform disorder-to-order transitions based on the binding of various targets [Tompa and Csermely, 2004]; Many well-known cancer proteins are IDPs (such as p53 and Myc); Many stress-related proteins are IDPs, and IDPs can rewire the protein interaction network to activate previously masked functions in response to stress; and IDPs are linked to epigenetic changes and phenotypic plasticity. By synthesizing these facts, it is proposed that stress can increase promiscuous interactions (through stress-response of IDPs) and rewire the protein interaction network, leading to state/phenotype switching.

Another recent example of applying the cancer attractor idea to conceptualize cancer is to consider cancer cells to behave as simple, minimal replicators. Based on the similarities between cancer and microbial populations, and the genomic instability of cancer cells, it was suggested that the loss of intercellular organization might involve a reduction of cellular complexity, resulting in a minimal set of intracellular components able to operate in a robust manner under noisy conditions [Solé RV *et al.*, 2014].

3.5.2 *Endogenous network hypothesis*

Ao's group promoted the endogenous network hypothesis to explain tumorigenesis. Aimed to provide a quantitative understanding, they tried to combine molecular biology, mathematics, engineering, physics under stochastic dynamics, and evolutionary adaptation. They considered cancer as an intrinsic robust state of the endogenous network that is not optimized for the interest of the whole organism.

According to them, molecular and cellular agents form a nonlinear, stochastic, and collective dynamical network, the endogenous molecular–cellular network. In normal situations, such network is able to autonomously control its operational functioning state. Stochasticity may accidentally cause a transition from one stable state to another. Under certain conditions such network is in a state not optimized for the interest of whole organism, resulting in sickness, though this state might be "normal" under other conditions. Based on these properties, they believe that the wiring rules among endogenous agents and their global dynamics interaction may be understood.

Interestingly, they have included many agents such as genes, pathways, molecular modules and networks; however, they did not pay enough attention to the ultimate importance of inheritance in evolution during their integration, as they believe that there are more important, and nearly autonomous, layers of mechanisms between genetic and environmental factors. Thus it is important for them to think beyond the genome to understand this new layer of mechanism [Ao *et al.,* 2010].

3.5.3 *Common challenges*

Despite offering new perspective/promise to cancer research, an increasing number of models/concepts based on systems biology (especially those based on the network regulation and cellular status changes) face some common challenges.

First, many systems biologists still apply reductionist ideology to study the system. They pay too much attention on the parts of the system such as gene mutations and pathways. "Systems biology is the science of discovering, modeling, understanding and ultimately engineering at the molecular level the dynamic relationships between the biological molecules that define living organisms" [Hood, 2014]. To date, many system approaches are discussing large-scale parts interactions without new frameworks with solid links to the holistic biological basis. For example, there is less interest to illustrate what bio-structure defines the network structure at the first place. Specifically in the field of cancer research, it is necessary to ask, should the complexity of cancer be explained mainly by molecular relationships? The usage of large data sets alone is not enough, and a new holistic framework of cancer is essential.

Second, how can evolution be integrated into their models? Even though most of them discuss the importance of evolution, they lack specific mechanisms. As we previously suggested, there are multiple levels of interaction between self-organization and evolution, and different levels of organization require different self-organization [Heng, 2009]. Furthermore, the pattern of evolution is different from the various levels of genetic organization.

Third, the importance of genome level variation has been ignored due to the above reasons. Biological inheritance is not just influenced by gene mutation and epigenetic variation, especially when considering the cellular population dynamics that occur during a pathological process, which drastically differ from a developmental process. Significantly, as we will discuss in Chapter 5 in more detail, genome-mediated stochastic macrocellular evolution rather than the clonal evolution holds the key for cancer evolution.

Fourth, many system biologists were trained as physicists and mathematicians and lack biological training, particularly in evolutionary biology. Some of their key biological knowledge is based on the generally accepted concepts derived from mainstream molecular cancer scientists who traditionally focus on gene or pathway level understandings. However, many of these "facts" are not solid. For example, pathway interaction is highly dynamic, and there is no fixed and clear cut pathway within the dynamic network, which makes it challenging to study. In addition, many pathway profiles of cancer cells are based on molecular average profiling, which drastically differs from the real world of tumor heterogeneity (see Chapter 6). Furthermore, the accumulation of gene level genetic change is not equal to genome level change, and there are no gradual changes between "normal physiological" and "abnormal pathological" status. Finally, it is clinical fact that most cancers display drastically altered karyotypes rather than display the same normal genome with system rewiring, and thus far systems biology has not taken karyotypic alterations into account, despite that karyotype defines the bio-system [Heng, 2009; see Chapter 5].

Ignorance of genome level information can directly mislead efforts of building correct models. Cancer populations are highly complex compared to normal cell populations. The fact that most cancer cells display high numbers of chromosomes (many more than the 46 chromosomes observed in a normal human cell), high frequencies of chromosomal alterations, and high degrees of variation among and within cell populations may not support the idea that cancer cells behave as simple, minimal replicators. The intercellular organization of cancer cells differs from normal cells, containing multiple different types of complexity rather than the proposed reduction of complexity. In fact, cancer genome complexity is clearly much higher than highly homogenous normal cell populations, as suggested by the high level of dynamics of RNA splicing, pathway switching and phenotypic adaptation. To appreciate this, scientists need to recognize and appreciate the highly dynamic genomes of the cancer cell population, which have been unfortunately ignored. One contributing factor is the misunderstanding of the terminology of the genome. For many, the genome is just a name for the sum of all genes or a collective name of network parts, but the genome is more than this (see Chapter 4).

Recently some system biologists have begun to appreciate the key differences between genome, gene and epigene and their contributions to somatic cell evolution. Sui Huang, for example, has recently published a model that links the overall fitness landscape and the epigenetic landscape of cancer cells [Huang, 2013]. This thoughtful model has recognized the importance of the multiple level landscape model of cancer [Heng *et al.*, 2011b; Chapter 5]. We are confident that this important integration will soon be followed by more systems biologists, as Huang's group is among the front-runners in the field.

3.6 Some Provocative Hypotheses Based on the Natural History of Evolution

Using evolutionary principles and methodologies to study cancer is no longer new. Following Nowell's seminal work a few decades ago describing cancer as a clonal evolutionary process, there has been an increase of research interest (Chapter 5). Traditionally, however, the focus of cancer evolutionary studies has been the identification of recurrent genetic changes such as common gene mutations that drive stepwise cancer evolution. Even though considering cancer as somatic cell evolution makes sense, progress has been limited. This is especially true in recent years, when various large scale –omics studies revealed a highly dynamic evolutionary process, questioning the stepwise pattern of cancer evolution [Horne *et al.*, 2015b]. It gradually became obvious that new evolutionary frameworks were needed to understand cancer evolution. In addition to tracing cancer evolution with various *in vitro* and animal models along with patient samples, simulation experiments were performed to illustrate how selective pressure from the microenvironment drives tumor morphology and phenotypic evolution (Anderson *et al.*, 2006). Recently, the theory of life history trade-offs and Peto's paradox were applied to explain cancer evolution [Aktipis *et al.*, 2013; Caulin and Maley, 2011]. It seems that evolutionary biologists are very interested in testing some well-known concepts/hypotheses with the hope that some existing evolutionary ideas can solve the mystery of cancer.

In addition to the genome-mediated two phases of cancer evolution (Chapter 5), cancer is also considered by some as a new species (Chapter 7), which fits well with both the separation of macro- and micro-evolutionary phases of cancer and the concept of using genome identity to define new species. There are also new hypotheses that search for the deep root of cancer from the life history point of view.

3.6.1 *Cancer represents a type of atavism*

Physicist Paul Davies and his colleague recently introduced the idea that, rather than cancer being a result of somatic cells driven out of control by gene mutations, cancers may be the phenomena of age-old atavisms [Davies, 2011]. They developed the following ideas/arguments to explain this concept: Since cancer is widespread among many species from mammals to flies, this suggests that its deep evolutionary roots can be traced back to the emergence of the multicellular life form. Most of the cancer subroutine evolved gradually over a period between 1.5 billion and 600 million years ago, shortly following the emergence of primitive multicellular life. Cancer shares some features of ancient unicellular life forms. These include limitless proliferation and favoring fermentation in low oxygen conditions.

The same cells (rather than evolved cells with different altered genomes) are responsible for different hallmarks of cancer including metastasis. Many behaviors of cancer cells mimic what happens during embryogenesis. The propensity of circulating cancer cells to invade other organs closely parallels how the immune system heals wounds.

Based on their synthesis, they have concluded that:

Cancer is a pre-programmed, well-organized and efficient survival response to a threatening cellular micro-environment. All cells have this pre-loaded ancient "cancer subroutine". This programmed cancer potential is not active under normal circumstances; however, it can be activated as part of the neoplasm's survival response.

If these ancient genes that guide the organism's basic body plan are re-activated and re-expressed in the adult form, cancer is the result. The reason evolution has not eliminated cancer over the many millions of years since its emergence is that part of the "cancer subroutine" plays a critical role in embryo development (many oncogenes are active during embryo development), wound healing and the immune system. The ancient toolkit would be a limited set of specific genes and therefore present a well-defined target for therapy.

While interesting, some basic biological factors need to be seriously considered. First, this idea is heavily influenced by gene-centric thinking, where some of the activated "ancient" genes would return the already evolved genome back to an ancient status. The similarities of asexual reproduction between human somatic cells and many microorganisms is due to the lack of sex-mediated genome level constraint (rather than a simple return to an ancient state), as most cancer cells display drastically altered genomes (Chapters 4 and 5). Similarly, the change of metabolic state in cancer is just an effective trick to be different from normal cells (providing evolutionary selective benefit) rather than atavisms. If we target the anaerobic metabolic features of cancer cells, surviving cells will become aerobic systems, as most cancer cells have normal mitochondria anyway. Gene-mediated microevolution and genome change mediated macroevolution must be distinguished (see Chapter 5). This also applies to distinguishing embryogenesis and cancer evolution where genome re-organization makes the key difference. Interestingly, for proper immune system function, genome level alterations within limited genome regions are crucial.

Second, it is useful to use the "genie in the bottle" metaphor to describe cancer to the general public, which can downplay the efforts of tracing causative factors by sequencing the cancer genome. However, it is also necessary to be aware that cancer is an evolutionary process. There may have been a genie's gene within the normal cell genome, but the key is to trigger genome alteration mediated somatic cell evolution, and so many other factors can function as triggers. Interestingly, the genie in a bottle analogy actually fits the genome theory, once the genie is out of the bottle you can't get it back in. While a gene could be the causative rubbing of the bottle to release the genie, once it is out you can't target the one who rubbed the bottle to constrain the genie. The reality is, even if the atavism of cancer proves to be true, researchers will still need to deal with how it occurs and determine the evolutionary process.

3.6.2 *The linkage between retrotransposon mediated genome evolution and cancer*

Transposable elements are repetitive mobile genetic elements that can relocate within a genome or be integrated into a new genome. The prevalence of transposable elements within a genome varies by species and the transposons and transposon-derived repetitive elements are responsible for 40–45% of the entire human genome, it is thus logical to illustrate their potential contribution to human cancer.

The dynamic behavior of retrotransposons has various impacts on the host genome [Muotri *et al.*, 2007]. Retrotransposition can destabilize the genome, reshaping genomic landscapes by insertional mutagenesis, deletions, and gene rearrangements. Transposons were involved in new gene formation during mammalian evolution, mainly by generating pseudogenes. In addition, relics of transposable elements can acquire a regulatory function.

Adam Wilkins recently proposed that cancers could be initiated by retrotransposon (RTN) activation through changes in the transcriptional regulation of nearby genes. By analyzing the nature of physiological stress(es) in RTN activation; the role of DNA demethylation in the initiation and propagation of new RTN states; and the connection between aging and cancer incidence in the context of activated RTNs and chromosomal aberrations, he made a strong case to investigate the contribution of RTN in cancer progression [Wilkins, 2010].

It should be pointed out that, according to the genome viewpoint, retrotransposons should have more global impact on the host genome rather than just nearby genes, as retrotransposons can potentially change the network structure through karyotypic alteration or simply through changing the topological relationship among genes. A closely related issue is how DNA parts of a foreign genome alter the landscape of the host genome. Clearly, only a small portion of the infected genome can be directly linked to cancer genes, but the majority could contribute to the elevated dynamics of the host genome, triggering cancer evolution.

3.6.3 *Human genome mismatches the fast changing environment including lifestyle*

From an evolutionary medicine point of view, it was suggested that many common and complex diseases including cancer are caused by mismatched living in a mismatched world. Since developmental plasticity (mediated by epigenetic regulation) is important for an individual's life, the risk of developing disease in later life is dependent on early life conditions. Interestingly, such influence can be detected for a few generations.

Despite various mechanisms for adaptation, a rapid environmental change can trigger the mismatch between the organism and the world they are living in. Such conflict will lead to disease conditions. Reports have concluded that the conditions of embryonic and fetal life influence the risk of developing later life obesity, cardiovascular disease, type two diabetes mellitus and osteoporosis [Gluckman *et al.*, 2011; 2013]. The linkage between obesity/diabetes and cancer is now obvious.

Interestingly, such idea fits well with the genome view of the somatic cell evolution, as genome stability is directly related to the environment. An unfit lifestyle can generate stress to the genome, triggering cancer evolution. As discussed in the next session, to adapt to unfit environments, variations are also required which can be linked to cancer as well (Chapter 5). However, even though important, many other factors (genetic elements included) can impact genome instability (Horne *et al.*, 2014].

3.6.4 *Cancer represents a trade-off of adaptation and survival*

Based on the observed high frequency of non-clonal chromosome aberrations both in normal tissue and particularly in cancer tissues, we finally realized that the advantage of genome change is that it increases system dynamics for increased evolutionary potential under stressful conditions. Such variation is important for providing functional composition, as it is necessary to change the genome to provide function. However, this important function is also an evolutionary trade-off, which can lead to unstable genomes and the potential of cancer formation [Horne *et al.*, 2014]. Similarly, even the immune system which protects against infection can cause cancer through friendly fire when there are errors or variations in adaptive VDJ recombination. Furthermore, the weapons of the immune system on cancer also can promote cancer evolution under certain conditions. There are multiple levels of evolutionary trade-offs, and the challenge is that it is hard to predict the direction of evolution.

In this sense, cancer represents a by-product of evolution that is essential for cellular adaptation (repair, functional composition, and providing diversity for increased fitness). For example, wound healing and cellular/tissue re-generation can significantly contribute to cancer. It is thus not reasonable to discuss why evolution has not yet eliminated cancer. In other words, we disagree with the statement that humanity is too young to develop the mechanism to get rid of cancer. The evolutionary process is not going to remove evolution itself. As long as life is dynamic, cancer remains a possibility. Furthermore, evolution does not aim for perfection. Rather, evolution is simply about survival by progressing in a changing world (more is discussed in Chapter 7).

The concept of the evolutionary trade-off (both at somatic cell and species levels) and refocusing on genome alterations in cancer and other diseases is of importance. It can explain why many common diseases are highly associated with aging, infection, lifestyle (including embryonic and fetal conditions), and overall environment (including social interaction), all of which are associated with increased genomic instability. Interestingly, the genetic variation within the human population also represents a trade-off! Some individuals pay the price for the species having higher degrees of heterogeneity and potential to survive [Horne *et al.*, 2014; Chapter 5].

3.7 Some Re-emerging Popular Concepts

Finally, we would like to briefly mention some re-emerging concepts. The limitations of the classic gene mutation theory of cancer have promoted the popularity of these concepts. In fact, most of them have been included into current gene mutation theory. Since these topics are constantly discussed in the field, no further details will be presented. Some obvious challenges for these concepts will be explained as they represent significant limitations and these are ultimately related to the rationale of pushing genome theory.

3.7.1 *Epigenetic alterations drive cancer*

The importance of epigenetic effects on cancer is now generally appreciated at different levels including gene expression, network regulation, changing cell status, cellular interactions, tissue microenvironment dynamics and the somatic cell evolutionary process [Esteller, 2008; Feinberg *et al.*, 2006; Feinberg, 2014; Jones and Baylin, 2002]. In fact, many different cancer theories such as most network-based cancer theories and TOFT, are highly influenced by epigenetics concepts. The epigenetic concept is not new. However, since the highly diverse gene mutations detected in cancer have challenged the gene mutation theory, increased researchers have switched their interest to the epigene. Conceptually, to search for the missing inheritability of cancer, an epigenetic mechanism becomes the easy choice. Furthermore, the availability of new methodologies for epigenetic profiling makes the field more popular. On one hand, chromosome and nuclear structure (high order structure of chromatin, subnuclear localization of key proteins, and bookmarking of genes by lineage commitment factors in mitosis) are linked to epigenetics. On the other hand, many key epigenetic mechanisms such as histone modification, DNA methylation, and non-coding RNA-mediated gene silencing are often deregulated in a variety of cancers [Heng *et al.*, 2009; Zaidi *et al.*, 2013; Chapter 5]. Importantly, epigenetic regulation stochasticity seems to be a general rule [Feinberg, 2014].

Based on the concept that various epigenetic modifications drive tumor cell heterogeneity with phenotype plasticity and stochasticity, a unifying model of cancer was proposed, where epigenetic dysregulation allows for rapid selection leading to tumor cell survival at the expense of the host [Timp and Feinberg, 2013].

Many big questions remain. In particular, what is the epigenetic evolutionary mechanism for cancer formation? While it is necessary not to ignore non-genetic regulation, one needs to appreciate the proportional contribution of epigenetics to cancer, as there are multiple levels of genetic and non-genetic heterogeneity involved. Since genome level dynamics and heterogeneity are overwhelming in the vast majority of cancers, and epigenetic mechanisms fine tune gene regulation, not define the system, the following questions need to be asked: What does heterogeneity at the epigenetic level offer within the context of global-scale, genome system heterogeneity? The chromatin configuration is now recognized as important for epigenetic regulation. How about drastic karyotype changes that determine new system inheritance? When there is a conflict between different levels of heterogeneity, which type of heterogeneity matters more in evolution?

Addressing these questions can highlight the limitations of epigenetics in cancer evolution. For example, despite that polyclonal epigenetic disruption of cancer progenitor cells might be important for cancer initiation [Feinberg *et al.*, 2006], and stress-induced IDP (epigenetic mediated) can nicely explain one of the major molecular mechanisms contributing to cellular state switching, the evolutionary process requires different levels of system change to impact those beyond epigenetics. This is especially the case for the genome level alterations that occur during the macrocellular phase of cancer evolution [Heng, 2009; Horne *et al*, 2014].

While more sensitive than gene mutation, the reversal of an epigenetic feature defines its role in cancer. If some changes mediated by epigenetics are dominant and are conserved in the evolutionary process, selection pressure will often switch them (or fix them) at the genetic or genomic level. Even though epigenetics are more important for developmental events, cancer evolution is more about pathological events rather than developmental events. In addition, epigenetic alterations should be more diverse than gene mutations, making prediction based on epigenetics more difficult. And finally, both gene mutations and genome alterations can be linked to the initiation factors of cancer, especially under higher stress conditions. Clearly, the two phases of cancer evolution and the association of epigenetic changes to the stepwise phase limit the contribution of epigenetics to cancer. Now the question is, if only a limited portion of epigenetic changes can lead to the cancer, what is the mechanism? An integrated model is urgently needed that reflects the real world contribution of different types of heterogeneity during cancer evolution [Heng *et al.*, 2009; 2013b; Huang, 2013].

3.7.2 *Cancer stem cells*

The cancer stem cell hypothesis has received much attention in the past ten years with the identification of sub-populations of cancer cells with stem cell markers and increased tumorigenicity within most cancer types. The general view of the cancer stem cell differs from the somatic cell evolutionary concept in that only certain progenitor cells in body can become cancer. These cells display self-renewal properties, and thus can have the chance to accumulate various gene mutations if clonal expansion is the pre-condition of cancer.

The cancer stem cell hypothesis is built on two key assumptions. First, stem cells in normal tissue or their immediate progeny are responsible for the origin of cancer, and while cancer stem cells produce cancer stem cells by self-renewal the bulk of their progeny are differentiated, similar to most somatic cells during the developmental process. Second, rare and distinct cancer stem cell subpopulations within cancers are responsible for drug resistance and cancer relapse [Sonnenschein *et al.*, 2014; Wicha *et al.*, 2006].

At the conceptual level, the cancer stem cell model was proposed as a competing model to the stochastic model of somatic cell evolution. Interestingly, current findings so far have failed to dismiss the stochastic model. In contrast, the multiple level of somatic cell stochastic model might just dismiss the stem cancer model (see Chapter 5). To explain some key limitations of the cancer stem cell hypothesis, we need to re-analyze its main assumptions to determine if they fit the facts.

First, the hierarchical organization of cells within normal tissue differs from cancer. When discussing the cancer stem cell, the relationship between normal stem cell and normal differentiated cells is often used to compare the relationship between cancer stem cell and bulk cancer cells. However, these processes are fundamentally different. Normal cell development is a well-regulated process, and the majority of cells share the same or very similar genome. In contrast, cancer progression is a stochastic process, where the key feature is genome variation and replacement.

Second, the generally accepted key assumption of somatic cell stepwise evolution (such as clonal expansion) only represents a small portion of the entire evolutionary process, and common genome alterations can only be detected within some exceptional cases. Self-renewal is not a necessary precondition for the majority of cases of cancer evolution (though it is essential for normal cell populations), as evidenced by the high genomic diversity of most tumors. By the way, the mechanism of cancer stem cell self-renewal has not been fully demonstrated. Karyotyping cell populations with stem cell markers is necessary to illustrate whether self-renewal can be observed at the genome level in cancer stem cells.

Third, a true stem cell and acquiring stem cell-like properties are very different. This distinction must be accounted for, especially when using different markers to identify cancer stem cells. Since the proportion of cells with these specific markers can change quickly in response to altering the environment and even culture conditions, not all cells with stem cell markers are indeed stem cells. This also indicates that a dynamic relationship exists between cells with and without some stem cell markers.

Fourth, there are many additional theoretical paradoxes, conflicting observations, and obvious limitations of the methodologies used to study cancer stem cells, including the use of immune-compromised mice to assess tumorigenicity. In simple linear laboratory models, the differential behavior of cell populations with certain markers might be obvious, but may not be in the non-linear reality of human tumors. If cancer stem cells are allegedly so powerful, why does the confirmation of their function require a very specific model? Considering the necessary host model specificity for cancer cell growth [Fillmore and Kuperwasser, 2007] and the well-known knowledge gap of mouse models (see Chapter 8), it is likely that cancer cells, regardless of stem cell marker expression status, will succumb after application of additional selective factors/constraints (e.g. immune system). As we discuss in Chapter 8, the strength of cancer lies in population heterogeneity and not in individual cells holding intrinsic features.

It is important to point out that some basic features of cancer stem cells have been recently challenged. For example, it was assumed that cancer stem cells are very rare and represent a tiny portion of the cancer cell population, but under specific host conditions, they can be plentiful [Quintana *et al.*, 2008]. Similarly, studies have shown that all colonies derived from randomly chosen single cells in mouse lung and breast cancer cell lines form tumors following allografting in histocompatible mice, suggesting the majority of malignant cells rather than rare cancer stem cells can sustain tumors under certain conditions [Yoo and Hatfield, 2008]. Furthermore, increased studies have shown that cancer stem cells are more sensitive to drug treatment under some typical conditions. Interestingly, based on the cancer stem cell assumption, computational simulation models predict that if selection acts at the cellular level, neoplasms should be primarily comprised of cancer stem cells, which is inconsistent with experimental observations where some neoplasms contain large fractions of non-stem cancer cells. One logical explanation is that the cancer stem cell hypothesis is incorrect [Sprouffske *et al.*, 2013]. Fifth, even if the cancer stem cell concept is correct, we still need to explain how the cancer stem cell leads to cancer. What is the evolutionary process, and how can we integrate the high volume of observed genetic variations? Recently, the cancer stem cell model has been linked to non-genetic heterogeneity [Kreso and Dick, 2014]. It seems that gene mutation and epigenetic regulation can unify the somatic cell evolutionary model of cancer and the cancer stem cell model. However, as we have stated in various places, again, genome level alterations have been ignored.

Finally, recent studies have demonstrated that non-cancer stem cells can be converted into cancer stem cells [Chaffer *et al.*, 2011; Gupta *et al.*, 2011]. Such dynamic relationship makes separating cancer stem cell and non-stem cell conceptually trivial. To solve the increasing paradox, efforts have been made to unify the somatic cell evolutionary model and cancer stem cell model with the microenvironment. It should be noted that cancer stem cell researchers initially pushed the difference between these two models in order to illustrate the importance and uniqueness of the cancer stem cell concept.

Now, by linking the cancer stem cell to mechanisms of non-genetic regulation, the transition between cancer stem cell and non-stem cell makes sense, as non-genetic changes can be reversible under certain conditions. It is now obvious that the multiple level of cancer landscape model is needed to integrate cells with and without stem cell markers [Huang, 2013; Chapter 5].

Even from a philosophical point of view, the dynamics of cancer stem cells and non-stem cancer cells make sense, as it is hard to imagine that some cell type would not follow the laws of evolution and biology.

In contrast, however, some researchers are still determined to push the cancer stem cell concept. Much of the hope of the stem cell theory rests on the idea that the seeds of cancer look, act and react differently than the average cancer cell. However, the seed of cancer is likely to be the genomic heterogeneity that somatic evolution selects upon. The outliers are different from the average, but they are stochastically driven and follow the laws of evolution (Chapters 5 and 6). Increased studies have linked cancer stem cell features to various factors, including different pathways, which all can be explained as increased evolutionary probabilities.

In fact, we and others have asked similar questions about the cancer stem cell concept for years. In our 2007 BioEssays paper, for example, we asked the following questions: Are cancer stem cells truly stem cells or cancer cells that have gained features of stem cells? How can the unlimited self-renewal feature be proven? What is the mechanism that causes stem cells to turn into cancer cells? These questions are still open for debate despite extensive research efforts [Heng, 2007]. Interestingly, we have suggested profiling the karyotypes of cancer stem cells and their karyotypic dynamics during the generation of progeny. Measurement of the heterogeneity of different loci in breast cancer tumors revealed no difference in the amount of heterogeneity in stem cells and non-stem cells, suggesting that cancer stem cells are genetically heterogeneous [Park *et al.*, 2010]. According to genome theory, cells that keep changing their genome are responsible for cancer progression. This hypothesis is directly supported by Park *et al.*'s and our own data. It would be interesting to investigate whether the cells with cancer stem cells markers are actually less or undifferentiated cells with high levels of dynamic, non-clonal chromosome aberrations.

3.7.3 *Infection*

Globally, close to 20% of cancers are related to infectious diseases. Such association rate is impacted by socioeconomic and medical conditions, as Africa is currently 2.5 times higher than the developed world (10%). Even though the virus-cancer link is well established, including human papillomavirus in cervical carcinoma, Epstein-Barr virus in B-cell lymphoproliferative disease and nasopharyngeal carcinoma, Kaposi's sarcoma herpesvirus for Kaposi's Sarcoma and primary effusion lymphomas, and hepatitis B and hepatitis C viruses for hepatocellular carcinoma, bacterial and parasitic infection may also increase the risk of cancer. A well-known example is *Helicobacter pylori*-induced gastric carcinoma.

According to Paul Ewald, an evolutionary biologist from the University of Louisville in Kentucky, the number of links between infection and cancer could be much higher [Ewald, 2009]. He proposed that infection is the major cause of most common and complex diseases including cancer. He estimated that by the year 2050, more than 80% of all human cancers would be linked to infection. Since this number was considered to be zero in 1975, he even thinks that, the number could reach as high as 95% [Grant, 2009]!

One reason Ewald focused on infections causing cancer is his disagreement with the cancer somatic gene mutation theory where genes dictate cancer susceptibility. This disagreement is based on his viewpoint that any disease-causing gene that reduces survival and reproduction would normally eliminate itself over a number of generations, and thus persistence of many common diseases requires contributing factors beyond genes.

Infection is an obvious link for cancer. First, infection is common and many diseases are caused by infection. Second, it is more straightforward to link specific pathogens to certain cancers than other diverse environmental factors (the molecular identification of a specific pathogen(s) in a bio-sample is more reliable). Third, there are clear links between viral infection and human cancer, and some of the molecular mechanisms are well established. For example, human papillomavirus (HPV) types 16 and 18 appear to play a critical role in anogenital malignancies. Mechanistically, E6, a HPV16 oncoprotein, targets p53 causing its degradation. Since p53 is a key tumor suppressor, the virus cancer connection seems obvious.

If only the above information can explain cancer, it would be reasonable to search for and to identify infections as the most important cause of cancers. However, the majority of cancer cases cannot be explained by infection, and pathogens that are considered cancer causing do not cause cancer in the majority of infected patients. For instance HPV infection occurs commonly within the population, but the majority of women do not get cervical cancer. This also fits the molecular data where at the very early stages, the virus is always detectable, but the cancer-associated mutations are detected at much lower frequencies (approximately 10%). Similarly hepatitis B infection rates are very high in Asia, but the majority of infected individuals will never get liver cancer (20–30% of adults who are chronically infected will develop cirrhosis and/or liver cancer). CMV infection is linked to lymphoid cancers, but CMV cannot be identified in a majority of these cases. Yes, *Helicobacter pylori* can induce gastric carcinoma. But individuals infected with *H. pylori* have a 1 to 2% risk of acquiring stomach cancer. Just like other types of system stress, infection contributes to cancer, but with limited certainty for diagnosis and prediction.

Different ideas have been suggested to explain the relationship between pathogens and cancer: (1) the pathogen is necessary, but not sufficient to cause cancer. In these cases other conditions like host genetic background, collaboration among different pathogens, or toxin exposure are necessary co-contributing factors; (2) some pathogens are favored by the presence of cancer. These pathogens do not cause cancer but are more prevalent in some cancer patients, with compromised tissue or/and immune system constraint; (3) the pathogen is a direct and key cause of the cancer, but the causality has yet to be understood.

If infectious agents are believed to be the root cause of cancer, we also need to explain the many cancer cases without pathogen involvement. In other words, what about the cases where infection is sufficient but not necessary? Like the gene theory of cancer, if infections are used as the explanation for all cancers, very soon the explanations will become complicated and individual explanations are likely to conflict with each other.

To shine new light on this issue, we need to apply a new strategy and study the host system's response. Since cancer represents a much more complex system than infectious diseases, we might need to not just focus on the linear causation between pathogen and disease phenotypes, but to explain the potential contribution of the pathogen in the process of cancer evolution. Within this new evolutionary framework, the pathogen-cancer connection could be explained as infection-mediated stress, which causes a stress response in the host. If the stress is high enough to require genome change in order to respond, the seeds of cancer evolution would be planted. This strategy would unify many other types of non-pathogen mediated stresses that contribute to cancer evolution [Heng *et al.*, 2013b; Horne *et al.*, 2014; Horne and Heng, 2014; Chapter 5]. In other words, the link between virus infection and cancer might not and should not be directly linked to a specific gene or pathway, but to a general system response. Thus, predicting the cancer potential of many infectious agents with low penetration and a less specific system response is less straightforward for predicting the potential for cancer evolution. Such estimation becomes more difficult when other non-infectious factors are also involved.

We have demonstrated that the HPV E6 protein, in addition to SV40 infection, leads to elevated stochastic chromosomal aberrations, leading to various molecular changes. Viral infection can lead to many diverse system responses as well. In addition to causing inflammation, they also can hit non-specific targets. For example, it was recently suggested that HPV infection can trigger an anti-viral attack (the viral genes), but at the same time, can also hit the cell's own DNA. Such friendly firing can promote cancer [Henderson *et al.*, 2014]. Of course, it also can stochastically trigger a large number of other molecular pathways illustrated by large scale gene expression studies.

The above analyses further enforce the challenge of cancer research based on individual molecular mechanisms. Single causative factors do not explain the vast majority of cancers. In fact, there are many contributing factors, and most of them are likely under the limits of detection. Just because exposures of certain chemicals or physical conditions are hard to monitor does not mean they are less important than infections. It is most likely that many different factors have a much higher association with cancer, but due to the complexity of cancer evolution, the value of high frequencies of association is reduced and each of these factors become less significant within heterogeneous patient populations.

Another interesting point is that overall cancer incidence is dynamically linked to many factors, and the dominant contributing factors can change due to environments and lifestyle. For example, although developed countries have lower rates of cancer-linked infection, the obesity-linked cancer rate is higher. If cancer represents the products of evolutionary trade-off, we perhaps will face this problem forever. Relatively speaking, more dominant factors may have changed over the course of history. Nevertheless, current efforts should be made to reduce the infection and cancer.

One last interesting point is that most infections can effectively trigger genome instability by inducing chromosome fragmentation (C-Frag) [Stevens *et al.*, 2007; 2010; 2011; Chapter 5]. Infection-associated inflammation can also induce C-Frag. In fact, infection-led inflammation plays an important role in cancer.

Together, infection represents an important contributing factor (but not the only key factor) in the complexity of cancer evolution. Increased types of cancer can be associated with infection, but infected individuals each have their own evolutionary process/stories.

3.7.4 *The Warburg effect and the mitochondria's contribution to cancer*

In recent years, the metabolic link to cancer has become a new cancer hallmark. Following increased understanding of the human mitochondria, it was hoped that tumors "addicted" to aerobic glycolysis represented the new Achilles' heel of cancer.

Otto Warburg initially observed in the 1920s that in contrast to normal cells, which mainly rely on mitochondrial oxidative phosphorylation to generate the energy required for cellular processes [Warburg, 1956], most cancer cells rely on aerobic glycolysis, a phenomenon termed "the Warburg effect." Though a high rate of glucose uptake by solid tumors is the basis for commonly used prognostic and diagnostic imaging using radio-labeled glucose analogs, and increasing numbers of studies illustrate many molecular details of the phenomenon, the mechanism of Warburg effect is not fully understood.

Given its importance in many vital cellular processes, such as the regulation of energy production, modulation of oxidation-reduction status, production of reactive oxygen species, control of cytosolic calcium levels, and involvement of apoptosis (Wallace, 2012; 2013), it is no surprise that mitochondria are involved directly or indirectly in essentially everything a cell does, as most biological processes can be linked to these vital functions of mitochondria. That is why mitochondria can be closely linked to cancer in a highly diverse way (see the evolutionary mechanism of cancer, in Chapter 5).

As the result of searching over two decades for molecular mechanisms that contribute to cancer, many mitochondria-related gene mutations have been identified and linked to cancer. Both somatic and germline mtDNA mutations have been identified for diverse types of cancers, including mutations in nuclear DNA-encoded mitochondrial enzymes. Factors in mitochondrial ROS production, and alterations in bioenergetics metabolism have all been linked to cancer [Vander Heiden *et al.*, 2009; Wallace, 2012].

Just like any other subfield of molecular cancer research, increased knowledge has made the picture more complicated, and conflicting information and results are overwhelming. It is now realized that the molecular linkage between parts of the mitochondria or metabolic process and cancer are not linear but multifaceted, highly heterogeneous, and highly dynamic [Dang, 2013]. The heterogeneity of metabolic profiles matches well with other levels of cancer heterogeneity. The same gene cannot only function differently in tumors (e.g. FOXO and MYC can either increase or decrease mitochondrial biogenesis, and p53 can enhance or inhibit oxidative phosphorylation in cancer) [Wallace, 2012], different cancer cell types display different bioenergetics variation and can become more or less glycolytic or oxidative depending on the stage of cancer progression and conditions. Furthermore, different subclones within a cancer can collaborate with each other despite different metabolic types, similar to cancer cells collaborating with normal cells. Cancer cells can constantly reprogram themselves to adapt to environments and alteration of growth conditions. For example, just to achieve necessary total ATP numbers, cells can change the ratio between glycolysis and oxidative phosphorylation, the ratio between catalysis of glucose and glutamine, or the ratio of catalysis between glucose/glutamine and fatty acids [Zheng, 2012]. This is in addition to the multiple metabolic phenotypes that are under evolutionary selection.

Due to such complexity, it might be less useful to continue illustrating each molecular connection, as there are so many, and more importantly, their interaction relationship is so dynamic and hard to predict. Even the Warburg effect itself is no longer that black and white. First, Warburg's interpretation of mitochondrial dysfunction is incorrect. In contrast to Warburg's hypothesis that cancer cells no longer rely on mitochondria, mitochondrial function is essential for cancer cell viability. Second, tumors do not consistently inhibit mitochondrial bioenergetics as Warburg suggested [Wallace, 2012]. Even if 100% of cancer cells use the same metabolic type, and mitochondrial dysfunction was the only causative factor, under evolutionary selection pressure new cell populations with different metabolic types would likely emerge. Evolution always finds its way. Thus, solely relying on targeting the Warburg effect to cure cancer represents yet another form of wishful thinking, just like previous ideas of targeting DNA replication, cell cycle, micro-tubulin, blood supply, etc.

Now, it is clear to us the real reason why many cancer cells use different metabolic types. It is simply the direct result of evolutionary selection that favors cancer just for being different compared to normal cells! As for somatic cell evolution, the generation of variation is the first pre-condition [Crespi and Summers, 2005; Heng, 2007; Merlo *et al.*, 2006]. Many of the other reasons could be secondary. Meanwhile, the heterogeneity of cancer also preserves the capability of switching back to previous states quickly. The real enemy that we are dealing with is highly complex and ever changing. Such explanation can solve many puzzles. For example, cancer cells use glycolytic metabolism prior to exposure to hypoxic conditions. Despite residing within the bloodstream with higher oxygen concentrations, leukemic cells are highly glycolytic. Likewise, lung tumors exhibit aerobic glycolysis even though they are in an oxygen rich environment. The reason is evolutionary selection favors the different phenotypes of the cancer among many potential features. Just to be different from normal cells is the main selection goal for cancer, despite the high cost such as less effective metabolic types. Based on the heterogeneous nature of cancer, we believe that even in hypoxic conditions, a proportion of cancer cells will retain the normal metabolic type.

With this understanding, the immediate task is to illustrate how mitochondrial alterations, as a subsystem of the cell, impact overall cancer evolution. To study this perspective, research focus needs to be on the system dynamics of the individual cell level and population behavior in the context of evolution, rather than one of the large numbers of the individual molecular linkages. The following questions/perspectives are some examples of how to directly address this issue:

(1) Future research should focus more on the package of evolutionary selection contributed by metabolic alterations, rather than on a specific feature of this package, such as proliferation. The evolvability of cancer involves many different phenotypes in a dynamic fashion. Although proliferation is a very important phenotype of cancer, it might not be a key phenotype in earlier stages. During the entire evolutionary process, the most important features are the formation and survival of new cell populations, as well as their ability for further evolution. The power of proliferation is often associated at the late stage.

(2) How does the energy level or metabolic type impact on the stability of both mitochondrial genome and nuclear genome? Does mitochondrial genome instability represent a common linkage similar to many molecular mechanisms and the nuclear genome? Our data and those of others have linked mitochondrial function to overall system stress and to elevated stochastic nuclear genome alterations.

(3) In the nuclear genome under high stress conditions, rapid and massive genome re-organization can occur to form chaotic genomes [Heng *et al.*, 2013a–b; Chapter 5]. Can similar phenomena be observed in the mitochondrial genome? Do high-level dynamics of the mitochondrial genome (fission and fusion) result in the emergence of new features without specific gene involvement like within nuclear genomes? How does mitochondrial heterogeneity and quantitative features of the cell population impact the emergent properties of diseases? Are mitochondrial genome dynamics also functionally important for short-term adaptation but also come with a cost later on?

(4) A key piece of information is that metabolic type variation, just like other key phenotypes of the cancer evolutionary selective package, is highly dynamic. Under evolutionary selection, many factors can be involved. The promise of targeting metabolics for curing cancer should not be overstated.

3.7.5 *Mutator phenotype*

In 1974, Larry Loeb *et al.* proposed that cancers exhibit a mutator phenotype. According to this hypothesis, the mutation rate of nonmalignant cells is insufficient to generate the large numbers of mutations that is observed in human cancers, and therefore a greatly increased mutation rate is required for cancer formation. Loeb proposed that the loss of function of one gene, such as a DNA repair gene, accelerates the mutation rate at other loci of the genome, increasing the likelihood of a tumor acquiring advantageous mutations.

Studies of DNA synthesis fidelity in normal cells and the possible cellular consequences resulting from mistakes in copying DNA led Loeb's group to suggest a unified theory that various carcinogenic agents affect DNA by causing errors in base pairing during DNA replication. Some of these errors would cause alterations in structural genes that code for enzymes that function in base selection, leading to an expanding number of errors in many other genes. Based on theoretical and experimental data analyses, they proposed that the infidelity of DNA replication might be responsible for tumorigenesis. Specifically, they suggested that DNA polymerases play an important role in base selection. "If mutations produce defective polymerases, the polymerases in turn will produce further genetic defects," they argued [Loeb *et al.*, 1974]. Mathematical models also indicate that mutator pathways are the most efficient paths to cancer and are most effective when they are expressed early in tumor progression. It was suggested for most cancers that require three or more driver mutations, a mutator phenotype may be inevitable (if a given cancer type only requires one or two mutations, such as inherited retinoblastoma, a mutator phenotype may not be necessary). During the 1990s, the linkage of colon cancer to mutations in mismatch repair (MMR) genes seemed to support the mutator phenotype hypothesis. The loss of function of MMR genes, such as MSH2 or MLH1, leads to changes throughout the genome including the coding region of some tumor suppressor genes, and more importantly, changes within repetitive, noncoding, microsatellite sequences [or microsatellite instability (MIN)]. MIN results in varied lengths of microsatellite "hotspots", which can cause frameshift mutations or affect spacing within DNA regulatory elements. The caveat is, however, only a small portion of tumors displayed MIN. For example MIN is linked to hereditary nonpolyposis colon cancer, but the majority of colon cancers are linked to chromosomal instability (CIN). Interestingly, despite a high frequency of mutations occurring within DNA repair genes in many cancer cell lines (over 50%), only a limited number of DNA repair mutations are detected in patient samples [Bodmer *et al.*, 2008]. Similarly, gene mutations can lead to aneuploidy or other forms of instability, which have been incorporated into the mutator type hypothesis, but once again gene mutations that are directly associated with aneuploidy and CIN are rare in human tumors. Furthermore, even if an individual has a DNA repair gene mutation, there is no certainty that the affected individual will get cancer.

A method to directly identify mutations revealed elevated frequencies of random nucleotide point mutations in non-repetitive sequences in cancer, and this phenomenon is referred to as point mutation instability (PIN). Identification of PIN followed the identification of MIN and CIN and is the best direct evidence supporting the mutator type hypothesis. Using the mutation detection method, the frequency of random point mutations in various normal tissues was found to be very low, $<10^{-8}$ per base pair. In contrast, on average, tumors contain 2.2×10^{-6} mutations per base pair, and an individual cancer cells contain ≈1,000–3,000 random mutations [Bartos *et al.*, 2007]. More recent whole genome sequencing of tumors revealed a wide range of mutation rates across tumor types, ranging from ~0.01 to ~1000 mutations per megabase [Lawrence *et al.*, 2013].

Recent studies in mice have shown that if DNA polymerase genes Pol δ or Pol ε are mutated and rendered error-prone, tumors occur in various tissues [Albertson *et al.*, 2009], seemingly providing further support for the mutator phenotype concept. However, similar to many other reports that link specific genes to cancer, while the linkage is there, the penetration of PIN in the cancer patient population is low.

Ironically, perhaps the strongest support for this hypothesis comes from cancer genome sequencing data. Cancer genome sequencing was initially designed to once and for all identify a limited number of specific driver mutations. In contrast to this original goal, cancer genome sequencing has identified an overwhelming number of diverse gene mutations (tens to hundreds of thousands of gene mutations) in each tumor type, and most of these mutations stochastically occur, making the genetic landscape of each tumor unique (see Chapter 4).The mutator phenotype hypothesis has been broadened from the original focus of polymerase mutations to include genes involved in microsatellite instability, point mutation instability, chromosomal instability, checkpoint instability, and epigenome maintenance [Loeb *et al.*, 2008]. Together, the mutator phenotype might explain different types of system instability (from point mutation instability to chromosomal instability) and their role in carcinogenesis. However, many important challenges remain.

First, the mutator hypothesis still relies on the framework of the gene mutation theory of cancer, as it considers cancer to result from progressive accumulation of large numbers of gene mutations during tumor progression. The mutator hypothesis does however differ from the current dominant paradigm that only a handful of driver mutations are important while a large number of mutations are just passengers. Fundamentally, this hypothesis focuses on point mutations, as these are easily measured and the vast majority of point mutations are neutral. The latter reason fits better within the Darwinian perspective that any changes must be small, and benefits of these mutations accumulate in a stepwise fashion. Though the mutator phenotype has been observed, its general mechanism is unknown. It is not known if it is caused by one gene, a combinatorial function of multiple genes or something else.

Second, the mutator hypothesis does not address the relative importance of mutations compared to that of genome alteration. For example, what is the biological meaning of a high rate of gene mutations that occurs in the background of a high amount of genome alterations? Cancer genome sequencing reveals that genomic alterations occur as or more commonly than common gene mutations. What then is the relationship between gene level mutation and overall genome level stability (not mainly caused by gene mutation rate)? Can the gene level mutator type result from genome instability? On the other hand, how does the mutator phenotype contribute to genome level changes, which are key for macrocellular cancer evolution? Without considering genome level changes, the implications of the mutator phenotype could be limited. For example, it is possible to target the genes that are responsible for the mutation rate. This could be useful for managing the growth of the clonal expansion phase, but more important is the stability at the genome level. Genome replacement represents the major force to drive macrocellular evolution, which overpowers genes for many challenging issues such as drug resistance.

Third, when does the mutator phenotype occur during cancer progression: earlier or later stages? In contrast to Loeb's view, the PIN phenotype has been suggested to occur relatively late in cancer progression, based on the observation that cancer genes can be selected over short periods of adaptation in *in vitro* culture and do not take the proposed decades of natural progression [Klein, 2006]. From a treatment perspective, does it matter when the mutator phenotype occurs? Also, does the mutator phenotype occur in cycles like the two phases of genome level evolution (see Chapter 5)?

Fourth, how does cellular growth rate heterogeneity (cancer cell growth rates are highly heterogeneous) [Abdallah *et al.*, 2013] impact the mutator phenotype? As cells with outlier growth rates (and other behaviors) often dictate population behavior, some subpopulations can quickly accumulate gene mutations by dividing rapidly without largely changing the size of the entire population (Chapter 6). Genome chaos, or highly complex genome rearrangement, (see Chapter 5) could also affect the mutation rate, as gene and genome level instabilities peak at similar times (unpublished observation). Even though much of the >200-fold mutation frequency increase in cancer cells compared to normal cells reflects inherent genetic instability [Bielas *et al.*, 2006], it is also possible that the stability is not reflective of the population. This is because a small portion of the cells could have drastically accumulated gene mutations as a result of rapid, localized division. It is known that even short term *in vitro* culture can increase the mutation rate, for example [Klein, 2006]. During the later stage of clonal expansion, individual cells from some dominant clones could have less mutation rate variation. But during the punctuated phase of cancer evolution, such variation could be very high.

Despite major differences between the mutator phenotype hypothesis and the genome theory, including that they mainly focus on different levels of genetic organization and have distinctive evolutionary mechanisms, these two concepts actually share many common features. They both focus on stochastic genetic changes instead of stepwise changes, they both emphasize the importance of instability and neither is supportive of further cancer genome sequencing in hopes of identifying a cancer specific pattern of targetable genetic alterations. As we will discuss in Chapter 4, the issue to further sequence or not sequence directly reflects the different viewpoints of cancer evolution and inheritance of the gene and genome theory.

3.7.6 *Heterogeneity and tumor society*

As a key feature of cancer, genetic and non-genetic heterogeneity (including epigenetic heterogeneity) are essential elements in cancer initiation, progression, metastasis and drug resistance. These multiple levels of heterogeneity provide population diversity, complexity, and robustness necessary for the evolutionary process of cancer. An in-depth discussion of the ultimate importance of heterogeneity is presented in

Chapters 5 and 6. Below is a synopsis of the history of the cancer heterogeneity field.

Studies of cancer heterogeneity began prior to the era of gene-based molecular cancer research. Early efforts focused on morphological and pathological heterogeneity [Foulds, 1954; 1975]. Beginning in the 1970s, multiple levels of heterogeneity including tumor histology, cellular morphology, karyotypic and other cytogenetic markers, growth rate, cell products, receptors, enzymes, immunological features, metastatic ability, and sensitivity to various therapeutic agents were described by multiple groups [Dexter *et al.*, 1978; Fidler and Hart, 1981; Heppner 1984; Heppner and Miller, 1998; Miller *et al.*, 1981; Woodruff, 1983; Welch *et al.*, 1984; 1989]. Unfortunately, much of this important knowledge has been ignored by molecular geneticists in favor identifying common patterns of gene mutations. Within the molecular paradigm of cancer it was necessary to eliminate heterogeneity, as it was considered "noise" that obfuscated molecular patterns. The various molecular methodologies developed during this time (including blotting, cloning, sequencing, etc.) and after compounded the problem by artificially averaging molecular profiles of studied populations. This process effectively hides heterogeneity, giving the false impression that average profiles accurately reflect the reality of cancer where the heterogeneity is the key.

Despite the effort to eliminate heterogeneity for the sake of simplifying data analysis, high levels of heterogeneity are overwhelmingly detected in most cancer cases. **Table 3-1** summarizes examples of some of the key features of the multiple levels of heterogeneity that exist in cancer cells. With the introduction of various high resolution –omics methods to characterize cancer, the types of genomic heterogeneity are growing rapidly. Furthermore, each individual type of heterogeneity can interact with each other to form an almost unlimited combination of heterogeneity (see Chapter 6). Interestingly, among different types, genome level heterogeneity is ultimately the most important for cancer, as it defines the macrocellular evolution, unifies the lower level of heterogeneity, and provides the overall basis of genetic heterogeneity to higher levels (for example, cellular network level and tissue level).

Table 3-1. Multiple levels of genetic, epigenetic, cellular, and environmental heterogeneity

Multiple genetic levels:

Gene/nucleotide level:

Nucleotide polymorphism

Various types of repeats (e.g., microsatellite shifts)

Spectrum of mutations (including conditional mutations)

Heterozygosity (allelic)

Splicing forms

Gene family members (paralogs)

Combinational effects of multiple genes and mutations

Genome/chromosome and sub-chromosome level:

Copy number variation, microdeletion/inversion

Loss of heterozygosity (LOH)

Chromosomal translocation/inversion/duplication

Defective mitotic figures

Chromosome fragmentation

Aneuploidy

Polyploidy

Combinational effects

Multiple epigenetic levels:

Chromatin folding and attachment to the nuclear matrix

Packaging of nucleosomes

Position of histone variants

Covalent modification of histone tails

DNA methylation

Non-coding RNAs

Cellular population level:

Growth/survival rate of individual cell

Subline interaction (Collaboration and competition)

Outliers (frequency and robustness)

Change of system status independent of epigenetic alteration

Environmental influence on the multiple levels of homeostasis

Tissue specificity

Physiological condition alteration (aging, immune, hormone, and metabolic levels)
Nutrition status
Different types of exposure stress
Variety of dosage, duration of the exposure
Differential impact on specific organs
Certain levels of stochastic response

Table 3-1 lists different types of heterogeneity. This explains why decades of efforts aimed to reduce "noise" (or heterogeneity) have largely failed, as there are so many of them (each aspect can be explained by a large number of genes and pathways), and they represent key features that define cancer (modified from Heng *et al.*, 2009).

Recently, there is a renewed interest in investigating cancer heterogeneity likely due to the following new findings: (1) Even cutting-edge molecular techniques like next generation sequencing have difficulty identifying common patterns at various levels of genetic organization [Abdallah *et al.*, 2014; Heng, 2007a; Wood *et al.*, 2007]. (2) Increasing clinical sample sizes has paradoxically revealed much higher amounts of heterogeneity than smaller scale studies. (3) Single cell analyses have shown that individual cells from the same tumor have vastly different profiles [Aubele *et al.*, 1999; Bartos *et al.*, 2007; Fujii *et al.,* 1996; Klein *et al.*, 2002]. Drastic changes are also observed between primary and metastatic cancer from the same patient. (4) The debate between the cancer stem cell model and the clonal evolution model has caused scientists to rethink the issue of heterogeneity [Campbell and Polyak, 2007]. (5) Epigenetic heterogeneity has been shown to be an important regulatory mechanism in cancer [Heng *et al.*, 2009] and the behaviors of nuclear topology, chromatin domains and non-coding RNA have gained increased attention. (6) Non-genetic regulators like chromatin domain change, non-coding RNAs, etc. display much higher degrees of heterogeneity than at the gene mutation level. As a result, many stress responses can be linked to non-genetic heterogeneity including epigenetic heterogeneity, and the predictability is very low.

Our group has pushed for a general embracement of the importance of genome level heterogeneity in cancer. We have tried to understand and identify the mechanism(s) of heterogeneity, the relationships of different types of heterogeneity, generation and maintenance of heterogeneity, and the dynamics of heterogeneity in the context of cancer evolution. The realization that genome alteration creates a new bio-system ushered in the departure from the gene-centric theory of cancer to the genome-centric evolutionary theory of cancer. Over a decade of study and synthesis resulted in the identification of genome instability as the common driver for the majority of cancers, and also shown that cancer evolution is most powerful during the macrocellular evolution phase. Furthermore, heterogeneity represents a new layer of complexity which is essential for cancer cell survival. This important new framework of cancer unifies the large number of diverse molecular mechanisms of cancer [Heng e*t al.*, 2013b; Horne *et al.*, 2014; for more details see Chapters 5 and 6].

It should be pointed out that while the importance of heterogeneity has now received increased appreciation following cancer genome sequencing projccts, most cancer biologists continue to focus on gene, epigene and network levels but ignore the genome level. For example, though some recent studies have measured genome level heterogeneity including aneuploidy and polyploidy, the explanations offered in these publications are largely based on gene dosage or interruption of specific cancer genes, without addressing genome based inheritance. Although these heterogeneity studies have primarily focused on gene level alterations, simple, single loci-based measures of intra-tumor heterogeneity are able to differentiate grades of breast tumors and are very promising [Almendro *et al.*, 2014; Marusyk *et al.*, 2012; Park *et al.*, 2010]. A major hurdle to using measures of heterogeneity to grade tumors and/or predict outcomes is that the Shannon and Simpson indexes (most common current methods of measuring heterogeneity) are based on measuring species diversity and are only capable of measuring heterogeneity of one specific locus or trait. In their current forms, these indexes cannot incorporate the genomic complexity of most tumors. A measure that can incorporate these complex data is urgently needed.

Importantly, only studying the genetic and non-genetic heterogeneity of individual cells is not enough. Cancer is a tissue/organ disease, and its progression occurs as a result of cellular ecosystem dynamics at the cell population level. As far back as the early 1980s, Gloria Heppner and others suggested that, similar to other mixed populations, the behavior of tumors may not be governed simply by the behavior of the most deviant members or by the average of all cells from an entire tumor. Instead, tumors should be considered *societies,* or ecosystems in which the various members (clones) interact to produce a group dynamic or emergent features that define the overall behavior [Fidler and Hart, 1981; Heppner, 1984; Heppner, 1989; Woodruff, 1983].

Interactions of various subpopulations have been constructed and studied both *in vitro* and *in vivo* to understand the behavior of the tumor society. Regardless of the diverse results, these experiments have forcefully demonstrated the importance of subpopulation interaction for cancer evolution and downplayed the importance of specific molecular mechanisms as explanations or predictors of tumor society behavior. These results demonstrate that most molecular mechanisms do not explain the highly heterogeneous emergent properties of the tumor society. As pointed out by Heppner, the mechanism of interaction of tumor cells seems to be highly dependent on the characteristics of the subpopulations in the mixture. Given the fact that most tumors are so heterogeneous that nearly every cell is a unique subunit, it is highly challenging to identify common molecular mechanisms responsible for the complexity of the tumor society. For example, prediction of a tumor's response to chemotherapy is not possible on the sole basis of knowledge about the sensitivity of its individual subpopulations [Heppner and Miller, 1989].

Heppner has insightfully stated:

"Although it may be that we have not yet done the proper experiment to identify a mechanism common to all interactions, perhaps we have missed the point because of the reductionist impulse to search for a mechanism, which usually implies a 'molecular' mechanism. Members of a society will interact; the mechanism is 'any way they can'."

Unfortunately, very few have understood her important message. When faced with the overwhelming heterogeneity in cancer, a common response is that proper experiments to overcome this heterogeneity have not yet been devised, and we must continue to dig to figure out these magical methods.

We later introduce the evolutionary mechanism of cancer concept in this book [Ye *et al.*, 2009a; Chapter 5], which shares Heppner's viewpoint and explains the relationship between the evolutionary process and individual molecular mechanisms of cancer. Such relationship is further discussed using the multiple level landscape cancer model [Heng *et al.*, 2013, See Chapter 5].

Recently, more advanced evolutionary experiments with various subpopulations of cells have been carried out to study population dynamics. To study the pattern of population interaction, cells labeled with different colors were used to trace different cell populations and produce quantitative data of evolutionary interactions [Merlo *et al.*, 2011]. A separate study that used gene mutations as clonal markers in mouse models of breast cancer showed that interclonal cooperation was essential for tumor maintenance [Cleary *et al.*, 2013], confirming the general conclusion that Heppner et al reached decades ago.

Further studies of tumor cell societies are needed in the light of cancer evolution. This issue has now been raised by some evolutionary biologists. Crespi *et al.* pointed out:

> *"Evolutionary and ecological approaches to understanding and fighting cancer have burgeoned over the last ten years. However, despite empirically based studies by Gloria Heppner in the 1980s and 1990s that described 'tumour cell societies' as key to cancer progression, none of this work has proceeded from basic, established inclusive-fitness principles for understanding and analysing interactions between more or less related entities..."*[Crespi *et al.*, 2014].

Clearly, it is the time to finish the job Heppner and others initiated.

Chapter 4

The Surprise Cancer Genome Landscape Revealed by Cancer Genome Sequencing Projects

4.1 Summary

Despite the rapid accumulation of challenges to the current gene mutation theory of cancer as illustrated in previous chapters, the vast majority of researchers firmly believe in the key causative role of common gene mutations. The biggest hope both for rescuing the gene mutation theory of cancer and translating the success of basic research into the clinic is the cancer genome sequencing project. Here we are. Following a decades' worth of sequencing efforts, we have massive genetic data coupled with surprise after surprise. Even though a large number of high profile publications have declared success by illustrating the power of sequencing, and sequencing everything has been pushed into the clinic in the name of precision medicine, the main goal of identifying commonly shared gene mutations has failed. Furthermore, newly discovered unmanageable complexity of cancer actually calls for a new conceptual framework. Yes, a genetic profile snapshot can be sequenced, but the highly dynamic landscape makes it less predictable. We do not even know the meaning of many gene mutations within the altered genome! To reveal many untold stories hidden in the ocean of data, the rationale, key findings and the clinical implications of the cancer genome sequencing project are discussed. In contrast to the high enthusiasm in the general public, we pay more attention to the new challenges ahead. Nevertheless, the sequencing efforts might finally end the practice of only focusing on gene mutations to understand and fight cancer.

4.2 Ten Years of Cancer Genome Sequencing: More Questions than Answers

Cancer genome sequencing (CGS) projects aimed at sequencing the entire genome or exomes of various types of cancer have now been underway for nearly a decade. Publication of new CGS reports occurs nearly every month in high profile journals. Media excitement inevitably follows every major publication and many of these news reports contained bold statements from experts stating that cancer will soon be a thing of the past. Decreasing genome sequencing costs and increased technical know-how have suddenly brought the idea of sequencing every cancer patient's tumor closer to reality and a number of sequencing based cancer diagnostic companies have recently started. Personal CGS seems sensible based on promising CGS results and the idea of personalized medicine, but is it?

While the deluge of data produced has led to excitement especially in current data driven research culture where more is always better, it has also led to new challenges including correctly interpreting large data sets that cause statistical issues and provide unavoidable opportunities for researchers to (often unconsciously) cherry pick genes of interest for one reason or another. Furthermore, there is increased confusion in the field as the prevalence of many key gene mutations is much lower, the mutational spectrum is more diverse, tumor heterogeneity is unbelievably higher, and stochastic genome reorganization is more common than originally expected. CGS findings in fact challenge the foundation of the somatic gene mutation theory of cancer (see Chapter 2), which CGS is supposed to validate. Few acknowledge or even see this big elephant in the room, unfortunately.

In light of this outcome it is time to ask and answer some poignant questions, which can broadly be split into questions affecting basic research, clinical research/outcomes, technological/methodological development, and overall policy. Questions related to basic research include: What is the impact of these unexpected CGS results on the current conceptual framework of cancer research? Has CGS validated the gene mutation theory of cancer as originally designed or does it paradoxically challenge the very foundation that many current strategies including CGS are based on? Do these key predictions of cancer gene theory including the stepwise somatic mutation paradigm hold up in light of newly available CGS data, and if not, is it time for a new conceptual framework? More specifically, does CGS focus on the correct level of genomic organization? What if the key driver of cancer is not gene mutations but genome level changes (such as the recently reported genome chaos)? With so many genes found to be mutated or differentially expressed in cancer, which ones should we prioritize? Newly identified rare mutations or the long known cancer genes? Many independent labs focus on the function of one gene or gene family, but the vast majority of these mutations are only implicated in a very small subset of cancer cases; should this research continue? How can the quantitative contributions from different gene mutations and genome alterations be calculated?

Questions related to clinical cancer practice include: Can CGS predict the unpredictability of the majority of cancers? Regardless of their price, are these highly diverse genomic profiles detected by CGS useful for diagnosis and treatment, especially if they consider the average genomic makeup of a tumor? If tumors are sequenced how do we make clinically relevant sense of the multiple levels of genetic alteration that occur? How do we decide between treatment options based on the sequence profile? In regards to the next generation sequencing (NGS)-based clinical tests that are now available, how can sensitivity and specificity problems be solved? For example many cancer genes are mutated in less than 1% of samples, while sensitivity and specificity of many NGS-based diagnostics is 99%. Unfortunately with these statistics if a gene is mutated at 1% frequency (or less), half of the mutations detected would be false positives. This results in making diagnosis a 50-50 guess as to whether that mutation truly exists in the profiled tumor.

From a technological/methodological perspective, questions include: Can we deal with the issue of tumor heterogeneity? Is single cell sequencing the answer, and if it is, how many single cells must be sequenced for each tumor? If single cell sequencing is used, should we focus on the results of each individual cell, or is the overall heterogeneity of the population more informative? How do we use a primary tumor profile to predict the metastasis and potential for drug resistance? Can we handle the large amounts of data from these analyses? Are more data always better? How can data from different levels of genetic organizations be incorporated into one analytical and predictive framework? Can the complexity of cancer profiles be transformed into approachable simplicity, and is simplicity reached through new cancer-based theories, or complicated algorithms? Finally, now that hundreds of the most common types of tumors have been sequenced, how many additional samples are needed?

From an overall policy aspect: With the deluge of sequence data, do we understand more now about cancer than we did prior to the sequencing era? There is little doubt that CGS has provided more information about cancer, but is it a successful or disappointing story based on the current framework of cancer? Ultimately, is this academically interesting project clinically useful as most tumor types have few genes mutated in above 10% of them and most targeting agents only work for typically months to about a year? Should funding priority be increased for characterizing gene mutations of high incidence within the patient population? How can technology development be balanced with theoretical synthesis to make sense of available data? How should we develop the necessary infrastructure and incentives to critically examine generally accepted theories, especially when they are clearly not working in clinical situations?

To answer these questions and more, we must review the current state of cancer genome sequencing including the rationales behind such efforts, the initial plans, the goals, and the current status of these efforts.

4.2.1 *Initial plans and goals of the various CGS groups*

The successful completion of sequencing the entire human genome during the Human Genome Project (HGP) generated new excitement in and about cancer research. Following the official announcement of the completion of the first draft of the HGP in 2000, the Clinton administration asserted that cancer and other common diseases would soon become history. The HGP cost billions and officially ran from its inception in 1987 to the finalization of the sequence of chromosome 1, the last to be sequenced, in 2006. Along the way an initial analysis was published in February 2001, and a "finished sequence" was then reported in 2003 [International Human Genome Sequencing Consortium., 2001; 2004; Venter *et al.*, 2001]. During the HGP, large sequencing centers capable of taking on the challenge of CGS were built. Since the HGP, sequencing had become more common place, with the costs to sequence entire genomes continually falling due to the development of new technologies. These advances set the stage for large scale efforts to sequencing the genomes of various cancers. With the technology and the infrastructure in place, CGS was the logical choice to scale up cancer research.

Conceptually, CGS projects were based on the central paradigm of cancer research which states that cancer is a gene disease resulting from a stepwise accumulation of gene mutations. Each tumor type is driven by a unique set of gene mutations differentiating tumor cells from normal cells. Identification of these driving mutations would lead to the understanding of the mechanism of cancer leading to new diagnostics and exploitable therapeutic targets. This line of reasoning is primarily based on the gene mutation theory of cancer, which was cemented by identification of the BCR-ABL fusion gene in chronic myelogenous leukemia (CML) that results in the expression of constitutively active ABL kinase domain which induces cellular proliferation [Horne *et al.*, 2013a]. Cloning and characterization of the BCR-ABL fusion gene led to the development of the drug imatinib, which inhibits the kinase activity of BCR-ABL, impeding proliferation and significantly improving five-year survival. If genetic alterations like those of BCR-ABL exist in other types of cancer, the large scale unbiased approach of CGS would surely find those causative genetic components.

4.2.2 *Key rationales behind cancer genome sequencing*

The central rationale and overall plan for CGS was laid out by Francis Collins and Anna Barker in their Scientific American Article [Collins and Barker, 2007]. As the director of the National Human Genome Research Institute and deputy director of the NCI, their perspective of the cancer genome project was more from policy maker point of view than researcher point of view. They argued that cancer genome sequencing is a logical extension of the HGP, which created a reference normal genome that could serve as a platform to understand the cancer genome. Since cancer is "a disease of genes," sequencing the entire cancer genome would identify key cancer genes much quicker than the fragmented "cottage industry scale" research of individual labs. A "systematic, collaborative and comprehensive exploration of the genomics of cancer" would complete a picture of all the genetic mutations and their pathways. Once a complete picture was obtained these genes or pathways could be therapeutically exploited. Reviewing of available cancer sequencing data coupled with the success story of imatinib in CML led to their conclusion that it was the time to start the grand project. They were careful to state "TCGA (The Cancer Genome Atlas) is only a foundation for the future of cancer research, and certainly not the entire house." Unfortunately, if this foundation is incorrect, the entire house of cancer research that is built on this foundation will never be sound. With the logistics of the project laid out, Collins and Barker compared the impact of CGS to that of the Lewis and Clark expedition. Although Lewis and Clark were able to document the grandeur that lay beyond the Mississippi, they never did succeed in finding a water route to the Pacific because one did not exist. With results of CGS pouring in without a defined genomic pattern of cancer, one could ask whether history is repeating itself. To promote CGS, Collins often likens finding the key cancer gene mutations to a man who had lost his keys on the street. The man was found looking for his keys under the only lit street light on the block as it was the only location where he could see. Sequencing the cancer genome, he reasons, is similar to lighting the entire street, making the search for the keys much easier. The key assumption to this analogy is that the keys were in fact lost on the street, within reach of the newly lit area. If the answer to cancer is not found within the sequence of the genome, no amount of light will find it.

4.2.3 *Initial goals, time frame, strategies and expected results of CGS*

The groups that formed to sequence various cancer genomes during the early to mid 2000's included The Cancer Genome Atlas (TCGA), the Sanger Institute's Cancer Genome Project (CGP), the International Cancer Genome Consortium (ICGC), and others. TCGA was born out of a work group of the National Cancer Advisory Board (NCAB) which promoted focusing the new found technology of genome sequencing on cancer. In 2005 the NCAB proposed an ambitious project of sequencing 250 tumors of 50 major cancer types for a total of approximately 12,500 tumors to be sequenced in an approximately 10-year time frame. The ICGC formed later as an answer to the US dominated cancer genome sequencing field. The ICGC later expanded the TCGA goal to sequence 500 tumors of 50 types for a total of 25,000 tumors [The International Cancer Genome Consortium, 2010].

Broad ranging goals outlined by TCGA include identification of a comprehensive list of the gene mutations that drive cancer, identification of the cellular pathways that underlie cancer, resolution of cancer into more homogeneous sub-groups, identification of markers for early detection, and improve selection of therapeutic targets, all of which would ultimately lead to faster and more efficient clinical trials with improved application of existing and new therapeutics. In order to reach these goals, TCGA (and others) would sequence entire cancer genomes, sequence and map epigenetic changes, determine both large scale and point copy number alterations (CNAs), identify structural chromosomal changes, and identify expression patterns in the major tumor types. A smaller scale proof of principle effort would be carried out where ~100 tumors of various types would be sequenced.

Development of a list of all significant cancer genes in order to determine the genetic landscape of each tumor type would elucidate the genetic basis of cancer. Determining this genetic landscape specific to individual cancer types (or subtypes) would help to better identify druggable targets that could be used effectively in the clinic. In this case, driver gene mutations would be those mutations that occurred in ≥5% of tumors of a given type. With this in mind, it was proposed that 250 tumors of each selected type (50 total to represent the spectrum of cancer) would provide a sufficient sample to definitively identify relevant driver genes for each type of cancer.

4.3 Measuring Success of the CGS Project

The worthiness of a scientific endeavor is typically judged by whether it is successful in achieving its original goals, even though unintended discoveries may be the most important results. A key to evaluating the success of CGS is to identify how success was initially defined. Surprisingly the NCAB report lists only a handful of conditions for the project to be successful (**Table 4-1**). For example, the identification of common gene mutation patterns is not mentioned, although it is clear that the goals are based on the gene mutation framework of cancer. It is possible that such a pattern is so frequently discussed that it is an obvious task and needs no mentioning, or perhaps, there was very little doubt that this goal would be achieved. The small number of clear goals and milestones associated with the creation of the TCGA gives a lot of latitude in defining success, leaving the project open-ended.

Table 4-1. TCGA goals spelled out in the original NCAB report

DNA biopsies should require only a small sample and give useful prognostic information.
The project should aim to propel the development of new, targeted treatments that are more effective and less toxic than conventional chemotherapy.
When coupled with long-term research on the biochemical pathways in which the discovered cancer genes participate, additional targets for cancer therapy will emerge.
A clear picture of the underlying defects of cancer cells will provide a coherent framework for drug development and testing.

4.3.1 *Early challenges to CGS*

The initiation of TCGA was broadly praised by many leaders and scientists within the cancer research community as it represented a new hope for the war on cancer. For example, Nobel Prize winner, then NIH director, and current head of the National Cancer Institute, Harold

Varmus claimed this project "could fundamentally change how we approach cancer." The NCI director at the time, Andrew von Eschenbach opined, "It will be an important step in our understanding of the genetic components of cancer and the genetic susceptibilities of people affected by cancer," [Rockoff, 2005]. However, support was not unanimous. A number of objections were raised to this project including fears that a "fishing expedition" project of this scale would divert funding, hurting hypothesis-driven individual investigators in the cancer field, and others questioned whether or not such project could wait for a maturation of sequencing and computational technologies [Weinberg, 2010]. Two other objections based on known cancer biology are worth mentioning. One is the vast amount and unimaginable heterogeneity in tumors, and the other is the key difference between primary tumors and metastatic cancers. Garth Anderson, a cancer researcher who is well aware of cancer heterogeneity, advised against focusing on finding "miracle" targets and then developing "miracle" drugs. "CGS," he said, is "putting a lot of effort into something that I don't think will accomplish much therapeutically" [Rockoff, 2005]. George Miklos, on the other hand, objected the project by questioning whether sequencing primary tumors would impact cancer mortality as metastases are what ultimately kill the majority of patients [Miklos, 2005]. Cancer mortality largely occurs as a result of metastases, which have already been shown to be genetically different than primary tumors. Sequencing only primary tumors would then miss the genetic changes that cause mortality. Miklos further argued that existing data do not support the idea of targeting networks, which was a commonly stated goal of policy makers and the main goal of the original NCAB report.

During initial discussions about the TCGA, we raised a major conceptual concern about the project. We had recently discovered that cancer progression is mainly driven by non-shared karyotypic (genome) alterations rather than shared gene mutations or clonal chromosome aberrations. Based on this new finding, we were not convinced that common gene mutations existed for most cancer types and thus we were skeptical that such patterns could be identified. We realized that genome alteration-mediated macroevolution rather than gene mutation-mediated microevolution drives cancer evolution, and each tumor may proceed in a unique course each time. These findings also provide conceptual explanations for both tumor heterogeneity and the discontinuities between primary tumors and metastases. Since the macro-evolutionary phase of cancer is stochastic, gene sequencing would be incapable of delivering a recurrent stepwise pattern of gene mutation in the majority of cancers. Through this finding and additional synthesis of the key limitations of the stepwise somatic gene mutation theory of cancer, we insisted that the gene mutation theory of cancer did not hold up in the face of critical analyses of available data [Heng, 2007a]. In fact an increasing amount of data supported that CGS would not result in identification of commonly shared mutational patterns for most cancer genes even though individual labs maintained a hyper-focus search for them. If key cancer gene mutations were not the common driving force of cancer progression, the project would not be very useful. It was obvious that a new paradigm of cancer was urgently needed and there is no solid basis for the cancer genome sequencing project (Chapter 5).

Our viewpoint piece was initially rejected by a number of journals starting in 2005 and was subsequently published in 2007 after multiple rounds of revision. Interestingly, during this period of debate, Bert Vogelstein's group published the first whole exome CGS studies [Sjoblom *et al.*, 2006; Wood *et al.*, 2007]. Their data forcefully support many of our points and predictions, namely little mutational overlap was found between tumors, and genetic landscapes were highly heterogeneous. Ironically it was the first CGS paper that finally convinced the editor to publish this piece following its initial rejection. Upon the acceptance of this paper, we have communicated with the NCI leadership about our concerns. Though this piece attracted their attention, their response was: (1) The project was not that expensive; (2) New strategies of large scale whole genome research were needed as single gene study-based research was not working; (3) They were confident that the pattern of gene mutation would eventually be identified with enough samples. We were also told that if no common pattern was found, the CGS data would actually support our new concept. In other words, who was right or wrong would only be proven by the TCGA.

4.3.2 *Have the initial concerns about CGS been addressed?*

Despite many high profile publications highlighting the achievement of cancer genome sequencing, few if any directly address many key concerns initially raised. Curiously, many of these concerns proved to be highly significant. For example, Miklos points out multiple problems with CGS projects in his article "The Human Cancer Genome Project — one more misstep in the war on cancer." These problems include a lack of a good track record in producing targeted therapeutics that work well from mutational data, a lack of mutational overlap in most cancers, and the preponderance of epigenetic and genome alterations that differ from single gene mutations in function. Now that CGS data are pouring in, Miklo's paper is often cited as an example of a nay saying prediction that is easily negated by the great progress CGS has made [Garraway and Lander, 2013]. However, CGS has not addressed many of these criticisms. Another key problem is that within the CGS community, there is little discussion of data that are inconsistent with the gene mutation theory the project is based on. Instead of admitting common mutations are rarer than expected, thereby challenging the somatic gene mutation theory of cancer, CGS publications instead highlight stories about new cancer genes found to be mutated in a small fraction of tumors. Left unchallenged, these nice but highly biased stories only encourage more sequencing.

Finally, CGS in fact does not solve the problems of targeted therapy. A key finding of CGS is that old well-known cancer genes such as TP53 are the most commonly mutated genes. Despite decades' worth of attempts to develop therapies to target these genes, there has not been much progress. In other words, CGS by and large has not identified new promising targets for therapy and does not provide a new perspective. Another finding associated with CGS that has gained a lot of fanfare is the finding of epigenetic change in cancer genomes, although epigenetic alterations have been associated with cancer for decades. Furthermore, like the gene mutation picture, we are only beginning to get a grasp on the complexity of genomic and epigenomic alterations. CGS-based diagnostics have been developed to look at mutations and copy number alterations of hundreds of cancer genes. While these products do a better job of matching subsets of patients (such as lung cancer patients with

EGFR mutations) to targeted and other therapies, so far these treatments only provide short-term responses. The long-term evolution of resistance is nearly universal and even in the best cases, this often happens within a year. Even given the benefit of the doubt, CGS is only an incremental improvement, allowing physicians to better match patients with temporarily effective therapies.

4.3.3 *A surprising, new cancer genome landscape is revealed by CGS results*

Even though many of the initial concerns about CGS have been proven true by Vogelstein's group, few in the community paid attention to their potential impact on the success of the ongoing TCGA project. In fact, it was stated that the unexpected heterogeneity could be addressed by simply sequencing more samples. As of early 2014, the TCGA, ICGC, the Broad Institute (a TCGA member), the Sanger Institute, and Chinese sequencing centers including BGI combined have sequenced over 30,000 tumors, exceeding the 25,000 cancer genomes that were expected to be finished by the end of 2012 [Garraway and Lander, 2013; Hudson *et al.*, 2010].

Indeed, CGS has changed the landscape of cancer research forever, but ironically, in an unexpectedly opposite direction of the initial goals. For most in the community, new and surprising findings include large variation of the mutation spectrum between various types of tumors and within single tumor types. For example, fewer than 20 gene mutations are thought to drive breast and colorectal cancers, while 450 genes are significantly mutated in lung squamous cell carcinoma. In addition to finding diversity in the genes that are mutated in specific tumor types, the mutation rate varies by more than 1000-fold between tumors and ranges in lung cancer from 0.1–100 mutations/Mb and is not constant throughout the genome [Chen *et al.*, 2010; Koren *et al.*, 2012]. Chromosomal alterations including copy number alterations and multiple rearrangements are commonplace as are extensive chromosomal changes including phenomena such as chromothripsis, chromoplexy, chromoanagenesis and kataegis. Collectively these phenomena were originally described as genome and/or karyotype chaos by advanced cytogenetic approaches [Heng *et al.*, 2006a; Heng, 2007b; Horne and

Heng, 2014; see Chapter 5]. These findings demonstrate the diversity within and between cancer types as well as demonstrate that there is no single cancer signature. In addition to the wide array of mutations and genome alterations present in most tumors, epigenetic change also occurs extensively [Heng *et al.*, 2009]. Broad changes to the epigenetic landscape of tumors occur commonly at both the histone and DNA modification levels. These changes drastically alter expression profiles [Stevens *et al.*, 2013a]. Taken together, the chromosome (genome) level changes combined with epigenetic alteration and gene mutations are very diverse and complex, and an ever-growing body of work shows that these alterations are highly heterogeneous even within single tumors. To give some detail to the extent of this multi-level change, meta-analysis of CGS results of some of the most sequenced tumor types (those where over 450 samples have been sequenced as of December 2013) is detailed below.

4.3.3.1 *Breast cancer*

Breast cancer was one of the first cancer types to be subjected to exome sequencing. Exciting results from an initial panel of 11 tumors and validation panel of another 24 tumors showed that there are 38–193 mutations per tumor. In this analysis a total of 139 putative cancer (CAN) genes were identified through bioinformatic analyses, approximately 10–20 per tumor [Sjoblom *et al.*, 2006; Wood *et al.*, 2007]. Interestingly, many genes frequently cited as breast cancer genes such as MYC, PIK3CA, CDKN1B, and PTEN were not identified in this study as cancer genes.

Thirty-six genes of the above CAN gene list that were found to be mutated in at least one tumor in a mutation screen and validated in at least one other tumor out of 24 in a validation screen in the above study were resequenced in a set of 96 additional tumors [Jiao *et al.*, 2012]. This analysis revealed few mutations in this "validated" set of cancer genes. In fact only 28 somatic mutations were found. Thirteen of these mutations were already known polymorphisms. In total, mutations were only found in 1/3 of the genes previously identified as breast cancer genes (ADAM12, CENTB1, CENTG1, DIP2C, GLI1, GRIN2D,

HDLBP, IKBKB, KPNA5, NFKB1, NOTCH1, and OTOF) [Jiao *et al.*, 2012]. NOTCH1 was the most prevalently mutated gene occurring in only 3.1% of tumors sampled. With such a low mutation frequency, it is not surprising that NOTCH1 was not found to be mutated in Sjoblom *et al.*'s study. These data underscore the paucity of recurrent mutations even in those genes thought to be cancer drivers.

The lack of overlapping mutated genes and pathways is even more evident when two different CGS studies are compared. A second CGS of 100 breast cancer tumors identified 40 driver mutations including *AKT2*, *ARID1B*, *CASP8*, *CDKN1B*, *MAP3K1*, *MAP3K13*, *NCOR1*, *SMARCD1* and *TBX3* [Stephens *et al.*, 2012]. Of these genes, the number of drivers per tumor ranged from one to six. Twenty-eight percent of tumors were linked to a single driver in this list. The drivers in this study varied between tumors, and different tumors had different mutation rates. Fifty-eight percent of tumors were linked to mutations in *TP53*, *PIK3CA*, *ERBB2*, *MYC*, *FGFR1/ZNF703*, *GATA3* or *CCND1*.

The data from either of these reports seem encouraging, however, only 2 genes (TP53 and PIK3CA) were found to be shared between the driver list of the Stephens study [Stephens *et al.*, 2012] and the list of CAN genes in the Wood study. The top 20 breast cancer gene mutations from nearly 1,000 breast tumors sequenced by the Sanger Institute show little overlap with the Wood data (only 3 overlapping genes) but a total of 16 of the 20 overlap with the Stephens data. Not surprisingly, these genes are overwhelmingly well-known breast cancer genes. Taken together, the mutational diversity of breast cancer is extensive. There is little mutational overlap between samples, and new driver events seem to be identified in each individual genome sequencing study. When overlap does occur, it happens typically in already known breast cancer genes, raising the question of whether CGS produces much in the way of new information. In addition, although overlapping genes can be identified, the frequency that they occur in each study varies widely. For example, the frequency of ERBB2 mutation identified by TCGA and by the Sanger Institute differs by over an order of magnitude, underscoring the wide variation of mutational landscapes. Although each individual report may make sense and make its data seem promising, it seems that CGS does not add much information to the big picture.

In addition to extreme variation in specific gene mutation signatures of individual breast tumors, localized hypermutation is often present. In a sequencing study of 21 breast cancers, 13/21 tumors were shown to have mutational hot spots, which varied by location and number in each tumor where they were present [Nik-Zainal *et al.*, 2012]. These mutational hotspots in some cases associated with inter-chromosomal translocations. The limited genomic architecture data that were presented indicates that breast tumors commonly have multiple chromosomal translocations that vary by tumor [Navin *et al.*, 2011]. In fact, single cell sequencing of breast tumor cells showed multiple chromosomal copy number alterations occurring during tumor evolution, and that cell-to-cell copy number heterogeneity is extremely high. Further single cell sequencing of a triple negative breast tumor and an ER-positive tumor supports our hypothesis that cancer evolution occurs in two distinct phases [Wang *et al.*, 2014]. The data in this paper suggest that chromosomal alterations occurred early in punctuated bursts, which was then followed by stable expansions in the formation of the tumor mass. These authors found that many mutations were present in <1% of cells, diversifying the phenotypes of the cells within the tumor. Even though only a few single cell-based studies have been reported, the data from these studies show that the genomes of single cells are highly diverse even within a single tumor, challenging the utility of any non-single cell-based sequencing effort.

4.3.3.2 *Colorectal cancer*

The first whole exome sequencing of colorectal cancer included 11 tumors that were initially sequenced and validation of those genes found to be mutated in a 24-tumor validation screen [Sjoblom *et al.,* 2006; Wood *et al.*, 2007]. The number of mutations per tumor varied from 49–111. A total of 139 CAN genes were identified with an average of 14 per tumor. Identified CAN genes included well-known colorectal cancer genes including APC, TP53 and RAS. Sequential mutation of these genes has been postulated to drive colorectal cancer progression for decades, serving as the key evidence for current stepwise somatic mutation theory of cancer [Vogelstein and Kinzler, 1993], yet in these full blown tumors it was more common for there to be no mutations in these CAN genes than for all 3 to be mutated. Additional sequencing of the top 40 CAN genes in an additional 96 samples only found 25 of these 40 genes to be mutated at least once, underscoring the mutational diversity of colorectal cancer.

A second report was released of colorectal exome sequencing of 224 tumors and matched normal tissue pairs [Cancer Genome Atlas Network, 2012]. Analysis using the MutSig algorithm found 32 genes to be significantly mutated including *APC*, *TP53*, *KRAS*, *PIK3CA*, *FBXW7*, *SMAD4*, *TCF7L2* and *NRAS*. The 32 cancer genes included 15 genes in cancers with a hypermutator phenotype and 17 genes in tumors without a hypermutator phenotype. Interestingly, 4 out of the 15 (27%) hypermutator cancer genes and 9 of the 17 (53%) genes in the non-hypermutator tumors were included in the Wood report CAN gene list. The majority of the genes shared by the two lists are already well-known colorectal cancer genes, including TP53, APC, KRAS, TGFBR2, FBXW7, and PIK3CA. Further analysis of these data revealed that much

Table 4-2. Genes from thc top 20 mutated colorectal tumor lists from TCGA and Sanger Institute that overlap

Gene	TCGA Frequency	Sanger Frequency
APC	58%	77%
TP53	44%	59%
TTN	34%	57%
KRAS	34%	33%
SYNE1	22%	34%
MUC16	18%	23%
LRP1B	17%	25%
FAT4	17%	24%
CSMD3	14%	28%
LRP2	13%	22%

like the Sjoblom report; it was less common for a colorectal tumor to have an APC, KRAS and TP53 mutation (13%) than no mutations in these genes (29%). In fact, out of all of the possible combinations of mutations (all three genes mutated, two genes mutated, etc.), having no mutations in these "putative cancer genes" was the most common state of a tumor.

In addition to these data, the Sanger Institute also sequenced 357 colon cancers and 118 rectal cancers, which shared 18 of the top 20 genes found to be mutated by the Sanger Institute. All together the Wood, TCGA and Sanger Institute studies only shared 6 of the top mutations (KRAS, APC, TP53, PIK3CA, FBX7W, and SMAD4), and 7 other genes of the top mutated gene list for each study are shared in 2 of the studies. Because different groups use different criteria for selecting their "cancer gene lists," we have also analyzed lists of the 20 most frequently mutated genes in colorectal cancer determined by both the Sanger Institute and TCGA. Out of these top 20 genes, 10 overlap between the two studies (**Table 4-2**). In general, genes from this list were found to be mutated more frequently by the Sanger Institute than by TCGA, but the frequency of mutation could vary wildly. Similar to the mutated genes mentioned just above, many of these genes are well-known cancer genes such as APC, TP53, and KRAS (**Fig. 4-1**). Thus CGS of colorectal cancers tells us two things we already know, the mutational landscape of each tumor is diverse and a handful of well-known genes are recurrently mutated but with much lower frequencies than previously thought.

There is often focus on identifying mutations in one broad type of genes in a given cancer. For example, kinase mutations have also been postulated to be major drivers of cancer progression. In another study, mismatch repair detection assays of 30 kinase genes revealed a total of 144 mutations were identified in 55 colorectal tumors. The most frequently mutated kinase, BRAF was mutated in 40% of tumors. In the first whole exome sequencing of colorectal cancer, BRAF was not found to be significantly mutated above the background mutation rate, in the second exome sequencing paper it was found mutated in 46% of hypermutator tumors, and was found at 20% in the COSMIC genome database [Bentivegna *et al.*, 2008]. Interestingly, the majority of kinase mutations (17/24) detected in these tumors however were mutated in <10% of tumors. Clearly, there are big differences between studies, which could reflect the heterogeneous nature of patient populations. In addition to the complication of patient heterogeneity, individual studies often present data in different manners and apply different mutational significance algorithms, complicating cross study analysis. In breast cancers, the story was completely different. Only eight kinases showed mutations and these were each only mutated once.

Not only do colorectal tumors contain a broad and diverse array of gene mutations, alteration of genome topology is also commonplace. Comparison of structural alterations in 97 tumors using WGS found that hypermutated tumors had fewer somatic copy number alterations regardless of whether the tumors were microsatellite unstable or not [Cancer Genome Atlas Network, 2012]. These results echo more recent reports finding tumors, in general, tend to have either mutations or copy number alterations [Ciriello *et al.*, 2013]. Ten large chromosomal arms were found to be significantly amplified while 12 were found to be significantly deleted. 8p and 20p were found to be both significantly amplified and deleted. A total of 250 inter-chromosomal translocations were identified (0–10 per tumor). Most of these translocations were unique to each tumor demonstrating a unique genome context of each tumor, although 3 tumors did share an 2;11 translocation which created a NAV2-TCF7L1 fusion gene [Cancer Genome Atlas Network, 2012]. In addition, current genomic profiling methods are largely based on the average of cell populations, making most Non Clonal Chromosome Aberrations or NCCAs undetectable (see Chapter 5–6), eliminating a major source of genomic diversity. In fact, single cell sequencing of colorectal tumors reveals a high diversity of genome alterations [Yu *et al.*, 2014].

Taken together, these results show that despite the well-known regular morphological progression of colorectal cancer (hyper-proliferation of intestinal crypts followed by development of polyps and growth of polyps into invasive tumors) and the hypothesized cascade of recurrent mutations, colorectal tumors exhibit mutational heterogeneity and do not follow the previously accepted linear genetic model (the

Vogelgram). This mutational heterogeneity underscores the point that there is no a common mutational signature of colorectal cancer.

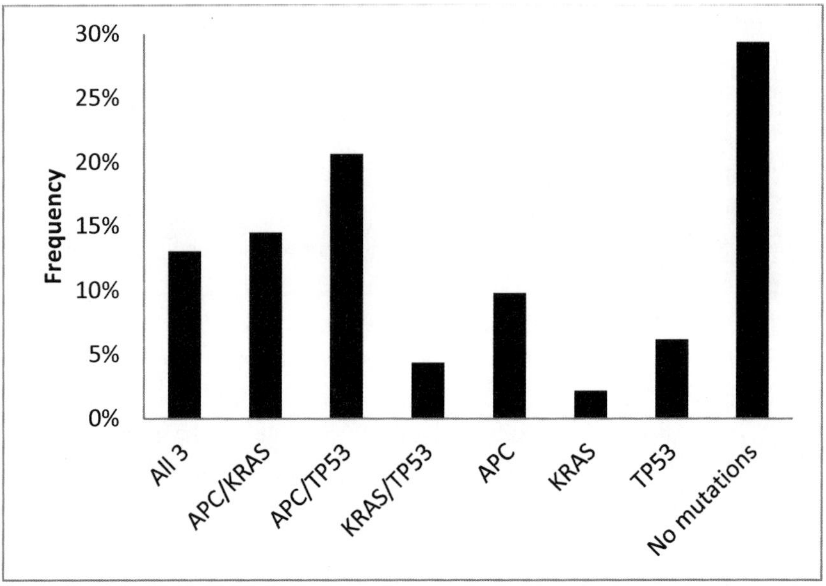

Fig. 4-1. Frequency of APC, KRAS, and TP53 gene mutations in colorectal cancers.

4.3.3.3 *Squamous cell lung cancer*

Squamous cell lung cancer is one form of lung cancer that is particularly well characterized. Although squamous cell lung cancer does not represent the entire spectrum of lung cancers, we have included it here due to its current level of characterization. Whole exome sequencing of 178 squamous cell lung carcinomas revealed 450 genes mutated significantly above the background mutation rate [Cancer Genome Atlas Research Network, 2012]. This list of significantly mutated genes surprisingly contained 101 olfactory receptors and 83 genes coding for proteins larger than 4,000 amino acids [Lawrence *et al.*, 2013]. In order to reduce the number of genes to an "approachable" number and eliminate genes that had no seeming connection to cancer, another algorithm was developed which again narrowed this list to 10 genes. Development of this algorithm was carried out because subsequent sequencing efforts have shown that the background mutation rate is not constant between tumors and varies by region in the genome. Taking a variable mutation rate into account, a second algorithm was employed to determine a significantly mutated gene list which only differed from the first by adding FBXW7 to the list. This list is impressive as at least 1 gene from the list is mutated in >90% of the 167 tumors represented in one major publication [Cancer Genome Atlas Network, 2012], and 80% of tumors contained a TP53 mutation. Interestingly, as of preparation of this chapter, 356 squamous cell lung carcinomas have now been sequenced and the data are publicly available. In the total list of 356 tumors, 46.6% of these tumors contain a mutation in any one of the significantly altered gene list, TP53 was found to be mutated in 41% of tumors, and only two tumors had mutations in six or more of the signature genes. The significant change in the presence of mutated genes from different publications suggests that lung squamous cell carcinomas are more heterogeneous than thought. Again, data presentations from different reports differ, as do the computational algorithms that are used to parse the data, making it difficult to decipher results.

In addition to extensive gene mutation, non-shared genomic rearrangement and large scale copy number change are common in squamous cell lung cancer. Tumors contained an average of 47 focal and 23 broad somatic copy number changes per tumor, where broad copy number change was defined as being 50% of the length of the chromosome arm or longer. These data underscore that extensive structural alterations are common in most tumors, especially if individual cells are sequenced rather than the population average.

4.3.3.4 *Prostate Cancer*

Exome sequencing of 23 prostate cancers derived from 16 tumors revealed 123 genes mutated in at least two tumors and only 23 in three or more tumors, underscoring the lack of mutational overlap in prostate cancer [Kumar *et al.*, 2011]. In an initial analysis, the only candidate gene mutation that was found was unsurprisingly TP53. Interestingly, three tumors had a hypermutator phenotype where the mutation rate was found to be ten times higher than other prostate tumors. Exome sequencing to 120X coverage revealed an average of 30 non-synonymous mutations per tumor [Grasso *et al.*, 2012; Barbieri *et al.*, 2012] while sequencing to 30–40X coverage only found an average of 16 mutations per tumor [Weischenfeldt *et al.*, 2013]. This has been attributed to genetic heterogeneity of prostate cancers and seems to be a likely assertion [Spans *et al.*, 2013].

WGS of 55 primary adenocarcinomas and matched controls revealed overwhelming genomic rearrangement (5,596 high confidence rearrangements) including the formation of previously identified fusion genes NRF1-BRAF and CRKL-MAPK1 and an average of 33 non-silent exonic mutations per tumor. Similarly, massively parallel sequencing of prostate tumor tissue and matched normal tissue from 7 individuals with high risk prostate cancer revealed an average of 90 genomic rearrangements per tumor [Baca *et al.*, 2013; Berger *et al.*, 2011]. These rearrangements formed a number of fusion genes including TMPRSS2 and disrupted many genes including TP53, PTEN and MAGI2 but the rearrangement pattern of each tumor was unique.

Despite the overwhelming frequency and complexity of genome level alterations (which change the genome context and the networks in which genes function; Chapter 5), there are still attempts to identify the sequential order of gene mutations. Authors of prostate CGS have offered a "consensus path" of mutations that occur in the following order, deletion of NKX3 or FOXP and fusion of TMPRSS2 and ERG, followed by CDKN1B and TP53 mutations and finally loss of PTEN [Baca *et al.*, 2013]. The authors go on to assert that, "these results imply that prostate cancer carcinogenesis favors the dysregulation of cancer genes in defined sequences, as has been suggested by (Vogelstein's) studies of developing tumors in colon cancer" [Vogelstein and Kinzler, 1993]. Not surprisingly, the frequency of any of these mutations is low. Of these genes, TP53 and PTEN stepwise mutation of ETS family genes followed by loss of PTEN only explains a minority of cases that have been sequenced [Chen *et al.*, 2013]. TP53, the most frequently mutated gene in prostate cancer is mutated in only ~15% of tumors and PTEN is mutated in 5% or less of tumors. The rest of the genes in this list do not even make the COSMIC list of the top 20 prostate cancer genes. How can a pattern that exists in far less than 10% of prostate cancers be considered a "consensus pathway"? Clearly, these types of analyses are biased by a flawed theoretical basis that says consensus pathways must be found to make any sense of cancer. Likening prostate cancer progression to colorectal cancer progression turns out to be a very good analogy; the biggest mutational pattern of each of these cancers is that there is no shared, stepwise pattern of mutation that can explain them. From an evolutionary point of view, even though there is convergence to the cancer phenotype, there is no simple fixed pattern of gene mutations or changes in gene expression that makes normal healthy somatic cells cancerous.

4.3.3.5 *Rhabdoid tumors – mutationless cancers*

Rhabdoid tumors are rare, highly aggressive tumors that primarily occur in infants and young children. Rhabdoid tumors occur in brain, kidney, striated muscle, and other soft tissues. Prior cytogenetic studies suggested deletion or point mutation of the SMARCB1 (INI-1) locus on chromosome 22 (22q11.23) occurred in over 90% of rhabdoid tumors, however whole genome sequencing of these tumors is changing that view [Cocce *et al.*, 2011; Lee *et al.*, 2012; White *et al.*, 1999]. In a whole exome sequencing-based study of 35 rhabdoid cancers (32 untreated tumors, and 3 treated tumors) from brain (20), kidney (3), and other soft tissue (9), revealed that SMARCB1 deletion occurred in 78% of tumors, while no SMARCB1 deletion was evident in 22% of tumors. Overall, SMARCB1 was altered in all tumors but one. However, the Sanger Institute has also reported rhabdoid sequencing results, and in these results, SMARCB1 was found to be altered in 41% of brain rhabdoid tumors, and in only 1 out of 15 skeletal muscle tumors (8%) [T.S. Institute, 2014]. Thus the Sanger Institute found SMARCB1 alterations in a total of 29% of examined tumors, far short of the 90% + of rhabdoid tumors they have been reported in. In addition to the SMARCB1 data, these two studies found a low mutation rate overall, and the majority of gene mutations aside from SMARCB1 occurred uniquely in one tumor. GARB2 was the only gene mutated in more than one tumor. Interestingly, Lee *et al.* found two tumors that harbored no mutations, even in SMARCB1 [Cocce *et al.*, 2011]. Sanger Institute results showed that 25% of CNS rhabdoid tumors and 31% of striated muscle rhabdoid tumors have no mutations. Chromosomal abnormalities in addition to 22q11.23 deletion have been described in cytogenetic studies of rhabdoid tumors, suggesting that CIN plays a role in rhabdoid progression. Taken together, these results demonstrate that gene mutations are not necessary for tumor formation.

4.4 Summary of CGS Results

In response to the many questions about CGS, Garraway and Lander have stated that, "In the end, however, the questions could only be answered with data." [Garraway and Lander, 2013]. So what do the data say, and more importantly how do these results support or challenge the stepwise mutational theory of cancer or the recently proposed genome theory? Though some of these points are discussed throughout this chapter, a brief summary is useful. In addition to the discussion below a synopsis of the CGS findings regarded to be important by Garraway and Lander along with our interpretation of the findings is presented in **Table 4-4**.

4.4.1 *Validation of known cancer genes*

In general, CGS has validated previous research that has repeatedly connected a small number of well-known genes to cancer such as TP53, P16/INK4A, MYC, BRCA1, and RAS. Unexpectedly these genes have been found to be mutated less frequently than previously thought. For example TP53, the most commonly mutated gene across tumors and frequently the most mutated gene within a single cancer type, is mutated in less than 50% of tumors of most types. A lower frequency of mutations of these genes than previously found reduces the impact and importance of each of these genes in cancer. In addition to these famous cancer genes, a limited number of "needle in a haystack" genes previously unknown to be mutated in cancer have been found such as

IDH1 and PIK3CA. While these genes are mutated at higher frequencies than randomly expected based on underlying mutation rates, they are typically mutated in less than 10% of tumors. With 20,000+ cancer genomes sequenced, it is not clear that there are few if any "key" needles to find. In fact, most had been identified before the sequencing era. Combinatorial mutation of a number of genes similarly does not provide a comprehensive picture of the progression of cancer. Interestingly, one of the key findings of CGS is the occurrence of "mutationless" tumors [Mack *et al.*, 2014; Parker *et al.*, 2014]. Though mutationless, these tumors do present chromosomal alterations. If gene mutations are the driving force of cancer, how can cancers exist without them? Clearly this finding challenges the notion that a handful of mutations drive all cancers. Surely, the confirmation of most previously known cancer genes is nice, but the challenge of using these gene mutations (or quantitative combination of them) for diagnosis and treatment remains. So far, CGS has failed to provide new strategies.

4.4.2 *Classifying driver and passenger mutations*

Prior to the cancer genome sequencing era, few reports addressed driver and passenger mutations, as only a limited number of gene mutations had been found and all were important. With identification of so many gene mutations, focus has shifted to the importance of distinguishing drivers from passengers. Distinguishing these represents central achievement of CGS.

Driver mutations are mutations that give selective advantages to cells by increasing survival or reproduction, leading to clonal expansion. However, as discussed in multiple places in this book, focusing solely on clonal expansion in cancer research is not beneficial. Identification of driver genes is a major goal of CGS because targeting these genes should slow or stop the tumors in which they are mutated. The NCAB formally proposed that sequencing 250 tumors of each type would be sufficient to identify mutated driver genes that occurred in 5% or more of tumors. However in most cancers, only a few genes have been found to be mutated in over 5% of tumors resulting in what has been called a "long-tail of gene mutations" [Garraway and Lander, 2013]. Now the bar for significance has been lowered to 1%, and the ICGC has proposed sequencing 500 tumors of each type in order to identify significant drivers. Following this effort, will the bar for significance be dropped to 0.2%, and will sequencing 1000 total tumors be required?

Not only are mutation profiles long tailed, but "driver" mutations often occur commonly in non-cancerous tissue. For example the BRAF V600E mutation, which is a poster child of the effort to identify and target recurrent driver mutations, occurs in over 80% of nevi (moles), and in lower frequencies in melanomas (**Fig. 4-2**) [Berger *et al.*, 2012; Pollock *et al.*, 2003]. While the presence of large numbers of nevi increases the risk of melanoma, the vast majority of nevi never become melanomas. Other cancer "driver" mutations occur in other healthy tissues that never become cancerous, raising the question: are mutations drivers of cancer? If they are, under what conditions, do these drivers really contribute to cancer (are genome level changes a necessary condition as suggested by the genome theory)?

In addition, passenger mutations are also important. Recent computational modeling based on CGS data shows that like genome change, driver mutations occur in two phases: a stepwise and stochastic phase [McFarland *et al.*, 2013]. In other words, the driver function of a specific mutated gene only lasts for a certain period of time before another mutation takes over the driver function. During the changeover from stepwise to stochastic phases, former passenger mutations can suddenly become drivers. The low penetrance and non-static function of most drivers strongly questions the basis of, and the ability to detect, drivers. Clearly, in the dynamic cancer selective landscape, distinguishing drivers and passengers is not particularly useful.

Furthermore, punctuated genome (karyotypic) alterations drive cancer evolution (for more, see Chapter 5) by defining the functional roles of many gene mutations in a stochastic manner. Drivers come and go based on the emergent properties of a given genome system. Drivers are not static and can therefore only be judged within the genome context in which they occur. We have previously reported this unique pattern of cancer evolution (see Chapter 5).

In conclusion, cancer evolution is a dynamic process with no fixed driver. For the process to occur, the perfect storm of a survivable but changeable genome must happen. Even if there are "drivers" that enable a certain genome to thrive in a given environment, environments are in constant flux and so are drivers. Mutations that occur in very low frequencies, or even mutations in genes that do not seem to be connected to cancer can be crucial to the survival of individual cells. These individual cells provide evolutionary flexibility to the population.

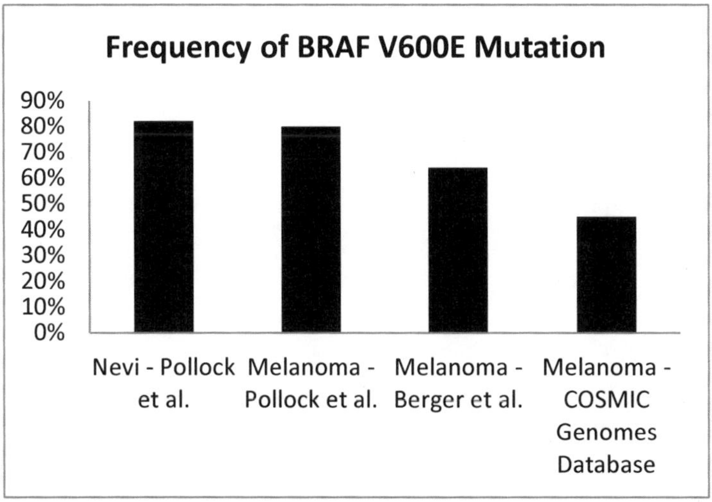

Fig. 4-2. The frequencies of BRAF V600E mutation in nevi and melanomas detected in various studies.

Targeting drivers in a heterogeneous and/or chromosomally unstable tumor is not likely to eliminate the tumor. Targeting drivers is largely only successful in linear models of progression where a driver's function persistently dominates. However, drivers if existent, are not static in most human tumors, especially during harsh treatment. Furthermore, drivers and passengers can easily be switched under crisis. Both drivers and passengers contribute to the evolutionary process of cancer, resulting in limited implications of identified drivers in single tumors. Identification of common drivers to explain a type of tumor is even more limited as each tumor progresses by unique runs of evolution.

4.4.3 *The mutational landscape is highly dynamic*

Perhaps one of the most important findings of CGS is the highly dynamic gene mutational landscape of cancer. Intra- and inter-tumor heterogeneity is extensive. The overwhelming mutational diversity should completely change the view of most researchers. It is quite apparent that there is no magic gene or pathway that will cure most cancers. Somehow, researchers that focus on a pathway or gene have convinced themselves to separate the facts of cancer evolution demonstrated by CGS from the strategies they are developing and wishing for. For example, despite the highly unpredictable nature of the cancer landscape, Eric Lander has recently lauded identification and targeting of specific genes/pathways saying, "But if you understand that this is a game of probability, and there is only a finite number of cancer cells and each has only a certain chance of mutating, and if we can put together two or three independent attacks on the cancer cell, we win." [Fallows, 2014]

Similarly, some system biologists have claimed that by identifying cancer specific network patterns, these patterns can be targeted with the hope that cancers do not have time to evolve new network patterns leading to resistance. They literally believe that although cells have evolved significantly to become cancerous, once the cancer pattern forms it will not continue to change under their treatment plan. In a sense, evolution will stop for them. This reasoning is the antithesis of evolutionary thinking. Life always finds its way. After all, we develop cancer despite all the mechanisms that exist to prevent it.

Revealing the cancer genome landscape is just the first step in better understanding cancer. The next step is to ask some "now what" questions. For example, with CGS demonstrating so thoroughly that the stepwise gene mutation theory as it stands cannot explain the results of CGS, should we then continue to cling to the stepwise theory of gene mutation accumulation or abandon it for a framework that does explain the findings of CGS? Some other questions include, knowing the highly dynamic nature of the multiple levels of the cancer genomic landscape, why should we still focus on gene mutations, especially since we have been doing so for the last few decades with no major breakthroughs?

Why should this same failing strategy to fight cancer continue to be used? As the picture is now clear, no matter how hard we endeavor to profile the diverse genomic landscape of cancer, these endeavors do not change the heterogeneous nature of cancer, they only reveal more and more diversity. Single cell sequencing will only demonstrate that intra-tumor heterogeneity is far greater than what was thought. Indeed, results of both single cell whole genome sequencing and single cell exome sequencing have shown tumors to be heterogeneous, especially at the exome level [Navin *et al.*, 2011; Patel *et al.*, 2014]. Similarly, measuring further layers of genetic or non-genetic organization such as non-coding RNA will only demonstrate that there is more heterogeneity, noise and complexity, and that these features are not stable.

The documentation of the highly heterogeneous cancer genomic landscape is not good news, but it is the inescapable truth. The glimmer of hope for a major contribution of CGS to understanding cancer is that the facts about cancer gene mutations (or the lack thereof) may finally promote much needed change in cancer research, both in conceptual frameworks and major strategic points of view.

4.4.4 *The mutational spectra of cancers*

CGS reveals many interesting insights about genetic changes in cancer, many of which have surprised those who believe gene mutation theory of cancer. One of the most surprising findings has been the diversity of the mutational spectrum of cancer. The mutation rate in cancer, once thought to be relatively consistent within each tumor type, varies by over 1000-fold in some cancers. Perhaps even more surprising, tumors have been found with few or no mutations [Lee *et al.*, 2012; Mack *et al.*, 2014; Parker *et al.*, 2014]. The diversity of the mutation rates and frequencies underscores the genetic variation within even a single tumor type. Such variation suggests a plethora of mechanisms that contribute to each type and likely even within a singular tumor. Surprisingly, despite the heterogeneity that is evident from single cell sequencing results, single cell sequencing has paradoxically shown that single cell-based measures of the mutation rate are significantly lower than bulk population-based measures [Wang *et al.*, 2014; Zong *et al.*, 2012].

Another interesting phenomenon detected by CGS that has stirred excitement is the finding that certain cancers have mutation patterns indicative of environmental exposures. For example, lung tumors from smokers with both lung adenocarcinomas and squamous cell carcinomas show a preponderance of C to A mutations and melanomas show a preponderance of C to T mutations due to ultraviolet radiation exposure. These findings have been presented with a lot of excitement and fanfare. It is indeed interesting that CGS in this case directly links known carcinogens with mutations throughout the genome, solidifying the etiological link between smoking, UV, and other exposures to cancer. Whether such excitement is warranted remains a question though. It has long been known that smoking and UV exposure tend to cause these mutations. Further, though CGS reveals a recurrent pattern of the types of mutations, these mutations do not hit the same genes in tumors, and actually affect a large number of genes, not to mention the karyotypic alterations induced by both agents. Knowing that UV exposure results in C to T mutations does not inform of the evolutionary process of cancer nor does not provide a treatment target. What clinical benefit does this finding produce? Similar to history, it is easy to look at a tumor and infer that a certain event caused the history of that tumor to proceed as it did, it is easier to document that a certain feature (mutation) happens in a given cancer type, but predicting the many possible routes to cancer in a given cell and/or population is difficult if not impossible, due to the nature of cancer evolution. As we better illustrate in Chapter 5, the evolutionary mechanism of cancer explains why individual pathways and/or mutation profiles provide limited predicting value in most sporadic cancers. There is still a gap between research characterizing the parts (the genes) and the actual process of cancer evolution in patients. Many "parlor tricks" have been performed where over-expression (or knockout) of a gene can cause tumors in a mouse model, and inhibition (or addition) of it causes rapid shrinking of the tumor. These simple linear models are wonderful in the laboratory, but in the clinic the same mutations that cause such strong phenotypes rarely occur. Similarly, the current brand of genomics is very useful from a research perspective but it has been much less clinically useful, likely due to differences in the way cancer occurs in models used in basic research and the way somatic evolution mediates disease progression in reality (see Chapter 5).

A good deal of the phenomena published by CGS consortiums is regarded by the authors as being interesting even if they are not medically relevant. Pure research curiosity does not cut it in the cancer field. Cancer affects nearly everyone on earth directly or indirectly. The promise of CGS was to once and for all enlighten the biology of cancer in order to eliminate this dreadful disease. Tens of billions of dollars are not allocated just for research curiosity. Interesting research does not equate to clinical usefulness.

4.4.5 *The importance of karyotype alterations*

Though long known to occur in cancer (see Chapter 3), high frequencies of often complex chromosome alterations including structural changes and aneuploidy have been confirmed by CGS to be pervasive in cancer. For years, molecular focused researchers favored gene-based studies of cancer for a number of reasons, including a relative under reporting of the commonality of chromosomal abnormalities. This was largely due to the assumption that only recurrent events are important, and most researchers today consider chromosomal changes that do not directly impact a specific gene to be insignificant. CGS data now showing the preponderance of chromosomal abnormalities in cancer should force many to wonder what the real causative factor of cancer is if there are so many chromosome aberrations and an overall lack of a mutational pattern. The finding that chromosome change is so common in cancer is perhaps one of the most important messages delivered by CGS. This finding directly challenges gene-based frameworks of cancer. In doing so, CGS brings about an immediate challenge, the need for a new cancer framework and research strategies to surround that framework. Should the genome (karyotypic) level prove to be the most important (and we strongly believe it will), further research is needed to develop platforms that can integrate the contributions of various lower levels of genetic/non-genetic change within a genome-based framework.

For example, a particularly interesting finding of CGS is the common observation of large scale genome rearrangements (chaotic genomes) including chromothripsis and chromoplexy. Although cytogenetic research has long shown these alterations to occur, they failed to catch the attention of the field (Chapter 5). Chaotic genomes have been long thought to be non-survivable cell culture artifacts, but they do in fact occur regularly in cancer. The "molecular confirmation" of these structures by CGS has led to their sudden popularity. Unfortunately, this finding has been approached with a traditional gene-centric point of view that focuses only on cancer genes directly impacted by the specific chromosomal alterations detected, as according to gene mutation theory dogma, any genetic changes (genome chaos included [Liu *et al.*, 2014]) are only significant when they mutate driver genes. Since many genetic and epigenetic and environmental factors can be linked to the massive genome-reorganization, and these aberrations could be responsible for the emergent properties of cancer even though they do not directly alter cancer genes, we caution against research that only focuses on linking these genome changes with yet another gene.

4.4.6 *Can pathways unify the mutational diversity?*

In response to the high diversity of the gene mutations discovered by CGS, a new explanation has surfaced in an attempt to turn disappointment into new promise. The explanation is that the inability to identify common gene mutations is not really disappointing, despite the fact that it clearly shows that CGS has failed to meet a central original goal, as a limited number of pathways can be tied to diverse gene mutations. These limited, core pathways, numbering from 10 to 100, are responsible for cancer, and represent a manageable number of targets for each cancer.

On the surface this synthesis seems reasonable but it is very problematic as illustrated by following points:

(1) Molecular pathways are defined by genes. The massive diversity of gene mutations will likely lead to a large number of diverse pathways, many of which are still unknown. In fact, of the identified gene mutations, only a small portion makes sense according to currently known pathways.

(2) Even a given gene mutation can be linked to many different pathways. For instance TP53 function is highly complex, it functions as a transcription factor, a signaling protein, a sensor of damage and stress, and a major driver of apoptosis and other forms of cell death [Attardi, 2005; Feng *et al.*, 2008; Fukasawa *et al.*, 1997; Godefroy *et al.*, 2004; Levine and Abrams, 2008; Olivier *et al.*, 2009; Sengupta and Harris, 2005]. Even specific functions of TP53 are highly complex. When functioning as a transcription factor, TP53 rapidly binds and unbinds DNA sequences to affect transcription [Luo *et al.*, 2004]. Because TP53 is such a central player in many basic biological functions, it is probably not surprising that it is overall the most mutated cancer gene. In addition, different mutations can have different impacts on the functions that TP53 performs. Despite this complexity many pathway solutions of gene ontology will often identify a TP53 pathway component of cancer. How in this complex highly non-linear system should the TP53 pathway be defined, and can it be defined as a static entity? In fact, many gene mutations are similar to TP53 in the regard that they are involved in many pathways. That is why there are so many "master regulators" in the literature. Some complex features of pathways are described in **Table 4-3**.

(3) Similar to gene penetrance, cancer pathway penetrance in specific tumor types is low, except for a few exceptions such as TP53, or various hormone signaling pathways. Estimates of the number of pathways involved in each tumor type vary by an order of magnitude from 10–100 pathways, further underscoring the complexity and diversity of cancer. Characterizing pathways by incorporating diverse gene mutations and/or changes in expression aids in the understanding of a proportion of gene mutations, but it does not solve the problem of highly diverse gene mutations and the pathways they define. Like cancer gene research where an ever growing number of cancer genes have been found, more cancer associated pathways will be found, and in particular more and more complicated pathway interactions will likely be discovered. Lowly penetrant pathways should not be expected to mechanistically explain cancer, especially given the high level of pathway complexity likely be discovered.

(4) Pathways are very dynamic. Large scale efforts to elucidate the individual pathways in cancer have been a central goal of cancer research. The current knowledge of pathways can be likened to an amalgamated snapshot. However, pathways are not this simple. Pathways change dynamically during cancer evolution and especially during treatment. Knowing the connections within some number of pathways differs from understanding the often stochastic dynamics of these pathways, especially under stress. Identification of the individual pathways that contribute to cancer progression may help to explain the relationship among some genes or mutations in well-defined conditions, but the real challenge is how to evaluate the contribution from different pathways, how phenotypes emerge from pathways, and how these contributions change during cancer evolution. Targeting a dominant pathway in a relatively homogenous tumor may work well initially and result in a promising short-term response. However, the dynamic nature of cancer means that altered genomes will evolve and with these, new pathways or the combination of new pathways will emerge. This difficulty is not changed by large scale data generated from CGS projects. Obviously such complexity requires a new conceptual framework for cancer research in order to deal with pathway switching within the evolutionary selective landscape (see Chapter 5).

Table 4-3. Summary of complex pathway phenomena

Pathway switching — The predominant pathway(s) that drive a specific tumor change. For example, loss of HER2 in a breast tumor that was previously HER2 positive.
Moonlighting — Many genes have alternative functions rather than their well studied functions, and can fill to make various pathways function when necessary [Copley, 2012, Chapter 5].
Network switching — Overall changes to the genetic network structure that occur as a result of genome reconfigurations.

(5) Cancer complexity challenges the idea of using limited pathways to fight cancer. The complexity of pathways is underscored by their "hairball" appearance even when changes over time are not considered [Weinberg, 2010]. Pathway focus in cancer tends to downplay the complexity of pathways as driver pathways pertaining to cancer tend to be more focused and smaller than the pathways created in databases [Leiserson *et al.*, 2013]. It is also often forgotten that pathways do not work in isolation and the linear pathways likely not continually have influence in the highly dynamic process of cancer evolution. Cancer researchers often reside in a world of linear causation and are not familiar with complexity science. For most, the concept of emergent properties (which apply to the pathways and the phenotypes) is alien. In physics solving the three-body problem is challenging, how can we solve the problem of 10–100 pathways where interactions are often switching?

(6) Why cancer researchers are confident to find the targetable pathways? One important addition is the confusion of normal physiological/developmental conditions and pathological conditions. Focusing on pathways in cancer has been attractive as pathway-based studies in normal physiological conditions have produced meaningful biological information. Many believe that if we can just find the differences between pathways in cancer and those in similar normal, healthy tissue that cancer can be solved. However, cancer is an macro-cellular evolutionary process where cancer genomes continuously change (for detailed discussion of macro- vs. microcellular evolution see Chapter 5), especially in response to treatment. In contrast, during most developmental processes, the genome remains the same. In this case, only gene mutations and epigenetic changes matter. In cancer, when the genome is altered, the new properties emerge through formation and elimination of many pathways. The genome level changes that occur during somatic cell evolution mean that cancer pathways and networks are an ever-evolving entity.

4.4.7 *Cancer is indeed an evolutionary process*

Prior to CGS, the view of cancer as an evolutionary process was not popular within the field. Who needs evolution if the gene mutation identification is the key? The results of CGS changed the viewpoints, as the key message of CGS is that cancer is an evolutionary process, where alterations (at the epigenetic, gene, pathway and genome levels) are constantly moving targets. Suddenly, cancer evolution has become a hot topic. However, a key challenge for this view is the definition of cancer evolution and its clinical implications. Most are not familiar with the details of cancer evolution, particularly how genome-mediated macroevolution is the key for cancer research (See Chapter 5).

If cancer formation and progression is an evolutionary process, we have to change the way we approach cancer. How can we continue to develop and apply a target specific strategy when subclones differentially expand and contract and genome makeup is constantly changing, making causational inferences difficult if not impossible? Research focusing on a specific gene or pathway should give way to more systematic studies that focus on tumor heterogeneity and evolution. Business as usual approaches will continue not to work just as they have in the past only worked in isolated experimental systems and not in the majority of human tumors.

The key challenge for the majority of these categories is that while they all can contribute to cancer evolution, each contribution is limited. The specific functional contribution of each specific change is not as important as its contribution to the total evolutionary potential of a tumor. No specific mechanism can drive the properties that allow a tumor to escape selection. Cancer evolution results in emergent properties derived from genome level changes. The ability for new properties to emerge is more important than changes or tweaks to the parts (i.e. gene mutations or epigenetic changes). Lower level genetic changes often co-exist and conflict with each other. In addition, once genome level alterations take place, specific alterations at the gene level are less important for the evolution of a population than genome changes. In the rare case of stable genomes, gene mutations can be targeted until any genome changes occur. However, targeting most events, including epigenetic events, can result in stress. This stress can temporarily reduce tumor growth, but many times it also paradoxically promotes cancer evolution, resulting in tumor progression and the promotion of drug resistance.

Table 4-4. Comparison of interpretations of CGS data based on gene mutation theory and genome theory

Finding	Interpretation based on the gene theory of cancer	Interpretation based on genome theory of cancer
Mutation rates are not static	Mutation rates vary by cancer type, by tumor, and even by genomic location (DNA sequences located in different areas of different chromosomes have different mutation rates). This finding was unexpected, as the original goal of CGS was to once and for all identify common gene mutations.	Mutation rates of cancer are heterogeneous, varying by orders of magnitude even within a given cancer type. Mutation rate heterogeneity should not be a surprise, as the mutational landscape of individual tumors is highly diverse. Predicted by the genome theory based on an expectation that gene mutations would be similar to chromosomal alterations that are by and large unique to individual tumors. Even when gene mutations are identified, the challenge is to make sense of a specific gene mutation within the altered genome context. Genome context alterations change the networks in which genes function. The meaning/function of a gene mutation is thus not static. This is important considering the high frequency of non-shared genome level alterations in most tumors.
Mutational spectra	Certain cancers tend to have mutational patterns. For example, melanomas have a mutational spectrum that suggesting UV exposure plays a role, while lung cancers have a different spectrum suggesting the impact of cigarette smoke exposure.	Yes, some mutational patterns exist, but 1) Many occur only in a small fraction of a given tumor type; 2) Patterns are only obvious when certain environmental conditions play a dominant role in tumor promotion (e.g. UV exposure and skin cancer). The contributing factors of most cancer types are highly diverse and do not lead to such a pattern. 3) These patterns may give insight into the cause of a given cancer, but they are by and large not clinically actionable, due to the constantly changing genomes of tumors (evolutionary dynamics).
Chromosomal gains and losses	Most cancers show both large-scale and focal gains and losses. Little discussion of these changes	Aneuploidy and gene amplification/deletion have been long known to occur in cancer, but only recurrent changes were thought to be important. CGS data underscores the vast genomic change that is highly

	except when translocations result in known gene mutations or chromosomal copy number variations change the abundance of a cancer gene.	heterogeneous even within the same tumor. Genome change alters the inheritance of the entire system, meaning that many gene mutations could be incidental.
Chromosomal fragmentation	Complex chromosomal changes occur in most cancer types. These changes are typically one-time events.	Large-scale genome change was previously ignored, as it was thought to be largely unsurvivable or an artifact of cell culture. CGS demonstrates that this is not the case, as chaotic genomes occur frequently in most tumor types. It is likely that large-scale genome change is underestimated by CGS as CGS is often based on a population average. Genome chaos (formation of chaotic genomes is a dynamic process that occurs over and over during cancer progression. Most of these large-scale rearrangements will be lost over time but every chaotic genome contributes to the evolution of the population. Non-clonal genome chaos is one type of NCCA. The chaotic genomes observed by CGS represent the stage of the evolutionary process at the time the sample was taken and challenge the stepwise model of cancer progression predicted by the gene theory.
Driver genes	CGS will identify and confirm driver gene mutations. These can include oncogenes and tumor suppressor genes. Identifying the true drivers would reveal new targets for therapy.	Driver gene mutations occur, but they occur in much lower frequencies than pre-sequencing era data predicted. Drivers that do exist are diverse even within a given tumor type. Many of these drivers are ultimately linked to system stability. Even when driver genes are identified, there is a gap between targeting the driver gene and targeting outcome. Within the oncogene class, there is a bias towards proliferation-related genes. Most of the tumors that have been studied are late-stage tumors. It is possible that late stages of tumor progression bias the data towards proliferation. Each tumor undergoes multiple rounds of evolution during its progression. Each round of evolution makes use of different drivers

		that are genome context-dependent. Therefore, there are no fixed drivers. Recent studies have also demonstrated that so-called "passengers" contribute significantly to cancer progression.
New cancer genes	Many cancer-associated genes have been found or have been associated with new cancers. Identification of these genes will lead to new therapeutic targets and tumor-specific treatments.	The vast majority (>90%) of these genes were already known to be associated with cancer. The mechanisms of most of these genes are already well characterized. Even though these genes are well known, and have been for decades, they have largely been found not to be targetable, which is likely due to heterogeneity and adaptive genome change that occurs as a response to stress. Any new cancer genes will not result in the long awaited "magic bullet" and will not change the current cancer research landscape.
Signal transduction pathways	Mutations or CNAs involved in many pathways have been elucidated. Pathways that are linear do not change, and because of this, cancer pathways can be targeted. Pathways and the network concept can unify the diversity of gene mutations in cancer. Gene mutation diversity can be reduced to a manageable pathway if the genes that are mutated are grouped into functional pathways.	Most of the pathways identified have been long associated with cancer, such as the EGFR pathway. Surprisingly, most of these pathway alterations have very low penetrance. Even in cases where pathways are identified and targetable, targeting is only effective for a short period of time. Single cell based studies that incorporate genome context information with pathway information will increasingly demonstrate that complex pathway switches occur each time the genome is altered. Pathways are defined by genes, and many of the same challenges of using genes to explain cancer still remain even if pathways are involved.
Metabolism	Newly discovered mutations and epigenetic changes in genes such as IDH1 and IDH2 shed light on the importance of metabolic	Most if not all cellular processes can be connected to cancer through a broad array of pathways. Changes to the genome cause the parts (genes) to be used in new ways. Even when these genes are found, however, their penetrance is low. Similar to other cancer hallmarks,

	changes in cancer.	metabolic changes simply provide increased competitive potential. These are highly dynamic features that are hard to target. It should be pointed out that while metabolic pathways are important to cancer progression, so are cell cycle pathways, stress response pathways, DNA repair pathways, etc.
Lineage survival oncogene transcription factors	Many transcription factors that are associated with terminal differentiation have been surprisingly amplified and overexpressed in cancers.	This finding is indicative of proteins gaining new function and pathway switching in cancer. Oncogenes can have tumor suppressor function and vice versa depending on the genome context that they occur in. Specific transcription factor changes have low penetrance and prediction power.
Epigenomics	Approximately 40 epigenomic regulating genes are altered in a wide range of cancer types. Mutation of these is a powerful way to rewire cellular circuitry. Histone modifications, nucleosomal remodeling, chromatin compaction, DNA methylation, and DNA hydroxymethylation are all epigenetic-related changes.	Epigenetic change is important as it leads to flexibility of the tumor system. Multiple known and unknown mechanisms contribute to an evolutionary mechanism. The key is that changes are not recurrent across tumors, underscoring the evolutionary mechanism of cancer. In addition, stochastic epigenetic change is common following many different treatments. While these changes are interesting, they are not specifically targetable, as when targeted, the results can go either way. Genome changes create new systems while epigenomic changes only modify the existing system. The primary importance of epigenomic change lies in the microevolutionary phase of cancer. Epigenetic alterations are highly dynamic. The effects of restoring a specific epigenetic state will not last long in the presence of continual genome level alterations.
Histone modifications	Genes that affect post-translational modification of histones such as histone methyl transferases, histone demethylases, histone acetyl transferases, and histone readers are altered in cancers.	Changes in histone modification allow the genome system flexibility and overall evolutionary plasticity. Due to this plasticity, few or no specific recurrent changes will be found.

Nucleosomal remodeling	Genes associated with nucleosomal remodeling such as parts of SWI/SNF complex or ARID1A are found to be mutated in cancer.	These mutations affect global gene expression and create evolutionary potential. Specific genes from this category are typically not highly penetrant except in a few specific cases of cancer. A great example of this occurs in rhabdoid tumors where SMARCB1, a chromatin remodeler, is the only gene found to recurrently mutated.
DNA methylation	CGS and associated genomic studies have identified some cancers as having a CpG island methylator phenotype, hypermethylation, and hypomethylation.	Overall changes to the methylation status of cancer genomes have been known for decades. Few if any specific even moderately penetrant patterns have been found, even with large-scale studies. This change allows for flexibility if it is highly dynamic within the population might be more important to the evolution of the tumor.
DNA hydroxyl methylation	A new type of DNA modification has recently been identified, 5-hydroxymethylcytosine.	It is not evident what this change causes. It has only been identified in a small number of hematological malignancies and in some myeloproliferative diseases. DNA methylation change in cancer has been known for decades, yet it has not resulted in better treatments. If hydroxymethylation was even as important as standard methylation, it would have been identified years ago.
RNA splicing	Multiple splicing related genes are mutated in various cancers at up to approximately 15% of samples.	Over 80% of non-cancerous myelodysplastic syndromes have splicosome gene mutations but do not become cancer.
Non-coding RNA	There are multiple changes in expression and mutations of non-coding (nc) RNAs in cancer. Identifying cancer specific nc-RNAs will lead to new cancer specific targets.	Multiple nc-RNAs change in each tumor. Similar to genes, nc-RNA alteration is highly diverse within individual cells and across tumors of the same type. Targeting specific nc-RNAs may result in transient tumor shrinkage, but it will not eliminate unstable cellular populations with high evolutionary potential.

4.5 Essays That Address Four Key Questions Related To CGS

With the above analyses, it is time to debate some important questions raised by CGS that require extensive attention. These issues represent some of the most important discoveries of CGS and also paradoxically challenge future sequencing efforts for understanding the cancer landscape as well as the basis of the gene mutation theory of cancer.

4.5.1 *How do sequencing data underscore the ultimate importance of genome level alterations?*

The initial goal of CGS was to identify gene mutations and other genetic alterations to find the key drivers of each cancer type. Paradoxically, highly diverse DNA profiles and the overwhelming presence of genome level alterations identified by CGS suggest that these "drivers" are hard to define and when found, they may not be as important as predicted.

Multiple levels of genetic alteration have been known to exist prior to cancer genome sequencing including DNA sequence mutation, focal copy number alteration, epigenetic change, and structural/numerical karyotypic change [Heng, 2007a]. However, the degree and frequency of genetic alterations have been significantly underestimated within and between tumors. There are a number of reasons for this underestimation, including: (1) Traditional molecular research focuses on the identification of recurrent mutations, writing off heterogeneity as noise. Most the initial successes identifying recurrent gene mutation profiles were based on research carried out in established cell lines or in blood cancers with more defined linear progression patterns, supporting stepwise theories of cancer evolution. (2) Karyotypic analysis of tumor samples often requires prolonged culture, which selects for the formation of dominant clones and reduces heterogeneity of samples. (3) Individual investigators often focus only on their favored level of genetic change (gene mutation researchers often dismiss genome changes). There are few if any comparative studies of the contributions of gene mutations and chromosomal alterations using a given model or tumor type. (4) Karyotypic data are considered to be of low resolution and genome alterations are viewed as byproducts of gene mutations. (5) Until recently, karyotypic alterations were difficult to identify from WGS. With methodological improvements, CGS has revealed a mind boggling degree of genome level alterations, far exceeding what was previously anticipated. Few proponents of CGS view genome alterations as important; for instance, only if they result in the production of a fusion gene. Many sequencing papers have pointed out that genome level alterations are observed, but aside from any specific genes changes (fusions and other changes) caused by these alterations, no specific discussion of their meaning is offered. Recently, it seems as though this situation may be starting to change with the identification of certain chaotic genomes including chromothripsis and chromoplexy; however, many of these studies still attempt to show that these alterations affect certain specific genes (see the genome chaos section in Chapter 5).

If genome level change occurs in nearly every tumor and large-scale changes occur commonly in multiple types of cancer, these changes cannot continue to be ignored. To explain the lack of identified common driver mutations, some have suggested switching focus to epigenetic factors, non-coding RNAs, or other unidentified mechanisms. But why not study the connection between cancer and these well-known, regularly occurring genome alterations, even if they seem stochastic and lack a specific recurrent pattern thus does not fit within the mainstream gene-centric theoretical framework? We already know that there is an overwhelming amount of diverse genetic changes across the multiple different levels of genetic organization, so why continue searching for common recurrent changes as magic bullet (through epigenetic or further gene-based studies)? Could it be that the framework of searching for highly specific patterns in such dynamic systems is not correct, that the pattern is not one that affects specific chromosomes, but rather the pattern is simply one of dynamic change? Changes in the concepts of both the gene and genome are needed. In particular, the true functions of specific genes and pathways within the genome-based evolutionary adaptive landscape must be addressed. As discussed below, only this genome-based framework can explain why most genome changes observed in cancer are not shared and why focusing on specific genes will not explain cancer in general.

An important consideration regarding genome alterations is how new systems are created through genome re-organization regardless of the involvement of gene mutations. It has been hypothesized that re-organization of genome topology creates new transcriptome-defined network structures [Heng *et al.*, 2011a; Stevens *et al.*, 2013a; 2014]. We and others have recently shown that the karyotypic changes indeed alter the transcriptome. Karyotypic alterations change the genome context, which in turn directly affects the global transcriptional profile and the networks that arise from it. It is well established that cancers have altered networks, but the connection of karyotypic alterations with transcriptome dynamics has become obvious. For example, exome sequencing has shown that the pathways utilized by individual tumors to progress from low to high grades are very different. In essence, every cell in a tumor has a unique karyotype, thus each cell has a unique set of drivers. The network structures within individual cells of a tumor are as heterogeneous as the karyotypic makeup of that tumor. Therefore, non-shared genome alterations provide cell-to-cell transcriptome variation which is essential to cancer evolution. That is the reason why only focusing on specific gene mutations or recurrent genomic change is not successful in the majority of cancers, as each tumor actually has hundreds or thousands of potential drivers. Furthermore, each tumor is a result of a unique evolutionary process where each different transcriptome provides diversity for selection to act upon (see Chapter 5).

The high degree of stochastic genome alterations that occur in cancer and their relationship with elevated transcriptome dynamics also explains why cancer is hard to define by a specific recurrent mutational and/or expression pattern, and why long-term positive tumor responses resulting from molecular targeting are rare at best. Somatic evolution does not stop with the formation of the tumor. Tumor cells are constantly evolving, meaning that transcriptomes and gene target(s) are also changing. Because of this evolution, cell populations are highly dynamic as well. Current CGS studies typically only profile a tumor at one stage due to a number of practical limitations, and the profile is mainly based on the average of a mixed cell population, which is a challenge for diagnosis and prediction.

The genomic diversity of tumors and their ongoing evolution also explain the difficulty of using cell population-generated "average" transcriptome measures to diagnose cancer. A major goal of transcriptomic profiling and later genome sequencing is to classify tumors into subgroups based on patterns such as mutational or epigenetic landscapes. Specific, more effective treatments could then be designed for these different subgroups. Therefore, examination of expression studies and the development of expression profiles may give some insight to the challenges that heterogeneity causes for CGS. Identifying patterns of transcriptomic change to better understand and create cancer subgroups initially seemed promising. However, despite advances in the technology, traditional pathological classifications such as grading are more predictive than expression profiles of tumors. Furthermore, expression profiles of randomly selected genes may be just as informative as cancer expression signatures [Lee *et al.*, 2012; Venet *et al.*, 2011]. Why do gene expression profiles not accurately diagnose cancer? We recently systematically analyzed karyotypic and transcriptomic change during immortalization in a well-defined *in vitro* model of immortalization. It had previously been shown that stochastic, punctuated karyotypic alterations drove the somatic evolutionary process eventually leading to immortalization. Following immortalization, karyotypic changes occur more slowly in a stepwise manner unless the population of cells is stressed, at which point punctuated evolution occurs. To determine the mechanistic basis of the evolutionary process, global gene expression was measured in concert with karyotypic change at defined points. Analysis of expression data revealed that there were no expression patterns that were consistent throughout the immortalization process. Pathways that contributed to the process constantly changed in concert with the genome context. Drastic changes in global gene expression were found when large changes in the karyotype occurred. Transcriptomic measurements in replicate samples showed increased variation when karyotypic heterogeneity was elevated. Together these data suggest that each unique karyotype in a population has a unique transcriptome profile. Thus, the overall level of transcriptome dynamics is a better indicator of cancer cell population behavior than specific fixed signatures. This conclusion is also supported by comparisons of single cells and cell population averages of cancers [Abdallah *et al.*, 2013; Patel *et al.*, 2014].

Single cell sequencing, though in its infancy, is a promising method to identify heterogeneity in tumors as methods focused on a single gene have shown that point mutations are heterogeneous within tumors; single cell exome sequencing has shown that single cell exomes of tumors are highly diverse [Bielas and Loeb, 2005; Patel *et al.*, 2014]. Single cell sequencing should give expanded insight into epigenetic diversity and karyotypic heterogeneity within tumors [Baca *et al.*, 2013; Chambwe *et al.*, 2014; Navin *et al.*, 2011; Yu *et al.*, 2014]. Perhaps the most promising aspect of single cell-based methods is to identify the degree of heterogeneity required for cancer cell populations to progress. Current methods of genome-based CGS still lack statistical power to identify individual point mutations, though single exome sequencing can capture some of these data [Patel *et al.*, 2014; Sandberg, 2014]. Genome sequencing costs have dropped to levels that were unimaginable just years ago, but single cell analyses of tumors are expensive as the cost of such endeavors is multiplied by the number of cells sequenced. Sequencing only 10 cells increases the expense by an order of magnitude! In addition to this, computational requirements for analysis and storage scale with the number of cells per sample. Furthermore, a challenging issue is the integration of single cell data into a population context. Population dynamics are more important, and focusing on the genome-based evolutionary potential is a better solution to predicting cancer than averaged gene mutation tumor profiles (see Chapter 6).

Applying sequencing information to determine more personalized treatments using primarily targeted therapies can result in a good initial treatment response by matching appropriate chemotherapeutics with gene mutations found in a tumor. For instance, identifying lung cancers with EGFR activating mutations in order to treat with erlotinib improves treatment outcome. However, the identification of tumor subtypes based on mutational landscapes has by and large not recapitulated the success of EGFR stratification in lung cancer. Furthermore, treatment benefits derived from stratification are temporary at best, as instability and the resultant heterogeneity increase the probability of the evolution of drug resistance. Due to the extensive heterogeneity of the mutational landscape of tumors, the application of mutational profiles for diagnosis and treatment is likely to be just as problematic as expression profiles. Again, paying more attention to genome level heterogeneity within the cellular population might shed light on strategies to avoid or reduce system chaos that would slow cancer evolution in the long run (see Genome Chaos section in Chapter 6).

4.5.2 *How do sequencing data support punctuated cancer evolution?*

The theoretic framework of CGS rests on the hypothesis that cancer evolution results from a stepwise accumulation of driver genetic changes. Unbiased whole genome screens could identify these specific recurrent patterns of changes in cancer genes, and cancers would therefore be susceptible to targeting these driver genes. However, current sequencing efforts have revealed an almost opposite picture. It is thus necessary to examine the theoretical framework of CGS.

Even though there are few formal discussions focusing on the mechanism of tumor evolution, it has been assumed by some that convergent evolution and/or parallel evolution should play a dominant role in cancer progression. This reasoning is primarily due to the fact that the final phenotype of cancer (out of control growth, invasion and other hallmarks of cancer) is the same for most cancers. These hallmarks then must be explained by specific recurrent genetic change. Some have also acknowledged that divergent evolution does occur in cancer. It was hoped that early molecular changes might be the same, and these divergent features somehow can be explained by convergent molecular pathways.

The evolutionary pattern of cancer has been traditionally studied using karyotypes especially in hematological malignancies. These cancers however have been proven to be exceptions as they tend to be highly clonal and show a stepwise pattern of evolution. In contrast, the majority of cancer types display diverse karyotypes. Despite the overwhelming presence of chromosomal change, NCCAs have been historically ignored. Observation of chromosomal change during the process of immortalization revealed that there are two phases of evolution. Significantly, one of these phases is dominated by punctuated evolution (see Chapter 5 and 6). Unfortunately the important message of this work has been written off, as some consider karyotypic studies to not be molecular enough, or that the experiment is a cell culture artifact. To some, it may be hard to imagine that such drastic changes could happen *in vivo*. Now that punctuated patterns of evolution have been illustrated by CGS to occur regularly in tumors *in vivo*, and the highly diverse genetic landscape of cancer can only be explained by punctuated cancer evolution, the field is beginning to shift towards embracing punctuated evolution in cancer.

The first sequencing confirmation of punctuated cancer evolution was reported in 2011. Using whole genome amplification on flow-sorted nuclei, Navin *et al.* studied tumor population structure and evolution in two human breast cancer cases. In contrast to gradual models of tumor progression, this report suggests that tumors grow by punctuated clonal expansions with few traceable persistent intermediates [Navin *et al.*, 2011]. Single cell sequencing revealed the presence of non-clonal genetic aberrations which have been largely missed by averaging the genomes of the population or ignored by previous studies. The confirmation of punctuated evolution by sequencing sent waves through the field about punctuated cancer evolution. Soon after, chromothripsis, a subtype of chaotic genome was discovered by WGS [Stephens *et al.*, 2011]. Chromothripsis is a process where tens to hundreds of chromosomal fragments are rearranged into a confined genomic space, often on one chromosome. It was initially identified in a patient with chronic lymphocytic leukemia. Subsequent analysis showed it to occur in 2–3% of most cancers but in ~25% of bone cancers. Bioinformatic analysis suggests that it occurs through one massive genomic rearrangement during a single catastrophic event in the cell's history, strongly supporting the concept of punctuated cancer evolution. In addition, single cell exomes of glioblastoma, a tumor noted for its genomic heterogeneity, have been shown to be very diverse. This finding is likely to extend to other cancers when single cell sequencing is carried out.

Recently, CGS of 57 prostate tumors illustrated the importance of rapid and massive genome re-organization by describing a new subtype of genome chaos (chromoplexy). The authors stated that chromoplexy resulted in a number of chromosomal alterations that occurred "in a highly interdependent manner" and contributed to punctuated evolution by affecting multiple genes in one fell swoop [Baca *et al.*, 2013]. Further sequencing efforts have identified other subtypes of genome chaos in cancer [Holland and Cleveland, 2012]. Recent exome sequencing of 23 primary low-grade gliomas and recurrent tumors from the same patients demonstrated punctuated evolution at the gene level. Significantly this work also showed that temozolomide-treated tumors used alternative pathways to progression and expressed a mutation phenotype [Johnson *et al.*, 2014]. Together, the concept of punctuated cancer evolution has now become a hot topic and is increasingly discussed in top journals. The occurrence of punctuated evolution in tumors is likely underestimated by current sequencing strategies which often focus on late stage tumors (where stepwise progression may be more common) and average the profiles of the tumor population. With the development of single cell sequencing methods, particularly those that focus on overall karyotypic compositions of individual cells, punctuated evolution will be connected to more and more cancer processes [Johnson *et al.*, 2014; Sottoriva *et al.*, 2015]. Most current methods of identifying punctuated evolution focus on DNA sequencing, thereby averaging population profiles and reducing the detected frequencies of punctuated evolution. In addition, our model which uniquely allows us to watch evolution in action has clearly shown that there are multiple cycles of stepwise and punctuated phase transitions, underscoring the importance of the observation window that is used to study cancer. Due to the dynamic nature of cancer, the punctuated phase of cancer evolution can be short, although it is very important for the overall evolution of a tumor (See Chapter 5).

Though it is encouraging that increasing sequencing data support the concept of punctuated cancer evolution, it should be pointed out that punctuated cancer evolution occurs and can be observed at a number of different genetic levels (gene mutations, copy number variations and karyotype changes). Our application of the punctuated concept to karyotypic studies of cancer progression was influenced by Gould and Eldredge's theory of punctuated equilibrium [Gould and Eldredge, 1993]. Gould and Eldredge postulated that most evolution is marked by long periods of evolutionary stability, which is punctuated by rare instances of rapid branching evolution. In contrast to their theory which centers on phenotypic change during organismal evolution, we applied this concept to illustrate how cancer genotype changes couple with phenotypes, and realized that karyotypic alterations are the key to understanding punctuated evolution in cancer. Even though gene level change may appear stepwise, genome level alterations are often punctuated during stages of macroevolution. As we will discuss further in Chapter 5, the genes as parts of the system are rearranged into unique karyotypes in response to stress, creating new genome defined systems. These systems arise quickly in a punctuated manner from ancestral karyotypes especially during genome chaos.

A final important issue needs to be emphasized here. The acknowledgment of the commonality of punctuated somatic cell evolution in cancer should cause a turning point in cancer research. The traditional concept of stepwise evolution must now be changed. Strategies and expectations of CGS built on the stepwise concept must also change. Ironically, though punctuated patterns of cancer evolution are confirmed in some of these studies, the same researchers that performed these important experiments still favor searching for stepwise gene mutation patterns despite proclaiming the importance of punctuated evolution. Such confusion reflects the challenge of understanding the deep meaning of punctuated cancer evolution and calls for a new framework of cancer research.

4.5.3 *Is there a (fixed) cancer genome?*

The HGP was an astounding success. Why has the current cancer genome project, with much more advanced technologies, become problematic? In fact, there is a major difference between the healthy, normal human genome and cancer genomes. The normal human genome is fixed; most cells have 46 chromosomes and share the same karyotype. The situation is drastically different from the genomes of cancer as cells in most tumors have highly heterogeneous karyotypes. Not only are the genomes of a given tumor heterogeneous, these genomes continue to change over time due to the effects of chromosomal instability and the evolutionary selection process. Note that, prior to the genome sequencing era, the somatic cells are thought to share the identical genome for a given individual. Now, increased multiple levels of genetic alterations are commonly detected from different somatic cell types, and diverse mammals have been found to carry distinct genomes in their cells, emphasizing the importance of the genetic chimerism at the somatic cell level [Giorgi, 2015; Chapter 5]. We have suggested that since most diseases occur at the somatic cell level, and there is an increased variation between germline and somatic cells, the heterogeneity in somatic cell is the key for many diseases, especially but not limited for cancer.

The assumption of CGS is that cancer cells are still a fundamentally human system that only functionally differs by a limited number of common and accumulated genetic and/or epigenetic alterations. The issue of overwhelming genome heterogeneity was not a main concern for early proponents of CGS, as according to the gene-centric concept, chromosomal changes are insignificant and consequential. Even now, with the knowledge that genome re-organization is a common phenomenon in cancers, cells within a single tumor differ drastically at the genome level, and the genomes of these cells continue to evolve over time, the field continues to deny the concept that there are no fixed average genomes within most cancers. As a result, many are sequencing a highly dynamic collection of variable genomes that are vastly different from the original human cells from which they are derived. Such a disconnection between theory and actual biology is traceable to the stepwise theory of somatic mutation in cancer and the average concept of the cancer cell population (more discussions in Chapter 6).

The implication of a lack of a fixed cancer genome is highly significant. First, it explains the large number of non-overlapping genetic changes that have been widely found across nearly every cancer type; second, it questions efforts to sequence more and more tumors in hopes of at last finding the long hoped for Achilles Heel of cancer. Merely cataloging lowly penetrant mutation after lowly penetrant mutation will not benefit the field. Some have started to point out it appears that we are nearing a genomic plateau in cancer [Vogelstein *et al.*, 2013]. In fact cataloging more lowly penetrant mutations and the fanfare that surrounds such announcements is often confusing to the research community. It leads many scientists who browse titles and abstracts to believe a long promised gene mutation pattern has been found. The problem here is that if it is generally believed that a pattern is found and everything in cancer research is sound, there is no incentive to search for a much needed new framework. Third, comprehensive cataloging of epigenetic changes similar to what has been done with gene mutations is similarly fraught with problems. Each epigenetic change (which is more plastic and its effects are already less predictable than a gene mutation) has a different meaning in different genome-defined systems. Because the genome system constantly changes in tumors these epigenetic changes also have ever-evolving meanings, reducing their predictive abilities. This is not to say epigenetic change has no importance. For instance, understanding epigenetic changes during developmental and other normal physiological processes is useful because these processes occur within a static genome. Findings about specific epigenetic changes might be true if found within the micro-evolutionary phase of cancer evolution, but they will only hold true in that phase and if the genome context of any tumors they are found in are similar.

Unfortunately, the misconception that specific cancer has a fixed genome is not only widely accepted, it also has generated much confusion and debate. Moreover, though many acknowledge that genome alterations are ubiquitous in cancer, they still regard the cancer genome as fixed. Recently, WGS of the HeLa "genome" generated a lot of publicity by raising some important ethical questions [Callaway, 2013]. HeLa cells, derived from a cervical cancer tumor, are the most widely used cell line in research. Unfortunately the cell line was started in the days before informed consent, and no permission for establishing the cell line was ever given by the donor, Henrietta Lacks. Publication of the sequence data caused an almost immediate uproar due to gene sequences possibly revealing genetic traits of Ms. Lack's family members including disease susceptibility. Interestingly the same data were already available in public databases, though it had not been assembled into a full genome.

In August of 2013, the NIH and the Lacks family reached an agreement about the sequence data. Realizing the scientific importance of the data but the sensitive nature of the data, the HeLa Genome Data

Use Agreement was written [Callaway, 2013; Hudson and Collins, 2013]. This agreement outlines acceptable use of the genome sequence and created a work group composed of representatives from the medical, ethical and scientific communities as well as two members of the Lacks family. However, despite a deal being brokered between the Lacks family, the NIH, and the cancer research community, there are still unresolved scientific issues surrounding the debate about the "HeLa genome." For example, is the genome of HeLa cells grown in labs today still Henrietta Lacks' genome? Are sequence data informative about the genetic relationship between (HeLa) cell lines and the original donor?

We published a short response in the journal *Nature*. We have argued that the HeLa genome is no longer Henrietta Lacks's personal genome. Although the two share some DNA sequences, the similarity ends there. Lacks's genome had the usual number of 46 normal chromosomes, whereas most HeLa cells have 70–90 chromosomes and more than 20 translocations, some of which are highly complex. These drastic changes have resulted from multiple cycles of genome reorganization that occurred during *in vivo* cancer progression and from the culture of these cells. Considering that chromosomes provide the genome identity and blueprint, it might even be argued that the HeLa genome is no longer a human genome [Heng, 2013a]. The reason we raised this important issue was not to callously discredit any concerns or to obfuscate the ethical quandary resulting from what would today be considered the theft of Henrietta Lacks's tissue (in fact, we fully believe that Henrietta Lacks should have been informed about what was being attempted with her tissue and those experiments should have only happened if she agreed to them), but to rather open debate as to what really constitutes a genome, namely is there is a fixed cancer genome, and whether cancer genomes still belong to the human genome. This discussion (below) illustrates why only profiling DNA sequences is not enough to explain genomic identity, and debating this issue is indeed important. To discuss this issue, we wrote following in a blog entry for Molecular Cytogenetics [Heng, 2013b]:

"Privacy and ethical concerns about the publication of the HeLa cell genome sequence have generated a great deal of commentary. In

addition to a number of pieces from well-known scientific journals, the New York Times has also joined this interesting discussion.

Today, regulations require informed consent to be obtained prior to establishing a cell line from an individual. How do we treat HeLa cells that originally were established 62 years ago and without a consent? Who should claim the ownership of the cell lines, the original patient or the researchers who experimentally manipulated and selected the specific cell lines? Perhaps both parties could claim the ownership, if any.

While interesting, debates such as this seem to miss one fundamental scientific issue. The present HeLa genome, or any other HeLa genomes, while derived from Henrietta Lacks's genome is not her personal genome. There is no doubt that the majority of DNA sequences are identical between HeLa cells and Lacks' original cells. However, HeLa cell genome and Lacks's genome show very little similarity. Lacks' genome originally contained 46 normal chromosomes while most HeLa cells have 70–90 chromosomes with over 20 translocations, some of which are highly complex involving multiple chromosomal rearrangements.

To fully appreciate this point, one has to understand and define what the genome is. The genome is not just a collection of genes, arranged like pearls in a string. While individual genes produce functional parts or proteins, the genome structure has a function defined by the karyotype that provides the blueprint that describes how these parts work together. In other words, the genome defines the network structure by providing the 'system inheritanc'. It is the karyotype rather than DNA sequence that defines the identity of the cell line as different derived cell lines could have similar sequences but different karyotypes (or genome context). This concept has recently been illustrated by re-interpretation the main function of sex, and the importance of the karyotype. It turns out, karyotype rather than specific genes define the identity of the species. In the case of the HeLa genome, the changes generated over the past few decades have occurred through multiple runs of genome re-organization during cancer evolution and initial cell culture experimentations. These changes not only differ greatly from Lacks's original genome but it can be argued that the HeLa genome is not actually a human genome anymore. In fact, some evolutionary biologists consider cancer cell such as HeLa cells to be their own unique species (see works of Julian Huxley and Leigh Van Valen). In recent years, an increase of investigators including our own group have proposed/supported the notion that many cancer cells are no longer 'human cells,' as judged by karyotypic changes (for more information, please check out our references on genome theory). Yes, many DNA sequences in cancer cells are still the same as compared to their original human cells, but if we are only judging by DNA sequence similarity, the same can be said to cells from other species. Most eukaryotic species are defined by their unique karyotypes (despite sharing similar genes), demonstrating the importance of the genome, which is to organize similar genes to define the inheritance of each species.

Despite the fact that cancer cells are derived from normal somatic cells, they form newly emergent systems. The process of cancer evolution is punctuated, and its products cannot reform the original genome structure. Without a fixed cancer genome, the DNA sequence and karyotype profiles of dynamic cancer cell populations are so divergent from their original hosts that there should be little concern for privacy or ownership issues."

A close collaborator, Root Gorelick responded with his concerns:

"I fully concur with Henry Heng that HeLa cells are not what they used to be. Extant HeLa cells have a very different genome from what Ms. Henrietta Lacks had. Genomes evolve. Or in terms that evolutionary biologists would find more appropriate, genomes develop insofar as changes are in a single generation, i.e. all mitosis and no meiosis. Genomes change in our own bodies while we are alive — see all the endoploidy (aka somatic polyploidy) throughout our tissues — and in cell culture lines that may be preserved after our

deaths. Henry Heng, as well as Harris Bernstein, have done more than most anybody to demonstrate this crucial difference between small-scale genetic sequence changes and large-scale chromosomal changes. Kudos!

My difference with Henry regards ownership. For me, extant HeLa cells belong to Ms. Lacks and her family. Extant HeLa cells are still her personal genome. I agree that this genome has changed, but development is something we expect of all individuals. Just because an individual changes, sometimes even drastically, does not make them a different person. We do not lose personal property or our personal identity when we get an education or get a tattoo or even when we get cancer. I wish that I could disown the genomes of my cancer cells, claiming that they are no longer my personal genome because of their large genomic changes, but I cannot. A similar argument applies to evolutionary change. Just because speciation occurs, even if via phyletic gradualism, does not disqualify any species from being part of the lineage. We expect change. While ancestral vertebrates incurred a pair of polyploid events, they are still part of the chordate lineage.

I admit that what constitutes 'personal' will vary from person to person. Similar sorts of debates rage about what constitutes an individual. These are normative matters. However, if genomic change is all that it takes to lose ownership, then most of my normal healthy liver and heart cells, which are highly endoploid, are no longer mine. That is a viewpoint I do not wish to adopt."

Our answer to Gorelick was:

"...One key disagreement regarding ownership might come from a differential appreciation of the genome's role in normal development and disease processes. Developmental processes differ from cancer evolution. When completed normally, development does not drastically alter the majority of a given person's genome but cancer progression does so extensively. This genome change is the main

challenge for cancer research as cancer macro-evolution is punctuated and results in high levels of cell to cell genomic heterogeneity while development is more or less stepwise and produces less genome level heterogeneity.

Another point here is that lineage and the genomic identity (species) are related but different. Under normal developmental conditions, cells share similar average genome and both the lineage and identity of the genome is clear. In cancer, however, despite the relationship of daughter cells and parental cells connected by lineage, these cells often drastically differ in terms of genome identity. Just because there is a clear lineage from DNA point of view, does not mean that the cells are the same from a genome point of view. We do not call other species Human even though they share a lineage with us.

Yes, for some tissues, like liver, polyploidy is common, possibly due to stress responses. It must be pointed out that massive chromosomal translocations commonly detected in cancer are rare in healthy tissues. We and others are actively investigating this subject. Finally, establishment of a cell line with a unique genome requires a lot of work and often some luck. Perhaps the scientist who created the cell line should be more responsible for the new genome identity if it is derived from experimentally defined artificial evolution?"

From the above discussions, it is clear that a defined genome system is important, and that an important, unique feature of cancer cell is that there is no fixed genome. As we will discuss this issue in Chapter 5, while normal tissue can contain cells that have abnormal genomes, normal cells dominate in these tissues. The static genome of healthy cells is why the original human genome sequencing project worked well.

4.5.4 *Are there enough CGS data to validate the gene theory of cancer or are more samples required?*

If overwhelming heterogeneity is already the rule rather than the exception now that tens of thousands of tumors have been sequenced, is there a need to sequence more samples? If cancer is a disease driven by dominant recurrent gene mutations, sequencing more samples is only logical if the addition of more data reduces the "noise" and makes drivers more apparent. Current CGS results clearly show that the more samples are sequenced, the more diverse the data are. In fact, in some cancer types such as squamous cell lung cancer or melanoma, addition of more samples reveals lower penetrance of mutations such as TP53 and BRAF than initial results (**Fig. 4-2**). What then is the rationale of sequencing more? Indeed as noted above, many prominent voices in the field have begun to advocate against sequencing any more samples for CGS because it will not tell us anything new about cancer [Vogelstein *et al.*, 2013]. Unfortunately this viewpoint is often offered without articulating why CGS sequencing does not work at the first place, and certainly without questioning the conceptual framework of the cancer gene mutation theory that serves as the conceptual basis of CGS. Initial phases of sequencing have been completed, where hundreds of tumors of the major types of cancer were successfully sequenced and yet few patterns have emerged for most cancer types.

Just a few years ago, focusing only on genes already identified as being mutated in an initial screen in additional samples was used to validate gene mutation patterns, similar to what was done for studies of single gene diseases. In fact, validating mutations by finding mutational overlap in tumors has been a focus of CGS-based research, but the lack of an overlapping gene mutation landscape between tumors challenges the value of validation. Even when mutations do overlap, the genome context that they occur in differs because most tumors are karyotypically unique. These large-scale genome alterations change the context of gene mutations, meaning that even overlapping mutations often have unique functions. The inability to identify recurrent mutational profiles after hundreds of tumors of a single type have been sequenced should raise a red flag as to whether the current theoretical basis of cancer is correct. As discussed in the previous section, if there is no fixed cancer genome, the increased number of different systems will not significantly increase our knowledge of cancer. Clearly, for diseases with highly diverse genetic landscapes like most cancers, increasing sample sizes without addressing the failing framework results in continually falling dividends.

Though genomic heterogeneity is beginning to gain traction, many do not understand its implications. Some suggest overcoming the lack of mutational overlap in tumors is to create thousands of cell lines in order to represent the entire mutational spectrum of each cancer type. A total 3,000 cell lines per cancer type has been suggested, for example. Unfortunately this line of reasoning ignores the fact that cancer is an evolutionary disease. Most tumors (and cell lines) do not stop evolving and continually produce heterogeneous populations. Generation of an almost unlimited number of cell lines from a single tumor is possible if selection conditions are right. Established cell lines continually change and end up varying widely. Even HeLa cells, the most commonly used cell line, while relatively stable compared to most other cell lines, vary at the genome level from lab to lab. Each genome variation creates a new cellular system, so many labs may be using the same cell line, but most are not studying the same system. This situation is much worse in most cell lines that are more genomically unstable. Instability can be retained *in vitro* for long periods of time. During this time, no dominant clonal population arises, and experimental results can thus vary from day to day. To get a true picture of the population complexity of these unstable cell lines, single cell sequencing is needed. Although at this point, the necessary number of cells that must be sequenced is unknown. Instead of establishing thousands of cell lines for each cancer type and blindly sequencing more and more samples, we must ask whether additional sequencing provides further understanding of cancer and clinically relevant information and whether these data justify continued CGS. If the assertion that thousands of additional tumors must be sequenced to gain a true picture of the mutational underpinnings of cancer is so wrong, where does it come from? The idea that this many tumors are needed is based on statistical models that are calculated based on now known mutation frequencies. Previous estimations of the frequencies of specific gene mutations were overestimations and resulted in calculations that suggested that hundreds of tumors should be sequenced for analyses to have sufficient power to detect what was important. These calculations are based on the assumption that recurrent gene mutations are the drivers of cancer. If however this model is not true (and based on an overwhelming amount of now available data, it is not!) the basis for suggestions that additional thousands samples are needed is not solid. The fact that almost each tumor represents a collaboration of unique systems with unique genome alterations that are constantly changing means that millions of tumors could be sequenced without finding the long sought after mutational pattern of cancer. Even when such patterns are identified, predictive values of such patterns are low, and signatures with good predictive value such as ER, PR and HER2 status in breast cancer have long been known. At this point, sequencing more samples no longer reduces the noise it only underscores the presence of heterogeneity. With the inclusion of instability, must we then develop and sequence thousands of sublines per tumor? How do we incorporate the different stages of cancer progression where stability is often changing? Should single cell sequencing be applied to all of these lines? If this trajectory continues, CGS will become a black hole, sucking up all available resources. The best case scenario that will (and already has) come about is that CGS will further illustrate "unexpectedly high" levels of diversity.

On the other hand, if we realistically accept the facts from by CGS (that there is no fixed cancer genome and that sequencing more samples will not change this fact), we then can finally focus efforts to battle cancer using a new framework after 100 years of searching for key cancer mutations. This new framework must be based on the evolutionary potential of a tumor rather than focusing on specific genetic patterns. From a purely research point of view, further characterization of as many samples as possible might be interesting (for example, the specific mutational spectra for some cancer types is interesting but it has little predictive value), and technically possible, but if the clinical implications of such work is limited at best, different research directions with new strategies are urgently needed.

4.6 Conclusion

Cancer genome projects were started with the hope that using powerful new technology, which allowed an unbiased approach to record all gene mutations in tumors, would once and for all identify the entire list of gene mutations that accumulate and importantly those that drive various cancers. Doing so would lead to the eradication of cancer, as the molecular mechanisms of cancer could be targeted and blocked. Surprisingly for many, these efforts have shown that there are few gene mutations shared in a given tumor type and the vast majority of mutations are individual tumor specific. The mutations that are shared in most cases are known cancer genes that were identified years or decades before CGS efforts.

Equally surprised, some normal tissues also display typical pattern of driver genes mutations similar to cancer [Martincorena *et al.*, 2015]. Therefore whether or not gene mutations are the cause of cancer becomes an obvious question.

It is worthy to mention that, recent increased GWAS (Genome-wide-Association-Studies) on large numbers of other common and complex diseases have shown the same general trend that the more patient samples are used, the more variants can be identified [Visscher *et al.*, 2012], yet, the increased diverse markers only paradoxically pose challenges for clinical implication due to heterogeneity-mediated complexity and reduced certainty for majority patients [Heng, 2008; 2010, 2013c].

Now, that CGS has revealed that the majority of cancers have a highly dynamic and complex landscape, and in some normal tissues, there are high levels of typical driver mutations, what should we do? These efforts do not support the theory of recurrent gene alterations leading to cancer. CGS has made it glaringly clear that this theory is no longer supported by the genetic facts of cancer. When a theory is no longer able to explain the facts that are uncovered by experimentation and observation, it is time to form a new theoretical framework that does explain the facts as they are known. It turns out that in contrast to the goal of finding a unified gene-based mechanism of cancer, CGS seems to support a general evolutionary mechanism of cancer. If there is to be any success derived from this expensive project, the gene-based framework of cancer must be replaced with a new framework. Fortunately, the genome evolutionary framework nicely fits available cancer data.

Chapter 5

Unraveling the Mystery of Cancer: Understanding Genome Variation Mediated Cancer Evolution

5.1 Summary

If we are to establish the correct framework for cancer research, we must understand how somatic cell evolution works and if there are unique principles which differ from organismal evolution. This chapter will introduce some important discoveries we made about a new pattern of cancer evolution: why the genome rather than individual genes defines the system inheritance, and how to use the genome-based cancer evolutionary theory to unify the cancer field and depart from the gene mutation theory of cancer. First, the long ignored biological significance of stochastic genome alterations needs to be visited as these genome level changes are constantly detected in experimental systems and patient samples. Second, genome alterations, a key feature of system dynamics, have been linked to tumorigenicity in various cancer models. By tracing the level and types of clonal and non-clonal chromosome aberrations that occur during active evolution, it was demonstrated that there are two distinctive phases of cancer evolution — the genome replacement mediated punctuated discontinuous phase and the gene mutation accumulation mediated stepwise gradual phase. The punctuated phase holds the key to understanding the stochasticity that occurs during cancer evolution, which is drastically different from the traditional model of gradual clonal evolution. Third, to explain the biological consequence of genome alterations in cancer evolution, the concept of genome context defined "system inheritance" is introduced to differentiate system inheritance from the gene defined "parts inheritance". This realization is a significant step in understanding how cancer evolution works, as inheritance is a key component for evolution. It also explains why chaotic genomes are common phenomena in most cancers. Genome chaos is an effective way to achieve macroevolution in somatic cells exposed to stress or under crisis. Further analyses have linked genome alterations to transcriptome dynamics, revealing the mechanism of how non-clonal genome alterations increase the evolutionary potential of a tumor. Fourth, the synthesis of the above findings within the context of the genome theory has led to the establishment of the evolutionary mechanism of cancer and its relationship to large numbers of diverse molecular mechanisms. Such a relationship explains the fundamental limitation of focusing on specific genes and pathways for treating cancer, and can be illustrated by the multiple level landscape model. Lastly, the general cause of cancer will be discussed within the context of biological heterogeneity, stress induced adaptation, and evolutionary trade-offs.

5.2 The Rationale of Rethinking/Redefining Cancer Evolution

Cancer evolution does not simply follow the pattern of somatic cell evolution defined by neo-Darwinism where the accumulation of gene mutations during clonal expansion is the key feature. Based on this important realization, it is necessary to redefine cancer evolution.

5.2.1 *A brief review of evolutionary studies of cancer*

The message of the first three chapters is loud and clear: a new conceptual framework is urgently needed for cancer research. But what should this framework look like?

The leadership of National Cancer Institute (NCI) has also been searching for the answers. In 2008, they organized three unusual think tank meetings under the title of "Physical Sciences-Based Frontiers in

Oncology." These meetings brought together top scientists from physics, physical chemistry, engineering, mathematics, cancer biology, and clinical oncology to collaboratively identify the most promising research questions and strategies in cancer research.

Four major focus areas emerged including the overall physics of carcinogenesis, the role of evolution and evolutionary theory in cancer, the complexity of cancer and information, and information theory in cancer. According to the NCI leadership, these priorities are the next logical step in moving the field forward [NCI Cancer Bulletin, 2008].

Understanding the evolution of complex cancer systems will aid in the development of a new framework of cancer research. Understanding the evolutionary process of cancer requires knowing how the cancer cell passes inheritance, as well as how complex systems behave under a high level of selection. Without evolutionary synthesis, we are left with only stacks of contradicting experimental results, case studies, and knowledge of the dissected parts of the system, all of which are confusing and hard to understand.

A bitter question for the cancer field is, however, why do cancer biologists need scientists from other fields to remind them of the ultimate importance of cancer evolution? Looking back, cancer researchers lost their insightfulness decades ago when reductionist approaches took root and began to treat the complex evolving system of cancer as isolated gene mutations and pathways.

Theoretically speaking, biologists should know evolution better than any other field does. There are only a few fundamental principles in biology, including the law of evolution and inheritance, the mechanisms of development, and non-linear complexity and emergent properties. Sadly, however, the reality is quite the opposite. While the principles of evolution are well taught in college classrooms, most cancer researchers do not apply them to their cancer research as this important concept seems unrelated to cancer. Furthermore, most researchers are not very familiar with the complexity theory, and instead separate parts and emergent properties of the whole system. The lack of this understanding led to the reductionist approach that permeates modern biology. Current bio-reductionists often continue to hold the belief that an understanding of the whole system can be obtained by characterizing all the individual parts. They often consider holistic approaches to study nature to be not mechanistic as they lack molecular details. All too often within the field, evolution is treated like a philosophy rather than a scientific tool essential for understanding cancer.

There is another practical reason for the lack of an evolutionary perspective in mainstream cancer biology. In the past half-century, even without this perspective, cancer biologists have made much progress by using molecular genetics. In particular, starting from the late 1970s, molecular-based cancer research identified many key cancer genes, seemingly validating the cancer gene mutation theory. The future of cancer research seemed brighter than ever as judged by increased power of molecular techniques. The coming of age of molecular biology left many cancer biologists so busy performing molecular experiments, they barely had any time to consider the big picture. The bottom line is: Who needs evolution theory when everything is working? The overall lack of reverence for cancer evolution is reflected in the extremely low frequency rate of the cancer literature mentioning cancer evolution. Even research reports of typical evolutionary problems like drug resistance rarely mention evolution [Aktipis *et al.*, 2011].

The evolutionary principle of cancer was directly borrowed from the current understanding of organismal evolution as key elements essential for the evolutionary process are shared by cancer (somatic cell) and organismal evolution. Both occur in complex biological systems that have the capability of evolving. Each has heritable population variation, and this variation affects reproduction and/or survival [Crespi and Summers, 2005; Heng, 2007a; Merlo *et al.*, 2006]. The key difference between cancer and organismal evolution is that it occurs at the cellular and individual levels, respectively. The evolutionary time scale is drastically different as well. Organismal evolution occurs over millennia, while cancer formation only needs years to decades. Finally, in the vast majority of cancer types, the success of cancer evolution reaches a limit with the death of the host.

The evolutionary view of cancer was proposed by Cairns and Nowell in the 1970s, mainly based on evolutionary studies of bacterial mutations and hematological malignancies [Cairns, 1975; Brash and Cairns, 2009;

Nowell, 1976]. The initial concept of cancer evolution focused on the process of clonal expansion that allows for the accumulation of gene mutations essential to cancer formation. If cancer evolved according to this model, clonal expansion of cells with accumulating mutations could be explained by a small number of mutations. Efforts focused on finding this elusive gene mutation set in each type of cancer.

Despite the initial success of identifying a number of gene mutations with high penetrance in certain patient populations [Rowley, 1988], which strongly supported evolutionary convergence of the cancer process, most subsequently identified gene mutations were only lowly penetrant in cancer patient populations. This finding does not fit the clonal expansion hypothesis.

There are a number of groups who consistently push cancer evolutionary studies, due to the realization that cancer is a highly complex process of somatic cell evolution that cannot be satisfactorily explained by individual molecular mechanisms or a combination thereof. For example, various aspects of cancer have been studied by Robert Gatenby's group including: competition between normal and cancer tissue, the contribution of microenvironments, the dynamics of the evolutionary landscape, and adaptive therapy [Gatenby, 1996; 2009; Gatenby *et al.*, 2009; Vincent and Gatenby, 2008].

Following the rapid expansion of gene based cancer research, Bert Vogelstein's group at John Hopkins introduced the importance of chromosomal instability in cancer using molecular terms. They have further synthesized their molecular genetic data within the cancer evolution concept. Specifically, they linked both chromosomal instability (CIN) and micro-satellite instability (MSI) to cancer evolution [Breivik, 2005; Cahill *et al.*, 1999; Calabrese *et al.*, 2004). Their influential studies have popularized the idea that an accumulation of small alterations (mutations) leads to big changes (cancer), and particularly, the evolutionary process of tumorigenesis occurs through multiple linear steps, which support the conceptual basis of the gene mutation theory of cancer [Hanahan and Weinberg, 2000; 2011; Huang *et al.*, 2009; Nowak *et al.*, 2006; Nowell, 1976; Vogelstein and Kinzler, 1993]. Unfortunately, their observation of the dynamics of stochastic chromosomal evolution (which is much more common in cancer than micro-satellite instability) was ignored by most cancer biologists who were more interested identifying the long-sought-after linear stepwise order of gene mutations that is thought to define each type of cancer.

Martin Nowak's group, in collaboration with Vogelstein, developed various mathematical frameworks to model different aspects of cancer evolution including the initiation and progression of cancer and the mechanism of drug resistance [Frank and Nowak, 2004; Michor *et al.*, 2003]. Importantly, they calculated that CIN could initiate colorectal tumorigenesis prior to inactivation of tumor suppressor genes [Nowak *et al.*, 2002]. Obviously, such an important message was not appreciated by gene-centric researchers who were insistent that chromosomal changes must be a later event than gene mutation. It should be pointed out that many collaborators and associates of Nowak's group including Steven Frank and Franziska Michor have done an outstanding job in pushing for a better understanding of cancer evolution.

Brian Reid's group used Barrett's esophagus as a neoplastic progression model to study how natural selection and clonal diversity drive cancer progression [Maley *et al.*, 2004a; 2006]. They also used agent-based computational models to show the promising strategies of benign cell boosters and chemosensitivity boosters [Maley *et al.*, 2004b].

Mitelman and colleagues have studied karyotype evolution in cancers by focusing on clonal karyotypic evolution of tumors and made efforts to establish the relationships between cytogenetic subgroups and karyotypical pathways. However, their analyses do not explain the overwhelming karyotypical diversity they have observed from clinical samples. They have identified over 40,000 different karyotypes in the 50,000 tumors they have analyzed, suggesting that no commonly shared, recurrent patterns of karyotypic change exist in the majority of cancer types and cases, particularly in solid tumors [Mitelman, 2006]. CIN and its associated NCCAs have been linked to this high heterogeneity [Gisselsson, 2003; Heng *et al.*, 2004a; 2006a].

A number of key features of cellular alteration or hallmarks of cancer have been analyzed in a stochastic multistep cancer model [Spencer *et al.*, 2006]. Simulations have been used to illustrate how tumors grow by a hybrid discrete continuum model [Anderson *et al.*, 2006]. The waiting time and transition from benign to metastatic phase has been studied based on newly identified numbers of common gene mutations [Beerenwinkel *et al.*, 2007; Jones *et al.*, 2008]. The invasion of cancer cells into the bone marrow microenvironment has also been studied using mathematical equations/modeling [Chen and Pienta, 2011]. Recently, a variety of evolutionary theories and ideas have been introduced and implemented in cancer research to understand the paradoxes of cancer through the evolutionary lens [Aktipis *et al.*, 2013; Caulin and Maley, 2011; Crespi and Summers, 2006; Crespi *et al.*, 2014].

There are many other exciting evolutionary studies in cancer as well. Interested readers should check out the following researchers: Daryl Shibata, Jose Costa, Larry Loeb, Christoph Klein, Mel Greaves, John Pepper, Charles Swanton, and Kornelia Polyak, to name a few. In addition, more recently in an attempt to make sense of cancer genome sequencing data, somatic cell evolution has become a hot topic. Increasing numbers of molecular biologists are using this terminology as evidenced by many high profile publications.

5.2.2 *The challenges of the current evolutionary theory of cancer*

Despite nearly 40 years of effort, the cancer evolution field has not yet changed the landscape of cancer research. This reality was clearly communicated in Robert Weinberg's recent view piece, which is critical of current cancer approaches [Weinberg, 2014]. He realizes that current cancer research needs a new framework, but in the piece, he did not mention the evolutionary framework at all.

Even within the cancer evolution field there are some fundamental questions or paradoxes that challenge common understandings of how cancer evolution really works. Cancer is easily explained by hand-waving, but specific explanations that hold up to rigorous questioning are few and far between. Unfortunately, there is no vigorous public discussion on this critical issue and no specific efforts to understand the paradoxes that challenge cancer evolution. Interestingly, these questions are frequently discussed during our private communication with some prominent thinkers in the evolution field.

On the surface, it seems incontrovertible that cancer evolution can be explained by Darwinian evolution. Most researchers seem to agree on the three essential factors that define cancer evolution (variation within the population, heritability of variation, and variation must affect cellular survival or reproduction). But at a deeper conceptual level, there are fundamental disagreements about the three elements, which lead to a sensitive question: can cancer evolution truly be explained by current Darwinian evolutionary principles?

Yes, variation is important, but what types of variation are most important? Gene mutations? Epigenetic alterations? Genome level changes? Something else such as microenvironments? Are all of these different levels of variation important and should they be considered equally in evolutionary studies, or are some levels of variation more important than others? What about if the levels have conflicting contributions? When we bring up these questions, a common response is it does not matter what types of variation are involved, it will not change the evolutionary dynamics under investigation. Is this true or false? We want to know the answer.

Yes, variation must be heritable, but what defines key inheritance for cancer cells: cancer specific genes/mutations, epigenetic states, or karyotype-defined genomes? What are the proportions of the genetic and non-genetic contributions to cancer evolution? Do different levels of inheritance cause different evolutionary patterns? If new types of inheritance exist, what is the impact of those on current evolutionary theory?

Yes, inheritable variation must affect survival or reproduction, but in which way? Does this occur in a stepwise manner where small advantages accumulate over many generations, or does it happen in a punctuated pattern where huge changes rapidly dominate within one or a few generations or even within hours? Is cancer evolution more closely related to the pattern of organismal evolution within species (micro-evolution) or above the species (macroevolution) level?

Furthermore, how do various genetic and non-genetic aberrations change the responses to tissue- and organ-defined environments (ecological factors)? What levels of change or combinations thereof are required to break the multiple levels of constraint in order to allow abnormal cells to form cancer? How can the evolutionary potential of different types of cells within and between cell populations (cancer cell populations can be highly diverse, consisting of cells with different tissue types, karyotypes or gene/epigene profiles) be compared and/or measured?

The answers to these questions are of ultimate importance, as any correct framework of cancer evolution must be able to answer these questions. Unfortunately, the current mainstream gene-centric cancer evolution theory does not satisfactorily address most of these questions. This has led to much confusion, as highly diverse multiple levels of genetic heterogeneity are observed in human tumors and traditional ideas about cancer evolution cannot explain the existence of such heterogeneity.

On one hand, increasing numbers of researchers now acknowledge the overwhelming existence of genome level changes, including genome chaos, within tumors. These large scale chromosomal changes differ from gene mutations and copy number variations. Their commonness in clinical samples would make it seem that they must be biologically relevant, but their biological relevance is difficult to comprehend within the current evolutionary frameworks. The key to Darwinian evolution is the accumulation of small changes over time. How then can positive selection result from large genomic rearrangements that change multiple biochemical processes and pathways within each cell, especially when the majority of them lead to cell death or loss of selection advantages?

The drastic stochasticity observed at the genome level within cancer also puzzles and even discomforts evolutionary biologists. Cancer evolution must be very different from organismal evolution, they often argue, as cancer is very bizarre. Cancer evolution more closely resembles a "hopeless monster" (given that most massive stochastic genome alterations are not survivable) rather than a "hopeful monster," as suggested by Richard Goldschmidt regarding organismal speciation resulting from genome level changes (as discussed later in this chapter, the transition from hopeless to hopeful is the function of sex). According to the modern synthesis that bridges the work of experimental geneticists and evolutionists, the genetic mechanisms are mainly built on gene-based inheritance with little appreciation given to genome level and non-genetic inheritance (see section 5.6). Given the status of current cancer genomic research which challenges the concept of the gene and the traditional definition of inheritance, the problem of the modern synthesis becomes obvious.

On the other hand, some evolutionary biologists just prefer to study evolutionary models without getting into details of the specific genetic mechanisms (if it is so hard to know what inheritance is in the first place). They reason that they can easily study evolution in nature without any detailed knowledge of genetics, as Darwinian dynamics only require a "mechanism of inheritance" in principle but otherwise are governed entirely by interactions between the environment and phenotypic properties of the organism. In fact, for over 100 years after Darwin's work, evolutionary biologists and ecologists developed an excellent understanding of the dynamics of life with no working knowledge of genetics. Similarly, genetics can be readily investigated without any knowledge or interest in evolution. This situation also can apply to cancer research, where a great advantage has been made at the basic research front without really considering how cancer evolution works.

In addition, from a practical point of view, cancer evolution concepts do not seem to directly help most molecular researchers. In the name of molecular mechanism research, the gap between basic research (focusing on "parts characterization") and clinical reality (driven by somatic cell evolution) continues to increase. Even though evolutionary theory has predicted and can explain this gap, it has not yet revealed the next magic molecular target. Without providing specific druggable targets, neither major funding agencies nor pharmaceutical companies will be excited enough to provide financial support for research projects, despite that evolutionary theory should guide current cancer research.

We strongly disagree with these viewpoints and reasoning. In order to address many of the previously mentioned paradoxes, we are convinced that we have to apply a new mechanism of inheritance and evolutionary principles to cancer, no matter how challenging. Perhaps, the current approach that only considers inheritance as a philosophical principle (without accounting for detailed mechanisms of inheritance) or as a gene-based principle has crippled evolutionary theory. We might in fact only understand gene-mediated evolution, which is drastically different from genome-mediated evolution. Such limitation might have caused the confusion between micro- and macroevolution.

The same limitation is also applicable to the field of genetics, which has failed to integrate evolution into its principles. After over a century of research, we now realize that many of the key mechanisms that are gene-based do not work! Challenging examples include the concept of the gene, missing inheritability, and the increase of identified diverse genetic and non-genetic elements that contribute to diseases. As we will discuss in Chapter 6, the knowledge of genetic parts characterization is extremely limited to explain or predict the clinical cases where the somatic cell evolution is based on the stochastic selection of a large number of different parts. Since such selection is above the individual gene, gene based inheritance no longer explains the majority of common and complex diseases. This situation can be considered as strong evidence for needed integration of genetics with somatic cell evolution. As without such integration, we will neither fully understand how genetics works in the context of evolutionary dynamics, nor how evolution works based on different types/levels of genetic/non-genetic organization.

This case also applies to current cancer research. For understanding cancer, analyzing genetic parts alone (such as gene mutations or individual pathways) is not enough, as somatic cell evolution involves inheritance above the gene level, and selective dynamics are mainly based on genome alterations during the evolutionary process rather than gene mutation information (for more information, see Chapter 6). However, cancer evolution is currently considered as highly gene mutation centric, and clonal expansion and diversification is the main explanation based on fitness/selection. It is now time to incorporate different types of inheritance to fully understand cancer evolution, and especially address the relationship between macro- and micro-cellular evolution regarding its genetic mechanisms, but how?

5.3 Searching for the Meaning of the Stochastic Genome Alteration in Cancer

Researchers have been deeply puzzled by the large amounts of seemingly random genome changes observed in the vast majority of cancers. At first glance, understanding why and how chromosomal aberrations contribute to cancer should be a top priority in cancer research. However, high interest in identifying various cancer genes and mutations has led to a lack of cancer studies focusing on chromosomal aberrations. According to the generally accepted view, chromosomal aberrations must be the result of gene mutations and are therefore late stage events. The only perceived benefit of chromosome based cancer cell analyses is identification of specific chromosomal patterns (primarily recurrent translocations) that directly affect a specific gene. Heterogeneous patterns of chromosomal aberrations are considered as insignificant "genetic noise". To many in the cancer field, the real "causative" factor(s) of cancer must fit Koch's Postulates. The drivers of cancer therefore must be gene mutations or alterations commonly identifiable in a given tumor type, and not found in healthy tissue. Based on Koch's Postulate, any factors that exhibit high heterogeneity cannot drive cancer. One of the biggest promises the gene mutation theory of cancer made was that there are a few key commonly shared cancer gene mutations that could be identified and targeted. Chromosomal cancer theory on the other hand was believed to be unable to deliver due to the high frequency of stochastic aberrations. To address this perceived issue, molecular cytogeneticists have developed more powerful methods to identify changes in specific chromosomal regions to increase the resolution of cytogenetic analyses in the hopes of identifying patterns of change, but this has only led to the realization that stochasticity is much more prevalent than the field predicted.

For example, spectral karyotyping (SKY) and combinatorial multi-fluor-FISH (m-FISH) were developed in the late 1990s as a method to accurately identify chromosomes especially complex translocations that are difficult to identify by traditional G-banding. By labeling DNA from each chromosome with a unique color or color combinations, the hybridization pattern along chromosome and among chromosomes can effectively reflect the chromosomal number or structural alterations

[Heng *et al.*, 2001a; 2003; Schrock *et al.*, 1996; Speicher *et al.*, 1996; Ye *et al.*, 2001; 2009a–b]. It was hoped that SKY would help identify a common signature of chromosomal alterations in each cancer type, but SKY has failed to identify common recurrent patterns of genome alterations in most cancers. In contrast, highly diverse genetic alterations are evident at the genome level, including many previously ignored forms of change, such as defective mitotic figures (DMFs), chromosome fragmentation (C-Frag), chaotic genomes and many unclassified chromosomal abnormalities that have been under the radar for decades [Heng *et al.*, 1988; 2004a; 2013a].

Something major did not make sense to us following SKY analyses of a large number of cell lines and cancer samples. No matter how careful we were, we could not find chromosomal alterations that were overtly common in a given type of cancer or across cancers. Instead we found a surprising and overwhelming amount of seemingly stochastic, non-clonal (non-shared) chromosomal aberrations. For decades, the failure of identifying common chromosomal aberrations in most cancer types was thought to be due to methodology limitations. But with new cutting edge technology in hand, like other researchers, we still failed in identifying a pattern. Why?

The common response to these findings is typically "well, you are not lucky enough" or "your samples must be anomalies" or "try harder and test more samples!" This approach in the past was successful, and many successful researchers found patterns and published these patterns in high profile scientific journals. Who can forget the story of the Philadelphia chromosome in CML?

The application of SKY and other molecular cytogenetic methods and the pervasiveness of stochastic NCCAs convinced us not to ignore these genome alterations [Heng *et al.*, 2004a]. We asked, "What if there are no commonly shared dominant genome alterations in most cancers, and the signature patterns observed from blood cancers are exceptions and not rules?" In this light, it is important to try to make sense of all these seemingly random data.

The data forced us to question mainstream reasoning. Questions changed from "Why is there no common pattern if cancer is caused by a stepwise accumulation of a few common gene mutations?" to "Why should there be a common pattern if the cancer process is highly dynamic, and many diverse genetic/environmental factors can contribute to cancer formation?" Changing the questions asked suddenly defined the mission we were on. Our own search for a pattern of chromosomal alterations that could explain cancer switched to a focus on understanding why and how stochastic genome alterations could contribute to cancer.

To answer this question, we were not only required to re-examine the majority of findings about cancer in the literature, but we also needed to develop a new concept and experimental system to analyze and comprehend the stochasticity observed in biological systems. This concept forced us to identify and address the limitations of molecular reductionism, which provides the foundation of the bulk of today's cancer research. This approach is designed to eliminate the "noise" of heterogeneity in order to ease the identification of patterns; such an approach unfortunately simplifies the reality of cancer and ignores the multiple levels of genetic and non-genetic heterogeneity, the very nature of cancer (See Chapter 6).

5.3.1 *"Noise" is the signal of system instability and a pre-condition for adaptation*

One major question we asked was, "If one considers biological systems to be complex with many pathways comprised of many parts, how can the behavior of the system be monitored without knowing the intricate, near infinite details of the parts?" We reasoned that instead of focusing on minute changes to specific parts, we would focus on identifying and measuring system stability, particularly when under stress in order to understand the system dynamics.

As illustrated in **Fig. 5-1**, according to the idea of system control, to determine whether or not the system is stable one can simply measure the level of variation and the time of recovery following the introduction of a stress condition to destabilize the system. Such measurement does not require pinpointing the specific factors that contribute to system

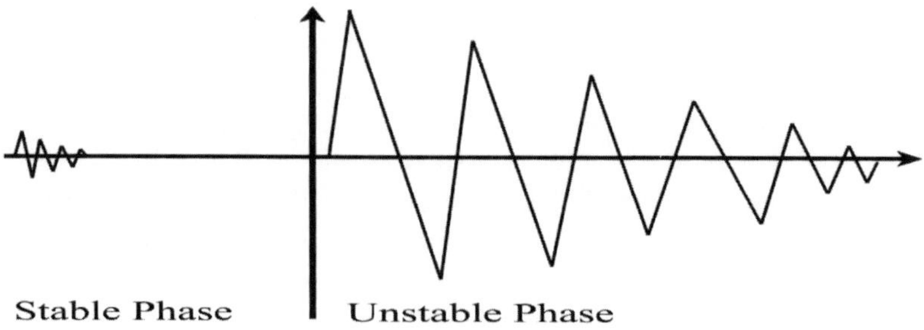

Fig. 5-1. System stability can be monitored by comparing the system response under a given stress.

In the stable systems, the stress induced system dynamics (stochastic alterations) are lower, and recovery time is shorter. In contrast, in unstable systems, the induced dynamics are high, and the recovery time is long. Thus, system instability can simply be measured using the level of stochastic genomic changes (providing the rationale to monitor NCCAs in cancer evolution).

stability. Thus the long ignored bio-noise is now an important tool to monitor system behavior.

Furthermore, for tumors consisting of heterogeneous cell populations, there are multiple sub-systems involved in progression, and "noise" within each of the subsystems is an essential precondition for evolutionary adaptation. Despite those researchers who consider "noise" as negative and try hard to eliminate it from their studies, "noise" is a "positive" pre-condition for somatic cell evolution. Thus, the high level of "noise" holds the key to understanding cancer.

The above realization also addresses a common question researchers often ask, "If there is no specific genetic alteration pattern, how can we understand the system?" The answer is to study and understand the system behavior in order to predict it. As the system status is more important than individual specific parts when the system is composed of so many parts with such complex interactions, the degree of dynamics of a given system can be measured by the level of "noise" within the system, and this noise defines the evolutionary potential.

5.3.2 *Which genomic level should we study?*

With the apparent importance of the stochastic genomic changes in bio-systems, an immediate follow-up question is what level we should monitor, as there are many levels of genomic organization including random gene mutations, copy number variations, chromosomal aberrations and epigenetic alterations?

The genome level seems to be the obvious choice to monitor the system itself rather than its parts. First, the genome is the highest level of genomic organization and genome alterations usually have much more significant impact than individual gene mutations. Second, according to the information theory, while lower levels of information might be easier to access, they contribute less to the overall control of the system. Finally, based on the science of complexity, understanding the parts often has limited contribution in understanding the system behavior, especially when nonlinear emergent properties are involved. We have also addressed this issue using the term "knowledge gap". More information about this gap can be found in [Heng, 2013c].

We thus realized that the tradition of molecular biology has been to push for higher resolution. However, while easier to study and comprehend, the lower levels that are illuminated by high-resolution studies have paradoxically reduced biological importance than higher levels of the system.

As illustrated in **Fig. 5-2**, there is a balance or trade-off between resolution and research importance. If the system behavior control is at the genome level, then pushing for higher resolution will likely jeopardize the importance of the research.

This paradoxical relationship is particularly applicable to all research of diseases that involve somatic cell evolution, especially cancer research. Selection pressure is applied to cancer cell populations primarily at the individual genome level rather than specific gene level. In contrast (as discussed in other chapters), in developmental studies for example, cellular differentiation gene regulatory changes by epigenetic mechanisms are the key drivers.

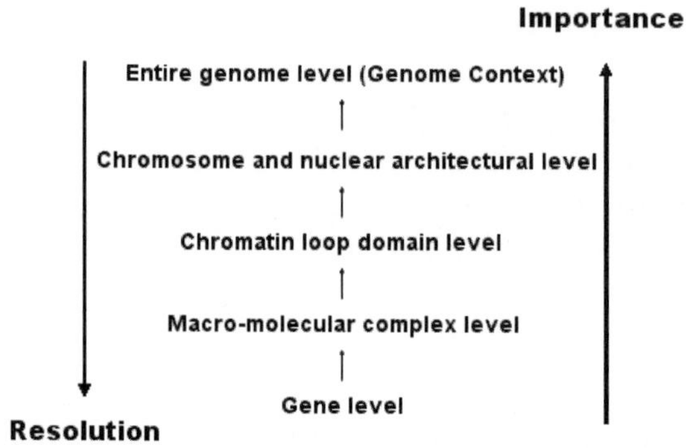

Fig. 5-2. The paradoxical relationship between high-resolution research and biological importance. Adopted from [Heng, 2013c] with permission.

This realization again raises an interesting point: we might not need to know every detailed aspect of the lower level parts of a system in order to understand the higher level of the system, especially during the evolutionary process. In other words, we might not need to know the mechanistic details of each genetic alteration to understand cancer in general, but this could be accomplished by focusing on the overall genome level stochastic changes. Interestingly, the holistic view of systems and complexity of diseases has gradually received increased attention [Sturmberg and Martin, 2013].

5.3.3 *Random genome alterations (e.g. NCCAs) are linked to genome instability*

While the above conceptual considerations may make sense, ample experimental evidence is needed to link the stochastic genome alteration to key behaviors of the cancer system in order to convince the cancer research field to no longer ignore "genetic noise". Since genome instability is the most common feature of cancer, the linkage between genome level stochastic changes and genome instability must be established. Note that although the presence of elevated genome level changes is defined as genome instability, most of the genome changes previously referred to as genomic instability were only recurrent changes, and stochastic changes were considered as noise.

To achieve this goal, the following approaches and evidences are needed: (1) Identify the important measurable "noise" at the genome level for cancer; (2) Link the level of change to known genetic instability; (3) Illustrate such universal link to various aspects of cancer.

5.3.3.1 *The definition and classification of NCCAs*

According to current cytogenetic standards set by the International System for Human Cytogenetic Nomenclature (ISCN), only clonal chromosomal aberrations (CCAs) are reportable cytogenetic anomalies [ISCN 2009]. CCAs are specific structural or numerical chromosome changes detected in at least 2 in 20–50 mitotic figures (occurring in 4–10% of analyzed cells). The majority of available literature reporting on cytogenetic changes in cancer does not report the frequency of specific CCAs; in the reports that do, the CCA frequency is usually greater than 30%. Therefore, NCCAs can be practically defined as occurring at a frequency of less than 4% in 50–100 mitotic figures.

Potential confusion of the term clonal stems from the complicated relationship between cell lineage and the gene/genome concept. The term "clonal" has two different meanings: one is related by descent and another is relation by similar genetic/genomic features. Even though a clear lineage may be traced based on historical relationship and/or shared short sequences of DNA, cells related by descent may not share the same genome. For example, parental and daughter cells are connected by descent, but cancer parental and daughter cells often have divergent genome identities (therefore they are not clonal). In contrast to the general assumption that genome-level changes are passed on to the daughter cells during cell division, parental cells with genomic instability rarely pass on the same genome. This leads to the unique feature of the cancer cell population, where the entire cell population of a tumor can be composed of unique non-clonal genomes.

The definition of clonality is often directly influenced by the level of analyses. Karyotypically clonal does not equate to DNA sequence level clonality. For example, two cells can be clonal judged by many gene sequences but be non-clonal at the karyotype level. Similarly, the cells can share the same karyotype but display different gene mutations. NCCAs can be classified into structural and numerical types. Increased new types of karyotype abnormalities have been documented [Heng *et al.*, 2004a; 2013a; Ye *et al.*, 2007]. Clearly, more attention is needed to further characterize the NCCAs.

Interestingly, overall NCCA frequencies can be elevated even in normal cell populations (tissues); although known to occur, few researchers actively study their impact. Most significantly, as we illustrate in the following section, elevated NCCAs are often associated with various diseases.

5.3.3.2 *Linking elevated NCCAs to unstable genomes*

We first linked NCCAs to genomic instability in collaboration with Dr. Alan Wang by using an ATM (ataxia telangiectasia mutated) and p21 double knockout mouse model. The ATM protein transduces DNA damage signals by activating downstream proteins that are responsible for cell cycle progression and DNA repair, while p21 is a cyclin-dependent kinase inhibitor that functions in the G1 checkpoint. ATM p21 double knockout mice spontaneously develop different types of cancers. In this model, elevated NCCA frequency was linked to lack of ATM and p21 that result in the inability of these mice to maintain genome integrity. Multiple tissues displayed increased levels of NCCAs as well as tumor formation. Elevated NCCAs are clearly associated with cancer mediated by ATM loss. These studies also linked tumor suppressors (like p21) to increased numerical NCCAs within the ATM-/- background [Shen *et al.*, 2005] and loss of p53 to structural and numerical NCCAs [Heng *et al.*, 2004a; 2006a].

We next linked expression of the HPV16 E6 onco-protein to increased NCCA frequency. The HPV16E6 onco-protein has its own defined molecular mechanisms, including suppression of p53 function. Previously, the majority of molecular characterizations of oncogenes (both HPV16 E6 and others) focused on regulation of the onco-protein itself and its upstream and downstream targets and their regulation. In contrast, our new studies focused on the behavior of the genome system as opposed to characterizing specific molecular pathways affected by the onco-proteins, which have already been extensively studied by other cancer researchers. A common result of onco-protein expression, regardless of the specific onco-protein, was again an increased level of NCCAs. This was unexpected. According to the gene theory of cancer, these onco-proteins are supposed to affect specific cancer pathways that drive specific events, such as increased cell proliferation, and not just generate "noise". A critical evaluation of the literature reveals that many oncogenes can be linked to elevated NCCAs — a point that most cancer biologists have failed to notice. Even the vaunted example of RAS gene-induced transformation is mainly linked to cancer progression due to its link with genome instability.

Lastly, we monitored NCCA frequencies in different cell culture models exposed to various carcinogens (mitomycin c, ethidium bromide, doxorubicin and others) that have been shown to induce genomic instability. The common result of these experiments was a significantly increased level of NCCAs irrespective of the type or target of the carcinogen.

Taken together, NCCA frequencies have been identified as a key feature of system instability which is essential to the cancer phenotype — elevated NCCAs can be linked to almost all oncogenes, tumor suppressor genes, dysfunctions of important pathways and any experimental manipulation that involves genome instability and cancer. These phenotypes include onco-virus infection, aging, inflammation, mitochondrial dysfunction, radiation, epigenetic dysfunction, and others. Stochastic genome alterations are not "noise" but rather drivers of cancer evolution as well as an indicator of the status of a system's stability [Heng *et al.*, 2006b, 2006c]. In addition, many of the NCCAs could be evolutionarily neutral at an earlier stage, and increased NCCA frequency results in increased evolutionary potential.

As illustrated in **Fig. 5-3**, most of the factors that can contribute to cancer also increase NCCAs. In this model, the NCCAs represent the evolutionary potential for cancer and the relationship between NCCAs and CCAs is the key.

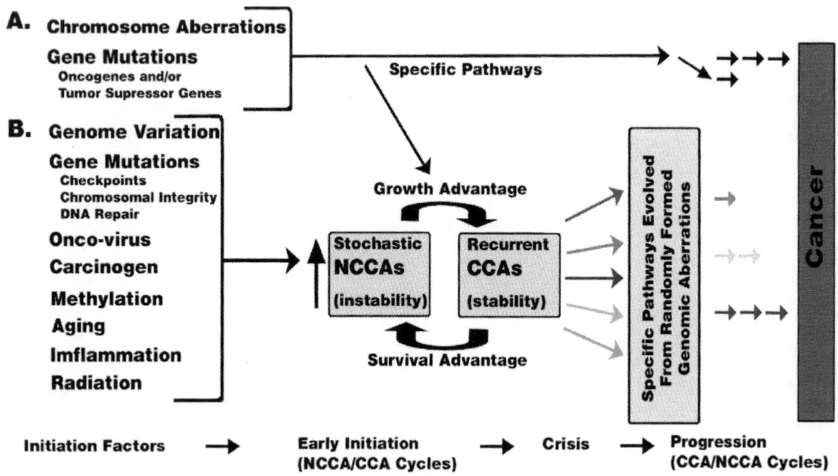

Fig. 5-3. A diagram illustrating the linkage between various factors that can increase NCCA frequency and cancer initiation/ progression.

Path A represents specific chromosomal aberrations and/or oncogene/tumor suppressor gene dominated pathways (suitable for the minority of cancer types). Path B represents a major pathway for cancer formation that involves genome variation, gene defects responsible for genome/chromosome integrity and other sporadic cancers that result from a variety of factors. In path B type cancers, the key-initiating factor is that increased genomic instability leads to elevated levels of NCCAs. NCCAs will then trigger specific pathways defined by oncogenes or tumor suppressor genes or other genes. These various combinations of genomic shuffling and the interplay that develops between NCCAs and CCAs then serves as a substrate for stochastic cancer evolution by providing the prerequisite system diversity needed to ensure survival. When cells pass the crisis stage, they reach immortal status (accomplishing initiation), after which additional NCCA/CCA cycles are needed (for further cancer progression) to reach the cancer phenotype. Various arrows with different intensities represent stochastically formed pathways. This figure was adapted from Heng *et al.*, 2006a with permission.

However, this information raises a fundamental question. If NCCAs are so common and important, why have cancer researchers, who study chromosomal involvement in cancer, only reported specific clonal chromosomal changes rather than stochastic ones they most likely observed?

5.4 Discovery and Validation of the Two Phases of Cancer Evolution

The surprising implication of the last section was that mainstream scientists might have cherry picked results to fit their reasoning, favoring the description of any patterns and avoiding "noise". But is this true, and if so, how did it happen?

5.4.1 *Reasons behind the cherry picking*

In many previous experiments, cancer biologists have not been able to link specific genetic or non-genetic changes or experimental manipulations to fixed patterns of genome alteration. For example, when experiments are repeated under the identical conditions, it is rare to find the same genome alterations as end products. This important fact cries out for attention, as these stochastic genome alterations provide the basis for cancer evolution and are more significant than genes for system control (See Section 5.6: parts vs system inheritance).

However, most publications (especially the most famous milestone papers) have promoted the idea of molecular specificity and causality. They often "successfully" identify a pattern of fixed gene mutations, specific pathway changes, identical chromosomal aberrations, or other common features of cancer. In most published data, the general conclusion typically supports a molecular, deterministic conclusion and overwhelming stochasticity is rarely, if ever, discussed. Now, knowing the importance of monitoring NCCAs in studying cancer, this cherry picking strategy might be the reason behind the drastic disconnection between our observations and those of others.

It further turns out that current paradigms and experimental approaches favor popular cherry picking strategies. Among these is the confusion regarding different system levels and misleading methodologies. Good examples of misleading technologies are experiments based on measuring some value of a population. Many current molecular methods homogenize the population and give an average measure of cells within that population. These analyses miss any elevation of NCCA frequencies resulting from specific experimental manipulations. Over time if these cells are continually cultured, NCCA frequencies lower and the cells typically become a clonal population under defined conditions. Molecular changes in these clonal populations can then be linked to the initial experiment and considered a common factor shared by the final clonal population. Experimental setups like this are common in the molecular biology field. However, when this process is repeated independently, a very different clonal population representing a different system is often observed, clearly challenging the molecular certainty others have concluded. Most molecular biologists monitor only the parts of the system (for example, a mutated p53 gene or protein) when repeating their experiments without considering that the system itself may be changing. However, each experimental run in a given system can result in altered genomes and create a completely different system.

In addition to homogenizing heterogeneous populations, many molecular biological approaches are very powerful and can identify the proverbial needle in the haystack. These key molecular methods including PCR and blotting can identify a single target among millions, providing the "positive" cherry picked data that can be based on the minority of the cells. In addition to the power of molecular methods to identify rare events, many well-defined experimental systems allow for apparent demonstration of molecular causality due to the artificial experimental systems that are used. For example, introduction of a powerful gene mutation perturbs the cellular system. Regardless of the source of the perturbation, greater than the normal range of system dynamics will be involved and if levels are sufficient these dynamics will reflect as increased NCCAs, random gene mutations or epigenetic changes. However, cancer biologists often measure only changes to known pathways associated with the mutated genes and ignore complicated responses of the overall system. Hence, reliance on Western blotting or other similar analyses can often pinpoint a causative relationship among some proteins in the same pathway. But when global gene expression is analyzed, many unexpected correlations emerge which can challenge the expected causality based on the linear system. In a sense, many linear experimental systems can artificially reduce heterogeneity and allow for illustration of a causative relationship that might not true in a real, heterogeneous system.

In summary, there are many avenues where the cherry-picking of results can occur. First, it can be done by ignoring general rules of typical cases and focusing only on exceptions (e.g. by using a very special model system or ignoring the overwhelming level of "noise" during analyses). Second, one can focus on the wrong level of analysis, ignoring the system level (e.g. genome alteration) and focusing just on the parts (e.g. specific gene mutations). Third, the overall evolutionary process can be ignored (observing within a specific time window without tracing the key evolutionary process). And finally, molecular averages can be chosen without considering individual cell variability and system heterogeneity (see Chapter 6).

5.4.2 *Watching evolution in action*

Stochastic genome alteration mediated cancer evolution affects large numbers of pathways, and each pathway can be cherry-picked by investigators focusing on different gene mutations or using different methodologies. In order to avoid cherry picking, there should be more focus on using experimental systems that allow cancer biologists to watch evolution in action. Specific knowledge of how NCCA and CCA dynamics really work in each run of cancer evolution and how to directly monitor this is a new area of promise for cancer biology. All too often cancer is thought of as a linear system and there is a clear genetic (part or pathway) separation between cancerous or normal cells. Therefore measurement showing a difference in some biological phenomenon (gene expression, oxidative damage, and others) in cancer cells and normal cells is a standard of identifying cancer associated changes. However, abnormal cells can use the same genetic parts but change the interaction landscape of the parts, resulting in emergent properties of a system with the same parts but a different topology. More importantly, it requires long periods of time to become cancer. During this time evolution and its interaction with self-organization occur. Watching this evolution in action has led to one of the most exciting and unexpected findings, the two phases of cancer evolution.

We used an *in vitro* immortalization model as an experimental system where different stages of linear progression could be monitored. In this model, fibroblast cell lines were established from patients with Li-Fraumeni Syndrome (germline p53 mutation) allowing cells to be analyzed at various stages of immortalization. Li-Fraumeni Syndrome is an extremely rare autosomal dominant hereditary disorder that drastically increases susceptibility to cancer through loss of function of one TP53 allele. In this model, fibroblasts go through a series of changes over time culminating in spontaneous immortalization at approximately passage 30.

By studying this model system, we soon realized that this system illustrates the point that there are no fixed pathways for immortalization. Each independent run of evolution in this system results in individual immortal cell lines displaying drastically different karyotypes (or genomes) and pathways. Although these cell lines evolved from the same progenitor cells, they all displayed diverse genomes and pathways once immortalization was attained, even under the same culture conditions. To explore this phenomenon, the entire process was followed/examined multiple times using cells from the same individual, and we investigated whether the same driving pathways were shared in different runs of immortalization.

There are obvious advantages of this model. First, this is a good model to watch somatic cell evolution in action, as the end product of evolution (cellular immortalization) is well defined, and the process to achieve spontaneous immortalization can be repeated simply by cell culture. Second, complex chromosomal changes often occur in this model allowing us the opportunity to study the characteristics of genome level dynamics prior to, during and after immortalization. Third, this model does not require any exogenous manipulations such as carcinogen exposure or genetic manipulation for evolution to occur. Fourth, both individual cells and the cell population can be traced at the genome level, and the details of clonal expansion as well as the dynamic relationship between common pattern and stochastic genome alteration can be studied in detail. In particular, this model allowed us to illustrate the pattern of karyotype evolution, showing whether evolution is driven by stepwise accumulating changes (changes shared across generations) or punctuated stochastic changes (no common changes are shared). Finally, the profiles of the entire process can be monitored both at the genome and gene level.

SKY analysis was used to trace the karyotypic dynamics throughout the immortalization process. SKY is an effective way to monitor gross genome alterations (both structural and numerical) with a high population resolution [Heng *et al.*, 2001a; 2003; Ye *et al.*, 2001; 2009b]. In SKY each chromosome is "painted" by a unique color, allowing for easy identification of translocations and aneuploidy. Each normal chromosome is labeled by one color while chromosomal translocations are identifiable by the presence of 2 or more colors. SKY is ideal to score large-scale stochastic genome alteration, which offers quantitative data as well.

The karyotypes of individual cells and cell populations were monitored throughout the entire process of immortalization in order to observe how karyotypes evolve during cancer progression. Karyotypes were analyzed at a total of 7 different time points before, during and after spontaneous immortalization. For each stage, 50 to 100 randomly selected mitotic cells were analyzed and all NCCAs and CCAs were scored. This analysis was used to determine which chromosomal changes were shared among sister cells and which were not shared. Those that were shared belonged to the CCA category and those not shared were NCCAs. The chromosomal changes that were able to pass between cell passages (to daughters, granddaughters and further generations) were analyzed. Alterations that were identified in early stages and remained at later stages were used as shared markers to trace evolution.

Although tedious, this analysis generated many surprising and important observations. First, no commonly shared CCAs were observed throughout the entire process of immortalization. In contrast to some types of blood cancers, clonal expansion is rather a limited phenomenon in this model. Most importantly, there are two distinct phases of karyotype evolution identified during the process (**Fig. 5-4**). A phase of punctuated evolution where NCCAs dominate the population and measured karyotypes change rapidly occurs prior to spontaneous immortalization. During this stage there is no stepwise accumulation of karyotypic change, and after passage 7, unique NCCAs are found in every cell (i.e. the NCCA frequency is 100%). Even at the very early passages (passage 7), there are multiple chromosomal changes, including NCCAs and CCAs, already present. However, all of the CCAs present in passage 7 are eliminated prior to passage 14! As the passages increase prior to immortalization, the number of CCAs in the populations falls. There is only one marker chromosome shared in more than one cell at passages 19 and 24, although the marker chromosome is different at each passage. During this time not only does each cell contain unique NCCAs (the NCCA frequency is 100%), the number of NCCAs per cell and the average complexity (the number of translocations in each NCCA) of NCCAs reaches its peak.

Post-immortalization stages, in contrast, display a clear stepwise pattern of evolution. Each measured passage following immortalization contains a handful of shared CCAs. Even at passage 236, nearly 200 passages after the first post immortalization passage where SKY was performed, the majority of cells within the population contain the same CCAs that were present at passage 54. During this post-immortalization time frame, new CCAs are slowly added to the CCAs that were present at passage 54. Some of the CCAs from passage 54 are also slowly lost over time. This slow gain and loss results in a stepwise pattern to chromosomal changes post immortalization.

Combining pre- and post-immortalization data revealed a pattern of karyotype evolution that drastically differs from the previous model of

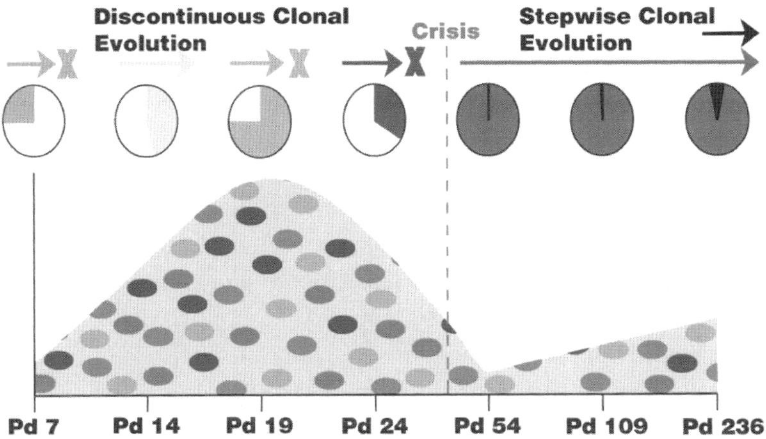

Fig. 5-4. Two phases of karyotypic evolution during the spontaneous immortalization process.

Karyotype evolution occurs in two phases, the punctuated and stepwise phases. In the punctuated evolutionary phase, the karyotypes are drastically altered without clear clonal evolution. Different shades within the pie diagrams represent different clonal patterns that were detected at each examined stage of the continuous cell culture. Within this phase, any clonal chromosomal aberration that occurs is short lived, and replaced by new clonal karyotypes. In the stepwise evolutionary phase, however, the clonal chromosomal aberrations are faithfully passed through to the later passages and displayed continuous patterns of clonal evolution. The shaded dots and the area underneath the curve represent the frequency of NCCAs detected from each corresponding stage. The higher the peak of the curve, the more instability can be detected from the cell population. The two phases are separated by the crisis event where one selected cell generates a stable cell population. Repeating this process multiple times reveals that all 'winning' immortalized clones contain distinct karyotypes. This supports Stephen Jay Gould's view that rewinding and replaying a tape of evolution would not generate the same fixed process and products (reproduced from Heng *et al.*, 2006a).

clonal evolution. Careful analysis of the pattern of karyotype evolution led to the following conclusions:

(1) There are two phases of karyotypic evolution, a punctuated or discontinuous phase where no dominant karyotypes are shared among cells or across generations, and a stepwise phase where common karyotypes are shared between cells and across generations.

(2) These two phases are inter-changeable depending on system stability. Phase changes from the stepwise phase require instability as evidenced by increased NCCA frequency. Therefore, instability is the mechanism of transition to punctuated evolution. Selection pressure and the presence of a cell with the capability of stability that is able to survive selection pressure is required for the phase transition from unstable to stable. For instance, in our immortalization model, the transition point when the punctuated phase changes to the stepwise phase co-maps with the cellular crisis stage where a large number of cells die. During this time likely only one cell emerges as a winner of evolution, and this stable cell will faithfully pass its genome into a large, immortalized cell population. In general, the highest level of NCCAs is detected within the punctuated phase while shared CCAs are detected during the stepwise phase. Introduction of stress (either internal or external) into the stepwise phase causes a switch to the punctuated phase. The stepwise phase is often longer. However, the punctuated phase can sometimes last for years (there are a few cell lines generated from drug treatments that have been cultured for over 2 years without the formation of clonal populations).

(3) We have performed multiple evolutionary experiments that start from the same progenitor population. In these experiments multiple runs of evolution occurred independently and showed unique progression of karyotypic change. Furthermore, many of the independent runs of evolution led to dead ends at various time points. Populations from successful evolutionary runs leading to spontaneous immortalization (the end result of evolution in this model) have distinct karyotypes. This experiment demonstrates a common pattern of cancer evolution (two phases) and the stochastic nature (winning karyotypes are unpredictable) of *in vitro* evolution [Heng *et al.*, 2006a]. These results explain the

overwhelming observation that no common genetic changes dominate most cancer types.

(4) The concept of punctuated karyotype changes differs from karyotypic diversification, a traditional definition of clonal evolution based on the classical Darwinian model of evolution that does not differentiate the gene and genome levels. Within the punctuated phase, new non-clonal chromosomal aberrations are generated nearly every cell division by stochastic genome re-organization. NCCA frequency at most stages is 100%. There is no traceable karyotypic intermediate between cells. Chaotic genomes with complicated translocations are also common during this phase (see the later genome chaos section). The formation of different genomes in fact interrupts the karyotype-defined inheritance between mother and daughter cells (see Section 5.6.2). In typically described diversification, however, the majority of cells are clonal across generations with traceable karyotype changes.

The concept of punctuated equilibrium was initially proposed by Niles Eldredge and Stephen Jay Gould to explain why most species exhibit minimal net evolutionary changes in phenotype for most of their geological history, and why key evolutionary changes happen rarely and rapidly on a geologic time scale [Eldredge and Gould, 1972]. Even though we borrowed the term punctuated evolution to describe the rapid and drastic genome-level changes that occur in contrast to stepwise evolution where the same genome is maintained, our definition of punctuated evolution focuses on the pattern of genotypic (karyotypic) evolution. Punctuated evolution in cancer is therefore a period of evolutionary change when the karyotypes within a population change drastically and rapidly.

The two phases of evolution form repetitive cycles, referred to as the NCCA/CCA cycle [Heng *et al.*, 2006b; 2006c; 2011b; Horne *et al.*, 2015a–b; Ye *et al.*, 2009a]. Each individual shift from one phase to another is known as a NCCA/CCA transition. NCCA/CCA transitions can be observed in all major transitions (selection events) including immortalization, transformation, metastasis, and drug resistance. Prior to each of these selection events, the frequency of NCCAs in the population increases until selection pressures result in a winning genome that then expands clonally. Any "successful" run of cancer evolution can be described as multiple rounds of the NCCA/CCA cycle. Interestingly, the punctuated phase is associated with survival and heterogeneity while the stepwise phase features growth and homogeneity. In addition, the punctuated and stepwise phases of evolution also occur at different levels of genetic organization as these two evolutionary phases represent a general evolutionary pattern. It should be pointed out that the punctuated phase can be long lived without a switch to the stepwise phase. For example, following genome chaos, some clones can remain in the punctuated phase for over a year.

5.4.3 *Validation of the two phases of evolution in various models*

The Li-Fraumeni cellular immortalization model involves p53 dysfunction. To illustrate the point that the two phases of karyotype evolution are not just limited to p53 function but represent a general pattern of karyotype evolution, we performed similar evolutionary experiments in various *in vitro* and *in vivo* systems. These systems included a spontaneous transformation model of mouse ovarian cells, a metastasis model and various drug resistant models [Abdallah *et al.*, 2014; Ye *et al.*, 2009a]. For example, spontaneous transformation of ovarian surface epithelial (MOSE) cells from wild-type mice can occur in continuous culture, even though there is no pre-existing loss of p53 function. Similar to the Li-Fraumeni model, transformation in MOSE cells is driven by chromosomal alterations that occur in two phases of cancer evolution [Lawrenson, 2010]. Multiple molecular mechanisms have been shown to be responsible for the transformation.

The commonality of punctuated evolution in cancer has also recently been confirmed by single cell level sequencing [Navin *et al.*, 2011]. It was found that tumors grow by punctuated clonal expansions with few persistent intermediates, in direct contrast to the stepwise clonal evolution model suggested by Nowell [Nowell, 1976]. Two points need to be clarified, however. First, the stochastic model and the two phases of cancer evolution published in 2006 is not a gradual model [Heng *et al.*, 2006a]. The described punctuated phase is clearly in opposition to the gradual model of cancer evolution. Furthermore, the stepwise phase accounts for only a small part of the process and is mechanistically less important for cancer progression than the punctuated phase. Second, while the conclusion was revealed by sequencing-based methods, they really only measured genome level change. It would be interesting to examine whether the punctuated pattern could also be observed at the DNA level, as DNA level continuity differs from genome level continuity [the same materials can be used to build the same type (gradual) or different types (punctuated) of architecture], and the discontinuity of the DNA level does not necessarily correlate with the discontinuity of the genome level (macroevolution). While genome alterations can occur within the stepwise phase, gene mutations may be discontinuous within this phase. However, because genome level organization is a higher level of organization the population would remain in the stepwise (micro-evolutionary) phase. It is likely that evolutionary continuity and discontinuity will be detected at both the genome and sequence levels by single cell sequencing.

In addition to detection of massive and complicated genome re-organization in a high percentage of tumors, two separate "mutational clocks" have been suggested to explain the cancer genome sequencing data [Fox and Loeb, 2014; Wang *et al.*, 2014]. Type one refers to the large-scale, structural changes in DNA (such as amplification and deletion of large blocks of DNA) which probably occur in punctuated bursts of evolution. Type two refers to the point mutations, which may accumulate more gradually, generating extensive subclonal diversity. The existence of two mutational clocks supports the concept of the two phases of cancer evolution.

5.4.4 *The significance of understanding the two phases of cancer evolution and the relationship between NCCAs and CCAs*

The discovery of the co-existence of two phases of evolution as one continuous process with the ability to switch back and forth in accordance with system instability is highly significant. Particularly important is the finding that within the punctuated phase, genome level replacement dominates. This knowledge explains many puzzling phenomena in cancer research. The following are two examples:

The inconsistency between results in various *in vitro* and animal models and in actual clinical situations has been a major challenge for cancer researchers. In light of this new understanding of the two phases of evolution, it seems that most of the model systems currently used in cancer research represents a single run of evolution. This evolution often occurs rapidly due to non-physiological drivers, and the majority of analyses focus only on the stepwise evolutionary phase due to the relative genome stability of these models. In contrast, tumors differ greatly as they each result from unique runs of cancer evolution occurring over years or decades. Furthermore, in many cancer models, the initial conditions can be carefully designed and/or controlled, which artificially introduces a dominant pathway, while in actual clinical cases there is no way to precisely control initial conditions. Most importantly, focusing on gene mutations and specific pathways has meant that the stochastic patterns of genome mediated macro-evolution have been missed. It is no surprise that experimental data and patient results are drastically different. Experimental systems are only similar to exceptional clinical cases (in terms of evolutionary phase and a more linearized progression) such as chronic phase CML [Horne *et al.*, 2013a].

The punctuated phase predominates during the acquisition of drug resistance. Traditionally, gene mutations were thought to be the central mechanism of drug resistance as is the case with bacteria. Knowing that drastic genome re-organization occurs during the punctuated phase, and that transition to the punctuated phase occurs following most drastic treatments, it is clear that such restructuring can generate totally new systems that do not rely on given gene mutations. During drug treatment, genome chaos can occur very quickly, generating new systems which display drug resistance. Experiments have shown that the resistant clones always display altered genomes, further supporting this explanation.

Interestingly, the ability of tumors in the stepwise phase to quickly switch to the punctuated phase following drug treatment also challenges the concept of cancer stem cells being responsible for drug resistance. Resistance occurs due to macroevolution and not because of a specific magic cell type that is somehow immune to treatment. It might be very informative to investigate whether elevated NCCAs are associated with some stem cell markers. It should be pointed out that even though some cancer cells may display a stem cell marker, it does not necessarily indicate that a cancer cell arose from a stem cell. Expression of the stem cell marker could be due to the dedifferentiation process that cancer cells go through. A major question is whether cancer stem cells have stable genomes. Although the jury is still out, it appears that cancer stem cells have similar levels of genetic heterogeneity compared to non-cancer stem cells [Almendro *et al.*, 2014]. This is a bit surprising, as it is likely that cells with increased cancer evolutionary potential coupled with stem cell markers should be more dynamic. This issue needs further study, as it is limited by using chromosome copy numbers to monitor overall genetic heterogeneity. Even if these cancer stem cells exist and we can identify them, targeting them will be problematic due to their instability and stochastic response to the treatment process. The focus on cancer stem cells has further delayed addressing the question of how drug resistance really occurs by somatic cell evolution.

The two phases of somatic cell evolution also offers new insight into evolutionary mechanisms and their implications for organismal evolutionary studies. These new ideas specifically challenge the current approach that focuses on tracing genes in evolutionary studies without incorporating information on the altered genome. For example, the pattern of cellular evolution might mimic organismal evolution, based on the notion that the genome defines a new system (when there is a new genome, there is a new system as illustrated by novel global expression profiles), and that genome replacement represents macroevolution (in the majority of species, the genome context is different). In somatic cell evolution, micro-evolution mainly involves gene mutations and epigenetic alterations, while macroevolution occurs primarily through genome level alterations. Macroevolution creates new systems (species), and micro-evolution modifies existing systems (species).

Together, the pattern of cancer evolution cannot simply be explained by Darwinian stepwise natural selection.

5.5 Linking Genome Heterogeneity to Tumorigenicity

NCCAs are important prerequisites for cancer evolution. As soon as they appear, competition begins between normal and altered genomes, especially when the NCCA frequency is high. Since different genetic or non-genetic aberrations can be linked to increasing numbers of NCCAs, various genetic factors can be linked to cancer evolutionary potential. To demonstrate this point, it is necessary to link tumorigenicity to the frequency of NCCAs, which serves as an index of genome diversity (rather than an individual molecular mechanism).

To this end, we compared tumorigenicity and genome diversity in five well-characterized cancer models. As illustrated in **Table 5-1**, each model had previously been linked to a specific molecular mechanism. Attempts to identify commonly shared features revealed that elevated NCCA frequencies were the key common feature regardless of the diverse molecular mechanisms. These experiments further demonstrated that genome population heterogeneity (particularly structural NCCA-mediated population heterogeneity) promotes tumorigenicity. As such, cancer represents a simple evolutionary story where system heterogeneity drives progression.

Experiments also illustrated that the successful control of cancer will likely follow evolutionary law. When identical cancer cells were injected into mice to form tumors and various treatments were later applied to reduce tumor size, the same treatment would often have different results in different animals. In some, the size of the tumor was drastically reduced while in others, the tumor size remained unchanged. Both the reduced and unaffected tumors were cultured for short terms and their karyotypes were analyzed with SKY. The tumors that shrank had low NCCA levels while the nonresponsive tumors had high levels of NCCA-mediated heterogeneity! The link between genome level heterogeneity and treatment response was also observed in various models and clinical cases [Chandrakasan *et al.*, 2011].

Table 5-1. Various molecular mechanisms are linked to an increase in NCCAs, the common feature of the evolutionary mechanism of cancer

Cell model	Previous findings (molecular mechanisms: features or identified pathways)	Current common findings
LNCaP	Increased microsatellite instability; gradually lost androgen response; Increased tumorigenicity	Increased frequencies of NCCAs; increased genome diversity
MCF10DCIS.com	Stromal–epithelial interaction; increasingly invasive phenotypes	Increased frequencies of NCCAs; increased genome diversity
MCF10-CSC	Increased ratio of BCL-xL/Bax; increased expression of PCNA, gadd45; increased tumorigenicity	Increased frequencies of NCCAs; increased tumorigenicity in vivo
MCF10-HoxA1	Activation of cdD1 and Bcl-2; increased tumorigenicity	Increased frequencies of MDFs; increased genome diversity
Mouse Ovarian	Change: cytoskeleton and focal adhesion complex, down: E-cadherin and connexin-43, increased tumorigenicity	Increased frequencies of NCCAs; increased genome diversity
MCF10-Rad6	Centrosome amplification, aneuploidy and transformation; benign hyperplastic lesions	Low level of structural NCCAs; aneuploidy
Myc-transgenic mice	Expression of A2 and E2F1; increased tumorigenicity	Increased frequencies of NCCAs

(Reproduced with permission from citation Ye *et al.*, 2009)

In conclusion, population heterogeneity at the genome level is the key to cancer formation and cancer response to treatment, both of which are indeed typical evolutionary stories at the level of somatic cell population [Heng *et al.*, 2009; 2011; Stevens *et al.*, 2011; Ye *et al.*, 2009a]. More examples can be found in Chapter 6.

5.6 Genome Alterations form New Genome Systems by Changing System Inheritance

Genome alterations are crucial to cancer progression, but what is the mechanism behind their importance? In particular, why are genome level changes so significant that they can overpower gene level changes? How do genome alterations contribute to somatic cell evolution and how does this differ from the gene's role? Is the relationship between genome and genes just a quantitative difference?

Inheritance is a key to Darwinian evolution. We therefore have to focus our attention on the fundamental differences between gene (parts) and genome (whole system) defined inheritance.

5.6.1 *Gene defined "Parts Inheritance"*

The gene is generally accepted as the basic unit (both physical and functional) of inheritance. Since the rediscovery of Mendelian inheritance in 1900, gene research has been the basis of modern genetics and molecular biology. We know how individual genes duplicate, mutate, code for proteins, are transcriptionally regulated, how proteins are translated and how and why post-transcriptional modifications occur. We know how gene mediated genetic information is passed through generations, how alleles spread throughout the population, and how certain alleles can cause inherited genetic diseases like hemophilia.

On the surface, over a century of gene research has seemingly figured out all the key aspects of biological inheritance. The rationale behind the human genome sequencing project was simple, all we need to know is the information about each unit of inheritance and then add the information of all the units together. If we know the sequences of all genes, we should be able to understand the blueprint of life and cure most if not all genetic diseases.

Following the celebration of the 10th anniversary of the completion of the human genome sequencing project, many questions were raised [Hayden, 2010a]. Why has the highly promised sequencing project failed to deliver? Where is the missing heritability? Is the conceptual basis of gene-mediated inheritance responsible for all these surprises? While most scientists are convinced that more data are needed to finally deliver the promised understanding of all genetic diseases, some have begun to ask hard questions. If the current genetic theory cannot explain reality, it is about time to re-examine our conceptual framework [Heng, 2007a; 2009; 2010].

The overwhelming commonality of NCCAs observed in various cancer models/samples forced us to rethink the genetic meaning of karyotypes. Why are there so many different genome level changes in cancer? What does this mean for inheritance and phenotype? What is the relationship and/or relative contribution of genome level changes and gene mutations to the cancer process and inheritance in general?

Let us first examine the relationship between phenotype and genome. It is well accepted that the phenotype can be explained by a combination of the genotype and environmental interactions. As for the genotype, the traditional view focuses on genes, and the phenotype often refers to a specific feature associated with a given gene (or set of genes) in discussions of genetic traits. In fact, there is limited if any discussion about any roles the karyotype plays in determining phenotype, as the chromosome only serves as the vehicle of genes. The karyotype is mostly used in the classification of species and to explain phenotypes of some well-known genetic disorders such as Down's syndrome (an extra copy of chromosome 21) or the Philadelphia chromosome in CML, where the phenotype and genotype are linked. However, as soon as any specific gene defects can be linked to an abnormal karyotype, the focus is shifted from the karyotypic change to the affected gene(s).

Genes contain information for making the functional parts of the cell, whether those parts are proteins or other functional nucleic acids. These parts are the basic functional units of the bio-system. Genes therefore can be thought of as the instruction manuals for producing the structural and functional parts of the cell. Therefore, the genes define the "parts inheritance". However, even with the completion of the human genome and knowledge of all the parts, many common and complex diseases remain unsolved. What are we missing?

5.6.2 *Genome context defines the "system inheritance"*

If genes define the "parts inheritance", does the blueprint of the biological system simply come from the parts manual? In other words, can parts inheritance also function as a blueprint for assembling the system? If not, where does this design blueprint stem from?

The instructions for making materials/parts are very different from the blueprint. A good corollary is that the instructions for producing the bricks and lumber used to build a structure do not say anything about the architectural design of the building. The blueprints of the building are a separate set of instructions from the parts manual. The concept of different types of coding systems is also analogous to computers. The instructions to make parts and the schematics of assembly are different, as are the programs that operate the computer. It is thus hard to imagine parts inheritance equal to the system inheritance in biology, especially knowing that despite species diversity within the biological kingdom, many organisms share similar genes. In construction, the same building materials can be used to construct different buildings with different architectures and functions. Similarly by changing the genome context, the same genes (parts) can be used to build a new system (species). Therefore, the blueprint of life is likely not coded by the genes (genetic parts or DNA sequences) themselves. So, what is the "hidden" blueprint of biology?

Prior to our cancer research, we studied the essential role the karyotype plays in the classification of species and in organizing high order chromatin structure for gene expression [Heng *et al.*, 1985; 1987; 1988; 1994; 1996; 2001b; 2004a–b]. We knew the karyotype was important, but the exact mechanism of its importance was not clear. Studies began revealing that changes in the karyotype resulted in large-scale changes to the organization of the three-dimensional structure of the nucleus. However, the disciplines of genetics and molecular biology have treated the genome as merely a bag containing all the genes, and the role of the genomic topology was not appreciated by most molecular biologists. Clearly, new concepts as well as experimental systems were needed to illustrate the function of the karyotype beyond representing a bag of genes for inheritance.

What is the role of the karyotype in inheritance? During our cancer research, the high frequency of NCCAs in cancer got our attention. We saw the same pattern of punctuated cellular evolution followed by selection and clonal expansion over and over. This pattern made us hypothesize that these NCCAs were important for evolution, but how do they function?

Since all NCCAs in our experiments are generated by the same parental cell population, they likely share the similar gene sequences (materials) within the population especially within a short time window of culture. The main difference between daughter cells with NCCAs and parental cells are the reorganized genomes in the daughter cells.

To connect the dots, we made the following thoughts, assumptions and predictions:

(1) The "system inheritance" is determined by the karyotype-defined blueprint, which differs from gene-mediated "parts inheritance".

(2) Karyotype alterations create new topological relationships among genes, which changes the network structure by changing the genetic interaction landscape.

(3) The important function of topological changes as a new type of inheritance can be explained by different ideas/facts. In general, different emergent properties can be achieved by similar genetic parts within different genomic topologies. Such topology is inherited information that is preserved by the karyotype. The karyotype provides the topological scaffold that determines the interaction among all genes (the order of genes within and among chromosomes in 3D structure), and thus the genetic interaction landscape. A gene's function is genomic context dependent. Topological changes to the genome can result in emergent properties of the system that can be explained by various molecular mechanisms.

This is a hard concept to accept as many biochemists have less appreciation for topology in bio-reaction. Nevertheless, increased data have forced people to consider this new concept. For example, many regulating proteins are present in limited numbers in each cell and their physical location within the cell matters; many proteins or RNAs interact with large numbers of other proteins with low or moderate affinity; a large number of proteins are sticky, all of which further reduce the molecular specificity of protein interactions. Physical geography changes in sub-cellular location of molecules caused by alterations of genome topology can bring molecules closer to low affinity substrates, increasing the probability of a specific interaction. That is the reason why many common regulating mechanisms such as most post-translational modifications can have diverse substrates. Diverse substrates mean that the molecule of interest has less specificity than previously expected. Molecular specificity between domains is the general principle of molecular interactions. In other words, interactions within cells are regarded as molecule A interacts with molecule B through specific domains. It is thus surprising that many proteins and other molecules have large numbers of different types of substrates. However, low interaction affinity can be overcome by changes to cellular topology. For example, when two molecules are brought closely together the interaction becomes reality despite a moderate affinity. On the other hand, two molecules with high affinity are improbable interactors if they are physically separated. Lower domain specificity results in more versatile parts, while cellular topology can control the specificity of an interaction (**Fig. 5-5**) (see more examples in Chapter 6). The interaction between "not so perfect molecular partners" is especially important under stress conditions, where the preferable interaction (according to molecular likelihood) is no longer impossible or not useful. The "fuzzy" interaction becomes important for cellular evolution (see Fuzzy inheritance in Chapter 6 for more information). The new layers of cellular topology mediated bio-specificity have so far been ignored by many *in vitro* protein interaction assays. Note that the rationale of discussing the role of topology in protein interaction is to support the importance of genomic topology for gene interaction, even though the

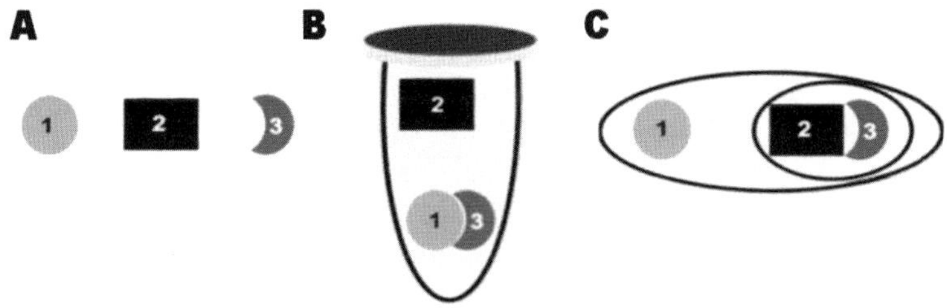

Fig. 5-5. Cellular topology can change the pattern of protein interactions.

We start with three proteins (A). Each protein has a unique shape. Interaction between proteins 1 and 3 has the highest affinity based on the parts' characteristics. In a test tube, where the proteins can freely interact with each other without cellular constraint, proteins 1 and 3 indeed are shown to preferentially interact (B). However, because of cellular topological constraints or localization in different organelles for example, proteins 2 and 3 actually interact within a cell (C).

mechanisms behind emergent genomes could be more complicated than protein interaction.

(4) The holistic view of molecular protein interaction in cells has emphasized the importance of intracellular compartmentalization of naturally occurring proteins in space and time. Most intracellular proteins can freely move in short distances, but usually live in defined physical spaces, limiting their interaction to proteins located within the same space at the same time [Feller and Lewitzky, 2012].

Clearly, genomic topology is essential as it can lead to changes of global gene expression and often leads to evolutionary innovation.

To conceptualize system inheritance, we proposed a model to illustrate the relationship between genome topology, the distribution of genes along the chromosomes and network structure (**Fig. 5-6**) [Heng, 2009; Heng *et al.*, 2011a].

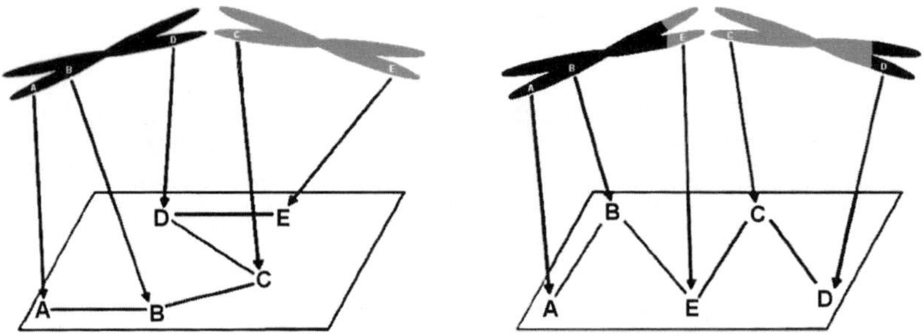

Fig. 5-6. This diagram explains how the genome context represents a new layer of information (blueprint) to determine the network structure (reproduced from Heng, 2013c).

Two chromosomes are drawn, representing all chromosomes within the cell and depicting the way in which genomic physical interactions affect networks. When the genome context alters (exampled by translocation) during a "genome re-organization" episode, the physical relationship between the same gene sets (genomic topology) changes and so does the pattern of the network (represented by the letters).

Accumulating evidence directly demonstrates or indirectly supports that genome alterations define a new bio-system. First, changing the genomic topology can create evolutionary novelty. An evolutionary experiment tracing *Escherichia coli* populations over thousands of generations demonstrates that a novel trait (aerobic citrate utilization) can evolve through rearrangement and amplification of a few pre-existing genes, rather than through invention of new genes or gene mutations [Blount *et al.*, 2012]. The observation that a novel trait can evolve through genomic re-organization raises an important point; *E. coli* has the building blocks (parts) necessary for the evolution of specific features, but realization of this capacity requires genome re-organization to recreate (not just rewire) the system. It is also recently demonstrated that large scale chromosomal rearrangements fuel rapid adaptation in yeast populations [Chang *et al.*, 2013].

Second, aneuploidy can overcome the dysfunction of a specific gene. Loss of function due to specific key gene deletion can be restored by a changing chromosome status, another example of how genome alterations lead to new features [Rancati *et al.*, 2008]. Haploid yeast contains only one copy of the MYO1 gene, encoding the myosin II protein which is essential for cytokinesis. Deletion of this gene is supposed to be lethal. Surprisingly, some MYO1 deficient cells were able to regain cytokinesis, not through restoration of the MYO1 gene, but rather by becoming extensively polyploid and/or aneuploid. Multiple novel mechanisms of cytokinesis emerged from the various types of genomic alterations that occurred. The evolutionary advantage of genome context changes is underscored by this experiment.

Third, karyotype changes are associated with large-scale transcriptome changes [Heng *et al.*, 2008; Stevens *et al.*, 2013; 2014; see Chapter 8]. These chromosomal alterations affect gene expression throughout the genome, and not just the genes that reside on the affected chromosomes. Chromosomal alterations do not merely cause a simple proportional change in gene expression due to gene dosage. In addition, biological replicates with high karyotypic diversity are less closely related than replicates with stable karyotypes. This indicates that each unique genome results in a unique transcriptome profile. This result is supported by recent single cell exome sequencing of glioma cells, which showed that gene expression of each cell is unique in these karyotypically diverse cells. Moreover, a single chromosomal change (either structural or numerical) can affect the expression of hundreds to thousands of genes throughout the entire genome far beyond genes only on the impacted chromosome. This viewpoint drastically differs from the traditional view that gene alterations are responsible for evolutionary transition points. Another good example of this is Down syndrome, where an extra copy of chromosome 21 leads to large number of altered genes' expression that is not explained by dosage alone, as the expression of many genes not organized on chromosome 21 is altered.

An interesting case is the drastic gene expression differences between males and females. Thousands of genes displayed sexual dimorphism in multiple tissues. Furthermore, the gene expression changes are not only limited to genes on the sex chromosomes, but on other autosomes as well [Yang *et al.*, 2006]. It has been suggested that sex-specific genetic architecture including genome topology contributes to human diseases [Ober *et al.*, 2008]. The important differences in gene expression may contribute to the fact that many diseases disproportionately affect one sex, and there are differential drug responses between males and females [Hayden, 2010b]. Since such difference cannot simply be explained by gene differences on the sex chromosomes, it is likely that the karyotype differences of male (XY) and female (XX) genomes are involved. Despite X-chromosome inactivation, and the relatively small number of genes (less than 2%) on sex chromosomes, the genome topological changes likely contribute to the molecular dimorphism as the genome context changes many genes' context.

Fourth, there are ample examples of speciation mediated by karyotype reconfiguration, supporting the importance of genome re-organization in the formation of new species. The best example is the divergence of the human and chimpanzee karyotypes. Chimps have 48 chromosomes while humans have 46 chromosomes. Karyotype/FISH mapping and genome sequencing demonstrated that human chromosome 2 was formed by the fusion of two chimp chromosomes. This fusion plus a few chromosomal inversions has changed the genomic topology between human and chimps, despite the fact that the gene sequence similarity between the two is over 98%!

Gene similarity and karyotypic divergence are common in closely related mammalian species. Current genome sequencing efforts have revealed that it is challenging to identify species based on specific gene differences (i.e. no speciation genes have been identified), despite great efforts to do so. In contrast, genome reorganization is very common in closely related species, and large numbers of euchromatic blocks are conserved during the genome re-organization process. For example, three very closely related species, the Indian muntjac, the Chinese muntjac, and the brown brocket deer have very different karyotypes [Indian muntjac (2n=6/7), Chinese muntjac (2n=46), brown brocket deer (2n = 70 + 3B)]. Reciprocal chromosome painting between these three deer species demonstrates that the karyotypic differences between Indian muntjac and Chinese muntjac can be explained by chromosome fusions and other rearrangement events. Interestingly, the euchromatic blocks of all chromosome arms of the brown brocket deer are apparently conserved in number and content in the Indian muntjac [Yang *et al.*, 1997].

Another good example is the human and mouse chromosomal relationship. Even though over 90% of genes are shared between human and mouse, there are over 250 chromosomal reorganization events that distinguish the two species' karyotypes. Chromosomal painting methods have elegantly illustrated the history of genome re-organization [Ferguson-Smith and Trifonov, 2007; Graphodatsky *et al.*, 2011].

Interestingly, sponges have 18,000 genes, compared with 22,000 human genes although sponges represent a much simpler life form. Clearly, the complexity of a species is not determined by how many genes it has, but rather by how the genes are arranged within the genome. In other words, in a given species with tens of thousands of genes, *de novo* invention of a new gene is likely not that important in the evolution of a species, as there already are enough genes to produce a high degree of complexity. Instead, evolution often builds on the raw materials by altering the usage of the same materials, such as by changing the architecture.

Fifth, due to the challenge of identifying common gene mutations in many common and complex diseases including cancer, the contribution of genome alterations in diseases has been increasingly discussed. In fact, linkage between genome alteration (nuclear and mt genomes) and diseases is common [Wallace, 2013; Heng *et al.*, 2013a]. This information emphasizes that, compared to gene mutation-defined parts defects, genome topology variation might significantly contribute to disease conditions as well as short-term somatic cell adaptations [Horne *et al.*, 2014]. Since many diseases involve pathological changes, it is reasonable to link them to altered cellular systems.

5.6.3 *How is system inheritance maintained?*

While the above syntheses support the hypothesis that there is a new layer of genome-defined inheritance, a crucial gap remained. How is system inheritance maintained in organismal systems and somatic cells?

Gene level information is maintained via semiconservative DNA replication. DNA replication has a low error rate leading to a low mutation rate (e.g. the mutation rate of phage T4 synthesis is 2.4×10^{-8}), ensuring the preservation of gene-mediated parts information. Semiconservative replication is also important for chromosomal maintenance. However, chromosomal aberrations frequently occur in various tissues particularly in cancer [Heng *et al.*, 2004a; 2013a]. It is estimated that, the *de novo* locus-specific rate of genomic rearrangement is at least 100- to 10,000-fold greater than point mutations [Lupski, 2007]. Obviously, there must be other mechanisms to ensure system inheritance.

In order to identify possible mechanisms of ensuring system inheritance fidelity, we compared the patterns of change that occur within two phases of karyotype evolution with the patterns of evolution that occur in sexual and asexual organisms. Shockingly, organismal evolution in asexual species mirrors the punctuated phase of cancer evolution, and sexual reproduction mirrors the stepwise phase [Gorelick and Heng, 2011; Heng, 2007c]. This comparison led us to recognize that there must be a mechanism to ensure genome stability during sexual reproduction, as species of sexually reproducing organisms preserve their karyotypic makeup with high fidelity. Stepwise Darwinian evolution (accumulation of microevolution) occurs only if system inheritance is preserved. If this hypothesis is correct, sexual reproduction should also function as a karyotype stabilizing force in contrast to the traditional concept, where the main function of sex is to introduce genetic diversity [Heng, 2007c]. We further suggested that sexual reproduction primarily functions to reduce genomic alterations despite its secondary function of mixing genes [Gorelick and Heng, 2011; Heng, 2007c; Wilkins and Holliday, 2009; Horne *et al.*, 2013b].

The function of sex is surprisingly linked to the separation between germline and somatic cells. Separation of the germline and somatic cells allows for high degrees of genomic variation to occur in somatic cells, which is essential for cellular adaptation within changed environments. Such variation, however, will not drastically change inheritance of the germ line. First, the germ line is more stable than somatic cells, and second, sexual reproduction functions to ensure the identity of the species by preserving the genome. The separation of germ line and somatic cell genome integrity is an important tool necessary for organismal evolution, as it allows flexibility (albeit with tradeoffs) in somatic cells but sex does not allow these somatic changes to be passed to the next generation. In other words, the organism's system inheritance is preserved by avoiding changes to a species-specific genome system identity (karyotype).

The lack of a sexual filter to eliminate genome variation allows somatic cell evolution to occur rapidly when cellular adaptation pressure is high. In this case the occurrence and accumulation of genome level changes sufficient for cancer can be accomplished within a time scale of decades. In the somatic cell situation a certain level of genome dynamics are needed to adapt to stress, however the trade-off is that the instability can also favor cancer progression. In a sense, the process of increasing adaptability can also decrease stability. Once genome stability is lost, the probability that rapid cancer progression will begin is high. Genome changes result in the formation of new systems; such changes are favored by cancer evolution, and alteration of system inheritance then becomes an effective means of cancer evolution. The requirement for genome alteration mediated somatic evolution in cancer progression is the reason why genome level changes are more dominant in cancer than gene level change.

5.6.4 *Why has system inheritance flown under the radar for so long?*

To address this question, we need to re-examine the relationship between the phenotype and the genotype. It is generally accepted that phenotype = genotype + environment. However, the genotype has traditionally been equated to gene mediated parts inheritance due to the failure to realize the key difference between gene and genome. For decades, the surprisingly low correlation between genotype and phenotype was explained by invoking environmental effects rather than searching for any overlooked inheritance. Distinguishing parts inheritance from system inheritance (the blueprint), and incorporating system inheritance into the equation reveals an additional mechanism/level of inheritance and rewrites the definition of the genotype (**Fig. 5-7**). In recent years, many investigators have tried to use the epigenetic layer of information to explain the inconsistencies between genotype and phenotype. While this approach is important, it also missed the true blueprint of the genotype. As some publications point out, epigenetic regulation layer is beyond the blueprint. Others state that epigenetic regulation modifies the blueprint. But, until our research, few have realized that the blueprint is contained within karyotype-defined system inheritance.

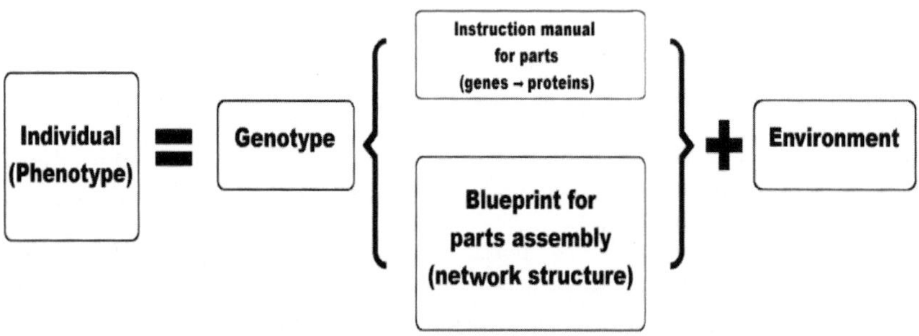

Fig. 5-7. The relationship between genotype, environment and phenotype, where the genome-defined blueprint has been ignored.

Realizing this was a real shock to us. The answer to how this inheritance could have been missed is rather interesting.

It turns out that for a century, most experimental systems used in traditional genetic research have artificially eliminated karyotype changes. By fixing two major components of the phenotype/genotype equation (using the same genome system, and reducing the environmental variation through precisely controlled experimental model systems), the genotype-phenotype equation became Phenotype = "parts inheritance", allowing geneticists to convincingly link some specific genes and mutations to phenotypic traits. Success in directly linking genes to phenotype in a limited number of exceptional cases hid the need to identify system inheritance. However, this practice no longer holds water in the case of cancer evolution, where stochastic genome alterations drive evolution for the majority of cancers, and genome context-defined system inheritance overshadows the contributions of individual parts. In most systems with pronounced genomic instability, the gene-phenotype relationship is much weaker than genome-phenotype relationship.

In summary: (1) Parts inheritance is relatively simple compared with system inheritance. It is often conserved between species, and its function can be enhanced or reduced, changed or eliminated by system inheritance. (2) Separating parts inheritance and system inheritance makes understanding why Mendelian genetics only holds true for a small portion of genes easier, and reveals the limits of applying Mendelian genetics to cancer where system inheritance dominates. (3) System inheritance plays a dominant role in macrocellular evolution while changes in parts inheritance function primarily in the stepwise phase.

5.7 Genome Chaos: Rapid Evolution by Creating New Systems Through Cancer Genome Reorganization

The link between the karyotype and system inheritance means that each NCCA represents altered system inheritance. Recent cancer genome sequencing efforts have shown that highly complex genome rearrangements occur commonly in cancer [Baca *et al.*, 2013; Stephens *et al.*, 2011]. The next questions are then, how are highly complex NCCAs so frequently produced in cancer progression, and what conditions promote normal cells to change their system inheritance during cancer evolution?

Our previous studies have linked to various physiological and pathological stresses to elevated NCCA levels [Heng *et al.*, 2006a–c; Stevens *et al.*, 2010; 2011; Stevens and Heng, 2013]. Meanwhile, highly complex genome rearrangements (chaotic genomes) are frequently observed at key evolutionary transitions, including immortalization to drug resistance, and are associated with the punctuated evolutionary phase of cancer. For example, genome chaos seems to be triggered during the punctuated phase of evolution that occurs in the MDAH041 *in vitro* immortalization model. We thus hypothesize that different cellular crises can produce extremely stressful conditions, leading to cell death and rapid genome re-organization. Alteration of system inheritance to form new systems is required for cells to survive under these highly stressful conditions.

This hypothesis is supported by a study where drug-induced genome chaos was identified as a mechanism of cellular survival strategy under crisis [Liu *et al.*, 2014]. Genome chaos, a process of complex, rapid genome re-organization, results in the formation of chaotic genomes (structural and numerical alike) (**Figs. 5-8, 5-9**). Although the majority of these chaotic genomes is unstable and will be eliminated by cell death, a small portion of them survive, and increase population diversity. Over time, stabilized genomes emerge from this diverse population following selection. This phenomenon was initially described by our cytogenetic analyses during the punctuated phase of cancer evolution, and was closely associated with chemotherapeutic drug treatment [Heng *et al.*, 2006a; 2007b]. For years, chaotic genomes have been ignored, as they do not fit the expectations of the stepwise cancer evolutionary model where small changes should accumulate in a stepwise fashion, partially because few believed that such massive changes would be survivable [Horne and Heng, 2014]. However, in recent years the common presence of large-scale genome re-organization in human tumors has been confirmed by whole-genome sequencing of cancer samples. Furthermore, the relevance extends far beyond cancer to other diseases or illness conditions [de Pagter *et al.*, 2015; McDermott *et al.*, 2015; Pellestor *et al.*, 2014]. Many names have been given to these highly rearranged genomes, including "chromothripsis", "chromoplexy", "chromoanasynthesis", "chromoanagenesis", "chromosome catastrophes", "neochromosomes" and "structural mutations" [Baca *et al.*, 2013; Crasta *et al.*, 2012; Forment *et al*, 2012; Garsed *et al.*, 2014; Holland and Cleveland, 2012; Inaki and Liu, 2012; Jones and Jallepalli, 2012; Liu *et al.*, 2011; Malhotra *et al.*, 2013; Meyerson and Pellman, 2011; Righolt and Mai, 2012; Setlur and Lee, 2012; Stephens *et al.*, 2011; Tubio and Estivill, 2011]. Now, this phenomenon must be important, as it was re-discovered by DNA sequencing projects, and most of the reports were published in top ranked journals.

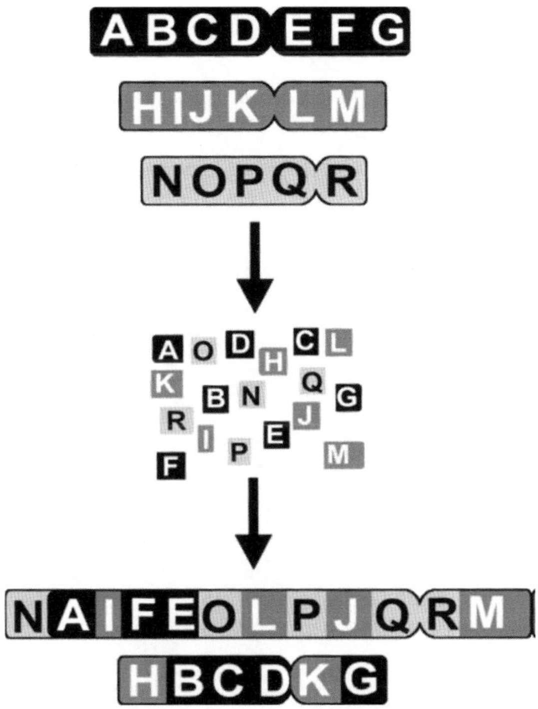

Fig. 5-8. The model of how genome re-organization occurs through genome chaos.

Three normal chromosomes are shown at the top. Each letter within the chromosomes represents a distinct region (or gene). Following exposure to sufficient levels of various stressors, chromosomes are partially cut apart by processes such as chromosome fragmentation [Heng *et al.*, 2004a; Stevens *et al.*, 2007, 2010; 2011; Stevens and Heng, 2013a–b] (middle). Following partial degradation of the genome, regions are recombined and rejoined likely through NHEJ, resulting in the chaotic genome illustrated at the bottom of the figure [Liu *et al.*, 2014]. The newly formed chimeric chromosomes can be a mixture of various chromosomal origins or occasionally from a single chromosome. The process of genome chaos represents an evolutionary process of genome fragmentation followed by reorganization of the genetic network [Heng, 2009]. This model is adapted from [Heng *et al.*, 2011b] with permission.

For numerical chaotic genomes, the total number of chromosomes for each cell can reach up to over 700. As for structural chaotic genomes, any type of known and unknown chromosomal aberrations can be observed [Heng *et al.*, 2013a]. **Figure 5-9** provides an example of giant chromosomes detected from chaotic genomes.

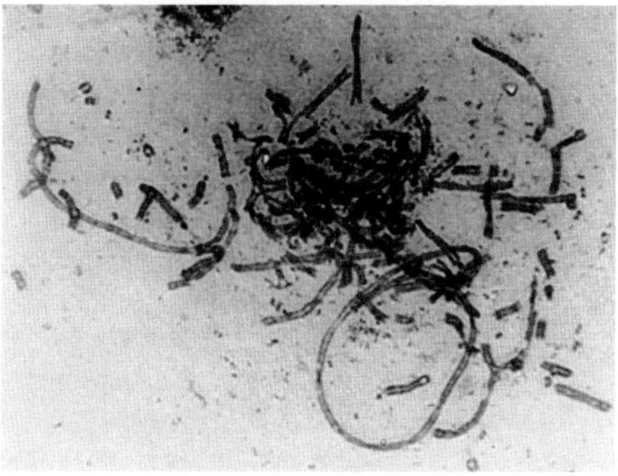

Fig. 5-9. An image of a chaotic genome.
Following chemical treatment, the mouse cell displays chaotic chromosomes. In this mitotic figure, there are normal sized chromosomes, chromosomal fragments, sticky chromosomes and many giant chromosomes. The size of the longest chromosome here is in the order of 20 times longer than a normal chromosome. All of these newly formed chromosomes are composed of different chromosomal fragments, which have altered the genomic topology among genes.

During the above mentioned study [Heng *et al.*, 2011b; Liu *et al.*, 2014; Stevens *et al.*, 2007; 2011], the entire process of genome chaos was monitored. Further synthesis has resulted in the following conclusions:

(1) Genome chaos is a survival strategy in response to high stress levels including various internal stresses (like crisis) or drug treatment. This process is frequently occurring both *in vitro* and *in vivo*. The first

phase is the induction of mitotic cell death. Recently discovered chromosome fragmentation (C-Frag) represents a major form of mitotic cell death [Stevens *et al.*, 2007].

(2) Both chromosome fragmentation (C-Frag) and non-homologous end joining (NHEJ) are involved in genome chaos. They serve to fragment the original genome and stitch it back together to form new genomes. This represents a macrocellular evolutionary process.

(3) The newly formed chaotic genomes with highly complex karyotypes are often unstable. Chaotic genomes will rapidly evolve during the first few weeks of time following induction. Over time stable clones arise at the cost of massive high frequencies of death.

(4) Each run of genome chaos will generate its own unique survivors with different karyotypes and transcriptomes [Heng *et al.*, 2008; Stevens *et al.*, 2013a; 2014; see Chapter 8]. The stochasticity of genome reorganization reduces pathway-based predictions of macroevolution.

(5) While many individual investigators have linked different molecular mechanisms to genome chaos, it is the level of stress that matters for genome chaos rather than specific molecular pathways. Many different molecular mechanisms can be linked to mitotic cell death. Genome alterations will occur regardless of which pathways trigger cell death. For example, different factors have been linked to the induction of C-Frag [Stevens *et al.*, 2011] but the key is the level of stress each factor is associated with (**Table 5-2**). This observation is highly significant, as it links specific molecular pathways to a common system response.

(6) Genome chaos likely happens much more frequently than it is detected in clinical samples. Genome chaos is often followed by more stable stepwise evolution where chaotic genomes have often been eliminated. These transitional chaotic genomes are frequently not detectable in these later stages of evolution.

(7) Different types of chaotic genomes can be observed from various samples. Even though chromothripsis has become a hot topic, it only represents a subtype of chaotic genomes. Different subtypes of chaotic genomes have been illustrated in *in vitro* models [Liu *et al.*, 2014]. This realization is important, as there are currently many

misunderstandings. For example, it was assumed that the mechanism of chromothripsis is different from other subtypes, as it only involves one or a few chromosomes. It is likely that when a few chromosomes are involved, there is a higher likelihood of survival. However, genome chaos is a highly dynamic process, and the end product could be drastically different from initial chaotic genomes. There is often a series of changes involved; thus, to claim that there is only one event associated with genome chaos is not correct.

(8) Genome chaos represents a new mechanism of rapid, adaptive, genome-based evolution that is essential for promoting rapid macroevolution, which leads to new genome-defined systems. This phenomenon may explain some unwanted consequences of cancer treatment.

Table 5-2. Various factors can lead to the same phenotype of C-Frag

Caffeine-induced uncoupling of the cell cycle
Chromosome-localized UV damage
Various gene mutations (AKT, p53)
Centrosome amplification
Measles vaccination
Calcium
Genomic instability
HSP inhibition
Oxidative
Temperature change
Unfolded protein response
Different drugs (MMC, Dox)

The discovery of genome chaos is an important piece of the puzzle needed to understand how somatic cell macroevolution works during crisis conditions such as chemotherapy, where emerging new genome systems fight for survival. Genome reorganization plays a key role in system transitions including immortalization, transformation, and metastasis [Heng, 2009; Heng *et al.*, 2011a]. Genome chaos should also play an important role in these steps.

The recognition of the importance of genome chaos including how genome systems can change to form new systems in a short period of time, and how drastically the network and phenotypes can switch, will ultimately help in understanding the relationship between genome reorganization and changes at all lower levels (gene, epigenetic or non-genetic) of the system. To illustrate this critical point, the concept of the multiple levels of the cancer landscape was recently introduced (see section 5.8.5) [Heng *et al.*, 2011b].

5.8 The Evolutionary Mechanism of Cancer

On one hand, so much progress has been made in identifying and characterizing the various molecular mechanisms that "cause" cancer. On the other hand, few of these findings have translated into improved cancer care. To solve this paradox, a common mechanism of cancer needs to be established. By putting all pieces of the puzzle together, it is clear that the evolutionary process of cancer is highly dynamic and cannot nor should not be explained by a specific molecular mechanism. It can only be explained through an evolutionary mechanism that unifies the individual and diverse mechanisms.

5.8.1 *Search for the general mechanism among diverse molecular mechanisms*

First, a large number of individual molecular mechanisms are involved with cancer evolution due to the multiple levels of heterogeneity that occur in cancer. Each new alteration can potentially generate new genetic landscapes by changing the combinations of genes and pathways. Second, the incidence of each individual molecular mechanism is limited in patient populations due to the large numbers of "cancer genes" and pathways. This is particularly true during the punctuated phase of evolution where every cell is different and the "winner" of each run of evolution is often unique and thus hard to predict. Third, any common mechanism would increase cancer evolutionary potential by destabilizing the system and increasing system heterogeneity rather than simply promoting proliferation and reducing cell death. Fourth, in cells displaying a different genome context (particularly when there are drastically different karyotypes), the biological function of many gene mutations might not be the same, as these cells have different gene interactive contexts. Molecular characterization based on fixed gene profiles or pathways alone is therefore likely to be misleading.

Similar to the statement that "all roads lead to Rome," many pathways or different molecular mechanisms can lead to cancer. To illustrate this, experiments and analyses using MCF-10A cells derived from the same individual were used to test whether different cancer-causing mechanisms could have the same result. The MCF-10AT line was established following introduction of an H-Ras mutation that caused transformation and subsequent tumor formation. In a separate experiment, cigarette smoke condensate, a well-known carcinogen, caused transformation and tumor formation despite the lack of an H-Ras mutation. In yet another experiment, by simply changing hormone regulation, cell transformation and tumor generation also occurred. These three experiments all involved discrete but different singular molecular mechanisms, but all triggered elevated NCCAs and increased population heterogeneity (the common mechanism driving cancer evolution). These analyses further suggest that by focusing on one of a large array of potential molecular mechanisms, it is possible to demonstrate that there are many effective ways to generate cancer in laboratory settings. Each of the altered pathways leads to transformation under given conditions, but because there are so many pathways that can lead to cancer progression, the importance of each is extremely limited. This is especially true in clinical settings where each case is unique and represents an independent run of cancer evolution that results in a unique genetic profile. In fact, the outcome of every road leading to Rome in cancer evolution can be illustrated very well by Sewall Wright's adaptive landscape model of genetic fitness, as there are many "roads" or means to form a tumor.

Analysis of these experiments also helps to explain why there are so many excellent publications that illustrate key cancer mechanisms but in all practicality the eloquent mechanisms described are not that important, as the same authors could publish many equally "important" papers centering on different mechanisms. Unfortunately, this is what many scientists have done throughout their careers. They have linked their favorite genes or pathways to different molecular mechanisms in a variety of experiments without realizing the limitations of each specific link. In fact, each identified link represents only one of many yet to be "discovered" mechanisms. Of course, biological processes like transcription regulation can be linked to almost all diseases and their responses to treatment and thus are very important, but the same is true for DNA replication, cell cycle regulation, cell death or mitochondrial function, endoplasmic reticulum (ER) stress, protein degradation, tissue organization, the immune system, individual lifestyles, and even a patient's social support.

Most molecular focused researchers are familiar with links between diverse biological processes and cancer, but many less can appreciate the importance of other factors when there is no clear specific molecular mechanism to explain the factors. These factors include lifestyle, social support, and even the mind-body interaction. On the surface, these factors do not seem to be linked to cancer, as most of us are looking for a specific molecular mechanism, but they are clearly linked to cancer. The lack of understanding of the workings of these factors is due to the fact that cancer is within a holistic system of the individual, and social interaction is part of the individual's life. For example, according to recent meta-analysis, stress is associated with poorer survival and higher cancer mortality. Breast cancer outcomes (five-year survival rate) can differ by as much as 25–28% in groups with and without family or social support, and depression increases mortality by 40% [Andersen *et al.*, 2008; Chida *et al.*, 2008]. Not surprising, psychological intervention can improve survival for breast cancer patients [Cao and During, 2012]. The drastic differences in outcomes between those with support and those without might be more important than any given gene mutation and are bigger than the differences between treatment and control groups in the majority of clinical trials of new cancer therapeutics.

Thus, specific molecular contributions to cancer are not all that special, and there are many other hidden contributors that are more important than some well-known gene mutations. The bottom line is there are many factors that can be linked to cancer.

If so many factors can be linked to cancer, what is the common link among them? How can we make sense of so many seemingly unrelated factors? In the case of the MCF-10A models, the common link is that they all shared elevated genomic instability. Regarding other factors such as an individual's social relationships, what is the common link?

Since cancer represents a somatic evolutionary process, and understanding evolution requires focus on the mechanisms of inheritable variations and the advantages of phenotype selection, it became obvious that we need to synthesize all of these diverse factors within the cancer evolutionary framework. These include the complex interactions between and among internal and external stresses, all molecular mechanisms, cellular adaptation, heterogeneity mediated genome dynamics, evolutionary potential, and emergent systems defined by the genome context.

This synthesis has led to a number of realizations:

(1) There are a number of two-way-interactive linkages, between stresses and molecular pathways/defects, between various molecular defects and genome instability mediated genome alterations, between genome evolution and constraints above the genome. There seems to be no clear general causative relationship within each pair, as the relationship is highly context-dependent, and there is no certainty regarding which partner will be impacted first. For example, drastic instability at lower levels can often impact higher levels. At the same time, when instability occurs at the higher levels of a system, increased instability can also be detected at the lower levels.

(2) Most genetic/non-genetic factors ultimately serve as system stresses regardless of the involved pathways. Our studies of the mechanism of chromosome fragmentation (C-Frag) induction demonstrate how multiple stresses affecting multiple molecular mechanisms can result in the same phenotype [Stevens *et al.*, 2011]. As long as each individual mechanism can generate sufficient cellular stress, mitotic cell death will occur. We demonstrated this by showing that various

molecular mechanisms and different types of cellular stresses could all lead to C-Frag (**Table 5-2**). Together, this study suggests that cell death is a diverse group of highly heterogeneous events that are linked to stress-induced system instability [Stevens *et al.*, 2013b]. Interestingly, cell death is also linked to both positive and negative evolutionary potential as it eliminates cells from the population, but can also increase the frequency of chaotic genomes and NCCAs.

The implication of this realization is highly significant, as it suggests that stress response is a system behavior with limited specificity. Even when there is initial molecular specificity in some experimental systems or defined pathways, any causation from this initial specificity will be lost in the long term due to the complexity of the cellular response or evolutionary process, where stress mediated stochastic genomic changes dominate.

It is thus reasonable to link all specific molecular mechanisms (genetic or environmental alike) to less specific stress response (see **Table 5-3**). The transition from molecular specificity to system behavior non-specificity becomes important in the context of cellular evolution. The question is: what bio-component serves as the transition linker?

(3) In addition to the linkage between cell death pathways and increased NCCAs (**Fig. 5-10**), almost all molecular/cellular defects can be linked to genome instability-mediated genome alterations, which in turn can be stochastically linked to different molecular defects. The important message is that most molecular/cellular mechanisms can be linked to CIN directly or indirectly [Heng *et al.*, 2013b]. Many well-known examples of molecular specificity in fact should be better explained by stress-mediated genome dynamics. The Holy Grail of molecular cancer biology has been to identify causative relationships by systematically testing the links between genes that are upstream and downstream in a pathway to cancer. However, our NCCA studies link many different molecular pathways or defects to a less specific event, NCCA elevation.

Table 5-3. Various internal and external factors can generate system stresses

Internal Stresses	External/ Environmental Stresses	Experimental Stresses
Aging/Telomere Shortening	Cellular Debris	Chemical Inhibitors
Calcium	Cellular-	Culture Conditions
Cell Cycle Control	Heterogeneity itself	Gene Knock Out/
Cell Growth/Differentiation	Chemical Exposure	Knock Down
Cell Death	Drug Treatment	Gene Overexpression
Cellular Interaction	Hormone	High Dosage Drug
Cellular re-generation	Lifestyle	Off-target Effects of
Chromosome Condensation	Metabolic/	Most Molecular-
Chromosome Segregation	Mechanical/	Manipulations
Cytoskeletal Aberrations	Mental Stress	
Development	Nutrition Status	
DNA Replication/Repair	Pollution	
Epigenetic Disregulation	Physical Stress	
Genome Alterations	Radiation Exposure	
Hypoxia	Viral/Bacterial	
Immune	Infection	
Inflammation	Wound Healing	
Mitochondrial Function		
Oncogenes		
Oxidative		
Post-Transcriptional		
Modification		
Protein Degradation		
Tissue Architecture		
Transcriptional Regulation		
Tumor Suppressors		
Unfolded Protein Response		

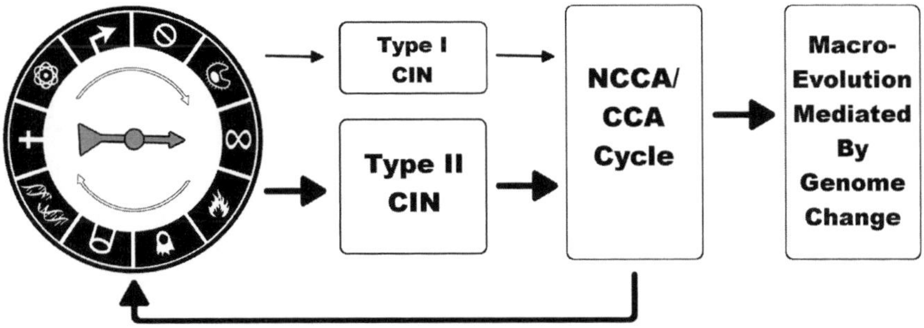

Fig. 5-10. Diagram illustrating the relationship among diverse individual molecular mechanisms of cancer and CIN-mediated macro/microcellular evolution.

The hallmarks of cancer were used to represent different pathways linked to cancer. Stress is the motor that turns the pathway wheel. Selection of a given pathway as a mechanism of cancer progression is represented by the arrow that selects a pathway based on probability. Individual pathways can directly compromise genome integrity (type I) or indirectly jeopardize genome integrity through general stress (type II). Both types I and II CIN are linked to elevated NCCA frequency (modified from [Heng *et al.*, 2013b]).

(4) There is no consistently dominant pathway in cancer patients and no dominant pathway throughout the entire evolutionary process of an individual tumor. Genome instability promotes the probability of evolutionary success rather than a fixed pathway.

As illustrated in **Fig. 5-11**, in a normal stable genome system, despite the occasional gene mutation within a specific pathway, overall system homeostasis results in no further somatic evolution towards cancer (a dead end for cancer evolution). However, when the genome system is unstable, genomic heterogeneity increases. The evolutionary potential increases due the increased number of pathways available within a population. Within the punctuated phase, genomic turnover occurs until a winning karyotype and resultant pathways are selected. Selection of winners appears to be stochastic, and it is difficult to predict which one

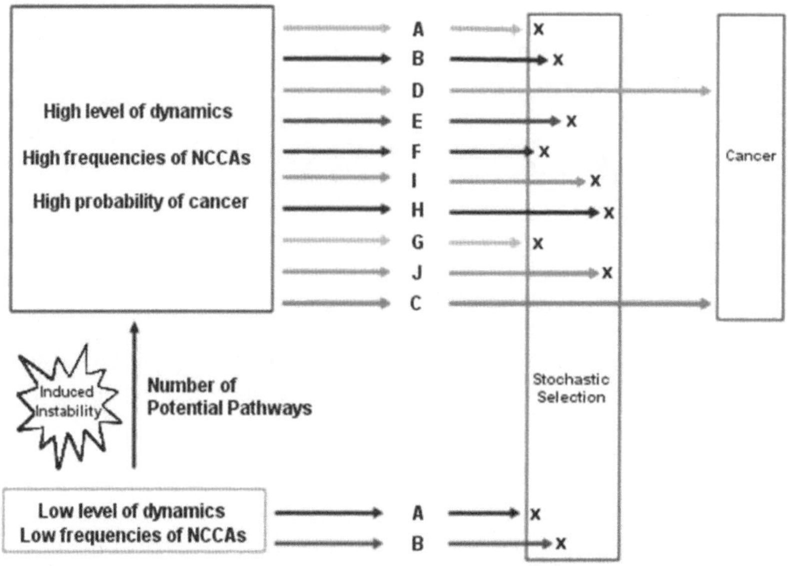

Fig. 5-11. The evolutionary mechanism of cancer and its relationship with molecular mechanisms.

Many types of stress linked to different molecular pathways can trigger system instability. Unstable systems can also activate different molecular pathways in response to system instability. The somatic selection process stochastically favors different packages of genome alteration, which become the driving force of cancer evolution. The lower left box represents a normal stable state that typically generates infrequent NCCAs that will likely go extinct under system homeostasis. With increased instability, higher frequencies of NCCAs occur, representing an increasing number of potential genome systems coupled with specific molecular pathways. Each arrow represents a given molecular pathway (a molecular mechanism). The increased number of pathways (represented by various colored arrows) increases the probability of evolution leading to cancer. A certain specific molecular signature may be identifiable in the end product of each cancer case, but it is extremely difficult to predict among all the potential pathways which one will be successful and how the driving pathways will fluctuate throughout the somatic evolutionary process (Reproduced with permission from [Ye *et al.*, 2009a]).

will be successful (both in individual patients and within a given tumor). When a genome is successfully selected, however, molecular methods can be used to identify gene mutations or pathways that are involved. This typically occurs at later stages of cancer evolution.

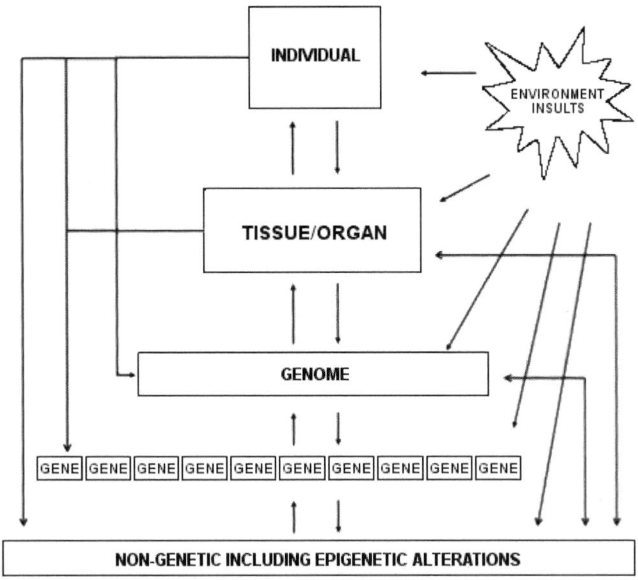

Fig. 5-12. Multiple level interaction model illustrating the complicated relationship among different stresses, various genetic/ non-genetic alterations and higher level constraints.

The relationship of cancer/genes/pathways/genome/tissue/individual and the environmental insults are illustrated. These multiple levels of complexity severely limit any predictions of cancer made based on gene mutation and epigenetic alteration. As illustrated by the arrows, lower level genetic alterations can impact higher levels. For example, gene mutation can impact the integrity of tissue organization, but changes of higher level constraint can allow lower level changes to become dominant. Cancer is a genome disease constrained by the probabilities contributed from multiple levels of systems. Understanding such a relationship is also important to cancer treatment, as focusing on only one molecular level could paradoxically reduce the system constraint from a higher level which could clearly affect cancer therapeutic outcomes (Modified from [Heng, 2007a]).

In addition to gene/epigene and genome heterogeneity below and/or at the genome level, there are multiple levels of constraint above the cellular level, including tissue specific fuzzy inheritance (see Chapter 6) to tissue architecture, the immune system, developmental/aging-related physiological changes, to an individuals' health status and even the social interactions of patients, especially family support. As illustrated in **Fig. 5-12**, all higher level constraints can alter cancer evolutionary potential. The presence of multiple levels of constraint is the reason that tumors are relatively rare despite the occurrence of so many genetic alterations over a person's lifetime. For most altered cells to become cancer, a perfect storm is often needed. Interestingly, reduction of higher level constraint can promote genome evolution and NCCA/CCA transition (also see **Fig. 5-4**).

5.8.2 *The simple relationship among stress, genome alteration and selection*

The above realizations illustrate the challenge of using specific molecular mechanisms to understand and treat cancer. The unpredictable nature of cancer evolution produces many possible molecular mechanisms that can be identified in patients, but few are dominant. Furthermore, each pathway is highly dynamic when subjected to medical treatment, as treatments change the genome, making pathways moving targets. Paradoxically, at the same time, it simplifies the common mechanism of cancer if we switch our approach from identifying an endless list of individual molecular mechanisms to understanding the evolutionary mechanism.

By synthesizing the above discussions and viewpoints together, the common mechanism of cancer can be unified as the evolutionary mechanism of cancer which is equal to or greater than the sum of all individual molecular mechanisms [Heng *et al.*, 2010; 2013b; Horne *et al.*, 2015; Ye *et al.*, 2009].

Evolutionary Mechanism = Σ Individual Molecular Mechanisms

The evolutionary mechanism itself can be explained by four simple components (**Fig. 5-13**):

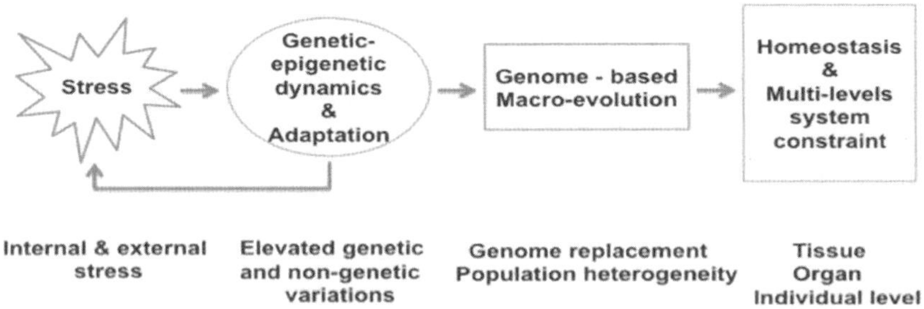

Fig. 5-13. Four key components of the evolutionary mechanism of cancer.

The overall causative factor of cancer is stress generated inheritable variation resulting from instability. The evolutionary process works to sort out the survivors. Although many diverse individual factors can trigger and influence this process, it can be distilled to two important aspects: (a) the key for somatic cell evolution leading to cancer is reaching the macro-evolutionary phase where the genome is changed; (b) as illustrated in Fig. 5-12, there are multiple system constraints that occur above the genome level. Sufficient macroevolution to break these constraints must occur for somatic cell evolution to progress to cancer.

First, the sources of stress are highly diverse factors that occur both internally and externally. All stresses that surpass the ability of traditional cellular stress responses to return the system to homeostasis will produce damage. These stresses can involve any major biological process and the damage resulting from these stresses can include genetic alterations, epigenetic alterations, or environmental factors (see **Table 5-4**). For example, elevated system instability can be a result of germline mutations, somatic mutations, bacterial/viral infections, environmental toxicity, nutrition deficiencies as well as aging and other disease processes. Time and environment interactions make the situation even more complex and further reduces the ability to predict this process based on specific types of and/or responses to stress, as the same inherited genetic factors may or may not result in the same phenotype depending on the environment. "Good" factors can be "bad" factors depending on the context in which they occur. The key is the combination of opportunities. Furthermore, the initial instability often can serve as an effective way for cellular adaptation by providing genetic variation. This mechanism confers a short term evolutionary advantage for somatic cellular function (even though at the same time, it also can provide opportunities that favor disease conditions by further disturbing the system homeostasis).

Second, there is a very complex relationship between the multiple levels of genetic/epigenetic interactions (**Fig. 5-12**). The first level of response to stress beyond standard stress pathways is at the genetic and epigenetic levels, where the epigenetic level is more sensitive than the gene level, and epigenetic changes are reversible. Most stresses do not cause significant damage. However, if strong enough, some will cause damage, resulting in epigenetic alterations then gene mutations. Some changes may induce significant instability further breaking the boundaries of the gene/epigenetic level and leading to further genetic/epigenetic alterations. These alterations can then serve as new stresses. Gene mutations and lately epigenetic changes during cancer evolution have been extensively studied. However, research has focused on specific mutations rather than incorporating the random mutations that occur during the process, and as previously noted these studies lack incorporation of any genome level changes, especially stochastic forms such as NCCAs.

Third, genome replacement-mediated macro-cellular evolution is a key to cancer evolution. If gene/epigenetic alterations do not return the system to homeostasis, genome level alterations must occur for cellular adaptation and survival. This creates population diversity and jumpstarts somatic macrocellular evolution. Only powerful macroevolution is capable of breaking down the multiple levels of system constraint that occur in the fourth step. While genomes constrain gene mutations, there are many system levels above the genome that can both enhance or constrain genome alterations despite the fact that genome context also influences the behaviors of these levels. Most of the genomes created during this stage will not ultimately be survivable and will be lost.

It is important to note that stochastic genome alterations happen regularly during cancer progression, even without major gene mutations. As discussed in Chapter 6, there is likely a tissue specific fuzzy inheritance that favors certain degrees of heterogeneity at various levels. In normal tissue, the level of genome heterogeneity is low, and even within the stepwise phase of cancer evolution, the increased level of heterogeneity does not alter the dominance of the average clonal population. This fuzzy inheritance-defined heterogeneity is essential for cellular adaptation. However, high levels of stress can often elevate the frequencies of NCCAs beyond their baseline fuzzy levels, drastically increasing the opportunities to form new CCAs following evolutionary selection.

Fourth, breakdown of multiple levels of constraint above the cell population. Any newly formed cell population with new genome systems will have to breakdown these constraints. These constraints include process constraints and geographic constraints. Process constraints include clonal over-growth (hypertrophy), immortalization, transformation, angiogenesis, tumor establishment, metastasis and challenge resistance including drug resistance. Geographic constraints include tissue and organ constraints. Other constraints include the immune system and mind-body interaction. Even though there are special relationships among different types of constraints, each level of constraint is very different and breaking it usually requires a different genome-defined system with novel features that are essential for breaking specific constraint. This might be the reason why there are multiple waves of NCCA/CCA cycles, and the "winners" of all the various stages display their unique genome context. For example, most newly formed genomes will not escape the constraint of the immune response, however, multiple waves of NCCA/CCA transitions make it likely that at least one will survive. The observation that primary tumors and metastatic cancer often display altered genomes supports this hypothesis, as new genomes are needed for such transition. It also explains why in many cancer types, the evolutionary process is a rather long process. Finally it explains why solid tumors display much more genome level diversity than many blood cancers.

5.8.3 *The basis of limited gene mutation specificity in most cancers*

The above discussions can effectively address one frequently asked question: how can stochastic cancer evolution explain the clinical fact that despite the huge diversity, there are higher frequencies of some cancer gene mutations in the patient population? Also, why is there an increased association between some cancer genes and specific cancer types and disease stages?

The pattern of cancer evolution and its selection environment/process is the basis for the increased frequencies of many genes. For example, the much higher p53 mutation frequencies observed in many cancers should be linked to the high level of genome alterations, despite that it could be linked to diverse molecular pathways, as stochastic genome alteration is the basis for creating new systems or pathways and switching among different pathways. This might be the reason p53 mutations can be detected in high frequencies across cancer types. However, due to its diverse nature and low specificity for pathway switching, it would be hard to target p53 mutations and achieve clinical success.

The higher frequencies of many oncogenes should be understood within the framework of the two phases of cancer evolution. Even though, as we have discussed, macrocellular evolution is the key for cancer formation, the phenotypic dominance of the formed cancer can greatly benefit from gene-mediated microevolution. In other words, cellular population expansion of already formed cancer cells can be achieved quickly with the help of various cancer genes that promote cell proliferation. This explanation is supported by the fact that oncogenes can be detected in much higher frequencies in many NCI cancer cell lines (where proliferation is a key selective feature) than in patient samples (where more features are involved in addition to proliferation). In addition, the "long trail" of gene mutations detected from the cancer genome project fits the explanation, as different oncogenes can be associated with needed micro-evolution. This analysis further suggests a possibility that many oncogene mutations are biased towards the later stage of clonal expansion. Yes, more typical oncogenes can be detected from formed cancers, though that they might not directly contribute to initial and intermediate processes. Similarly, for metastasis, it is the genome alterations that lead to the initial success of metastasis, but some common genes can help the already metastasized cells become more dominant. Yes, these genes can be identified, however they might tell less about how initial metastasis occurs.

As for some tissue-specific distributions of cancer gene mutations, the selection pressure from specific tissue types and cellular stages can favor certain genes. But again, these contributing factors are less dominant when compared with genome level changes (explaining why the frequencies of these oncogenes are only around a few percent), and can be easily changed under new stresses (this is why they cannot serve as ideal drug targets, especially for the long term).

Based on this prediction, it is easier to achieve better treatment responses when cancers present dominant gene mutations rather than highly diverse genomes. It is thus easier to treat some blood and childhood cancers than cancers with high levels of genome instability.

5.8.4 *Challenges for making clinical predictions*

If understanding these four components of cancer evolution seems simple, the practical implications are rather complicated and hard to handle. The combinational number of all molecular mechanisms is huge and it is extremely difficult to predict which one will be dominant for which patients and when. Examples of large numbers of individual molecular mechanisms are listed in **Fig. 5-10**, **Table 5-3** and **Table 5-1**.

The complexity of cancer evolution leads to another paradox, the insignificance of the significant. No matter how significant a gene mutation appears to be, further unbiased studies will result in reduced significance, due to the complex nature of cancer evolution. This sad reality has been illustrated in many clinical observations and underscored by current cancer genome sequencing efforts. Sequencing of additional tumors invariably reduces the frequency that a given gene mutation is observed. Now it is easy to understand why there is such a huge gap between basic research and clinical practice.

The significance of the insignificant also explains the limitations of some of the most famous cancer genes. As mentioned in Chapter 2, the key experiment illustrated that defined genetic elements (the ectopic expression of the telomerase catalytic subunit, hTERT, in combination with oncogenes SV-40 large-T and H-ras) can create human tumor cells [Hahn *et al.*, 1999], demonstrating that disruption of intracellular pathways suffices to cause cancer. The same experiment in human diploid fibroblasts however does not always produce transformation. More interestingly, when transformation occurs, chromosomal changes are always necessary [Akagi *et al.*, 2003; Heng, 2007a].

Finally, the evolutionary mechanism of cancer solves one of the key confusing questions regarding the importance of massive genetic alterations (especially gross genome level changes) that are nearly universally detected in cancers. Traditionally, gross genome changes have been thought to be non-causative factors resulting as a consequence of key gene mutations, since most of the genome level changes seem to occur stochastically among patients. After illustrating that genomic variation is essential for cancer evolution to occur, arguments arose regarding their importance and whether this represents "only correlation" but not "causation". This argument seems to make sense based on the current molecular approach of pinpointing a molecular mechanism (such as the identification of gene mutations) in the search for a causative factor. However, it does not make sense from a clinical perspective, as so many diverse individual mechanisms (e.g. gene mutations) exist. Most of the identified molecular mechanisms thus far have failed to provide useful clinical applications as they lack commonality. Now, the importance of diverse genome variation only makes sense under the evolutionary mechanism of cancer. As to the question of defining which genomic changes are "the" common causative drivers, this invokes a broader question of evolution in general: What is the common causative factor of evolution? If natural selection is the answer, then genetic variation is a precondition, and no individual specific genetic alteration universally matters (there is no common gene mutation that defines evolution). The same standard should be used in cancer research. In cancer evolutionary studies, each tumor represents an independent successful run of evolution.

In macroevolution, genome replacement is the common driver. The key to cancer evolution success is the establishment of a new genome-defined system. However, despite their ultimate importance, it is extremely difficult to predict which genome alteration will be the driver of a specific cancer case when so many could potentially be the driver for that tumor. In agreement with this, the degree of stochastic genome level alterations correlates well with tumorigenicity in most solid tumor types. Thus focusing on end products of evolution rather than dissecting an unlimited number of genetic components is a better approach in terms of predicting the evolutionary trajectory of a tumor.

In conclusion, for cancer evolution to be successful, elevated levels of inheritable genetic variations are required, and genome level change is the most dominant form of variation compared to any gene mutation or epigenetic alteration. With increased evolutionary potential, cancer will find its way. Thus, fighting cancer is not a simple issue of stopping a given pathway or a few pathways, but the issue of how to reduce cancer evolutionary potential.

5.8.5 *Multiple level landscape model*

The evolutionary mechanism of cancer can solve many paradoxes: the significance of individual gene mutations in basic research and their decreased significance in clinical implications; the importance of the characterization for a specific pathway and decreased importance of using such information to predict the direction of cellular evolution; the precise understanding of gene and epigene has limited contribution to understand the genome alteration mediated cancer evolution. It is thus necessary to build a model to articulate the relationship among different pathways within cellular selection and among different types of genetic/non-genetic organization. To illustrate these relationships, we have proposed the multiple level landscape model of cancer evolution [Heng *et al*., 2011b; 2013b].

The concept of a multiple level landscape is similar to Sewall Wright's adaptive landscape model of genetic fitness [Wright, 1932], except that there are two fundamental genetic levels within the landscape. These levels include the Waddington's epigenetic model, which fits into the local landscape (gene level or epigenetic change) as well as bigger "mountains" that describe the global landscapes (genome alterations) (**Fig. 5-14**).

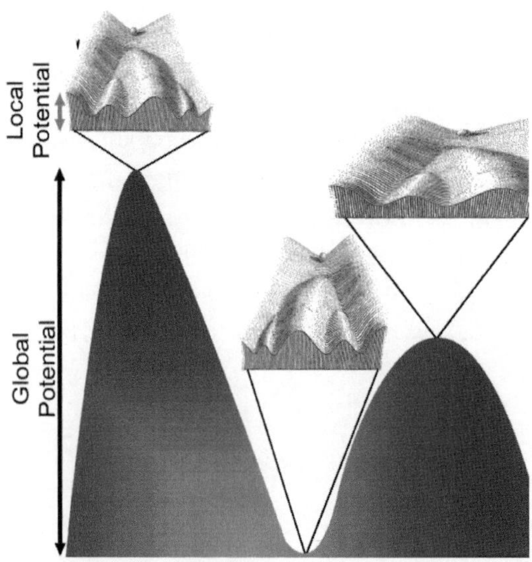

Fig. 5-14. The multiple level landscape model of cancer.

The genetic landscape of cancer can be broken down into two levels of evolutionary potential. Local potential refers to adaptation potential provided primarily by gene-level or non-genetic changes. While important to many biological processes such as development, local adaptive landscapes do not typically drive the evolutionary process of cancer. The global potential of the evolutionary landscape of cancer is derived primarily by genome level changes that drive macroevolution. It should be noted that local landscapes are found throughout the global landscape. Due to the different mechanisms involved in micro- and macroevolution, changing the global landscape is not achievable by simply accumulating local landscapes (Reproduced with permission from [Heng *et al.*, 2011b]).

As illustrated in **Fig. 5-14**, the local and global landscapes are very different in terms of the energy barriers and the degree of stress needed to change the landscape. The major "mountains" of the global landscape are high enough to prevent local attractor transition between mountains. In contrast, crossing small local hills requires only low energy potential, which can be achieved by the same genome with different gene or epigenetic states. Note that in applying this model to understanding organismal evolution, perhaps the most important element that distinguishes it from Wright's model is that different genomes are prevented from crossbreeding while similar genomes allow it. Wright's model best fits microevolution, where interactions between gene mutations, epigenetic regulation, selection, environment, population size and crossbreeding have an influence. However, to cross global landscapes, such micro-evolutionary interactions are not enough. Genome change is required to form a new species (see Chapter 7).

As a result, in order to understand the local landscape, one needs to understand how gene mutations and epigenetic regulation impact the gene regulatory network. In contrast, to understand the global landscape, one has to focus on the formation of new systems created by genome reshuffling. Factors that influence the global landscape include the frequency of altered genomes, the number of genome re-organization events in each cell, the ratio of NCCAs and CCAs, and the degree of complexity of altered genomes. The ability of a system to respond to stress is dependent on its ability to rewire an existing network. If the stress level is moderate enough to allow rewiring within a system, no genome change is required. However, when the stress exceeds the system's boundary of tolerance, that system will no longer be viable. That is the reason why many cells are eliminated under stress despite the rewiring ability of networks (the limitation of micro-adaptation).

The current adaptive landscape model (without integration of the genome level) offers only extremely limited insight into cancer as it does not incorporate genome-mediated macroevolution. In addition to our proposed fitness landscape that separates the gene and genome levels, we propose a new "survival" landscape that is separate from the traditional fitness landscape. Within the traditional fitness landscape, fitness can accumulate over generations. In contrast, accumulating gene changes do not fundamentally alter the survival landscape. The survival landscape occurs immediately when a population of cells is exposed to a stress. Cells are either able to survive the stress or not. For example, if cancer cells are exposed to high dosage of a theoretical treatment, most cancer cells will be eliminated within a short time, and will have no chance for gene level changes to accumulate over many generations.

The multiple levels of the cancer landscape model also explain why only focusing on a handful of oncogene-defined pathways will not provide reliable information that is relevant for clinical application. The evolutionary adaptive landscape allows for unlimited combinations of genetic/epigenetic changes (through evolutionary potential) that can change the system status leading to cancer. In reality, the simple branching concept of evolution is convenient and easier to understand but unfortunately is not real. In fact, the concept of evolutionary branching or trees only reflects alterations to the patterns of dominant pathways in most experimental systems and limited exceptional cancer types rather than the actual possible pathways in most patients due to genome/environmental heterogeneity. Since each tumor represents an independent run of multiple cycles of somatic evolution, the presence of a few recurrent dominant pathways in human cancers is rare especially when the genome is unstable. More common dominant pathways are often observed in blood cancers and in some pediatric tumors that are noted for their genomic stability. Interestingly, the relationship between adaptive landscapes and evolutionary branching can be used to illustrate the relationship between NCCAs and CCAs. NCCAs represent all possible genome landscape changes, while CCAs represent successfully selected changes. Thus, the dynamic NCCA/CCA interactions reflect the relationship of evolutionary potential and the reality of the evolutionary success in limited runs of evolution. Recently, the multiple level landscape model has received further support from some system biologists [Huang, 2013].

5.8.6 *Summary of the genome theory of cancer evolution*

To conclude this chapter, a brief summary of the genome theory of cancer evolution is listed in **Table 5-4**. The genome theory emphasizes the importance of selection at the genome package level for macro-cellular evolution. This concept is ready to be integrated with the multiple levels of homeostasis, as there are both cooperation and conflict among them, and not all levels contribute equally to disease progression. To achieve higher level of homeostasis, lower levels can sometimes be sacrificed or negatively impacted, or *vice versa*. For example: 1) while biodefense mechanisms could act to retain homeostasis, they could also lead to cancer (e.g. an enzyme crucial for fighting off malaria infection is linked to lymphoma; wound healing promotes cellular proliferation and increased genetic diversity); 2) there is a fine balance between cancer and autoimmune disease; 3) and targeting cancer cells to restore tissue homeostasis can damage the immune system and a higher level of homeostasis. The relationship between genome alterations and transcriptome dynamics illustrates the molecular basis of package-level selection and the importance of the genome, as it is the level of dynamics (evolutionary potential) that matters rather than specific gene expression (more see Chapter 8). Note that some key concepts regarding the mechanism of cancer heterogeneity (such as fuzzy inheritance and the importance of outliers in cancer evolution) are discussed in Chapter 6. Additional information (such as multiple levels of emergent properties, or genome as a unit of evolution) can be found in the following publications [Gorelick and Heng, 2011; Heng, 2009; Heng *et al.*, 2011a; Horne *et al.*, 2014; 2015a]. The detailed illustration of the relationship between genome alterations, transcriptome dynamics and evolutionary potential can be found in section 8.8.1.

Table 5-4. Brief summary of the genome theory of cancer evolution

(i) The genome is not just the collection of an organism's DNA. The genome organizes the interactive relationship among genes using genomic topology. The physical relationship among genes serves as a platform for emergent properties of genomic function. The genome rather than individual genes defines the "system inheritance," or genetic blueprint, while individual genes only provide "parts inheritance," or the instructions for making parts. The same gene content can form different genomes with different genetic network structures by re-organizing the genomic topology via karyotypic alteration. For most cancers, the key genetic information responsible for system evolution is achieved by new genome formation rather than the result of a few specific gene mutations.
(ii) The mechanism of passing genetic information differs from the classical gene-centric concept with genetic precision. (1) System inheritance is maintained by safeguarding the genome context or karyotype. For sexual organisms, the genome identity is effectively preserved by the function of sex. Individual gene mutations will not change system inheritance. In contrast, for asexual organisms, genome alteration is overwhelming. (2) While the core genome is shared by most individuals of a given species, there are multiple levels of genetic and non-genetic variations generated by fuzzy inheritance. These variations are not just generated by mistake, but guaranteed by fuzzy inheritance, as certain levels of variation (but not specific changes) are programmed for cellular adaptation. (3) The separation of germline and somatic cell provides unique mechanisms both for short-term, individual level adaptation (by increasing somatic cell diversity) and long-term species-specific genome stability (through germline preservation). To increase the capability of adaptation, the somatic cell can be highly plastic through utilization of fuzzy inheritance. For the conservation of a given species, the germline displays minimal changes, and the reproductive process further eliminates the vast majority of genome level changes to retain the identity of the species. Based on this mechanism, many diseases or illnesses will be contributed by somatic cell variations.
(iii) The pattern of cancer evolution differs from neo-Darwinian evolution due to different genetic principles and constraint involved (e.g. without normal constraint, cancer represents an extreme case of utilizing fuzzy inheritance to generate necessary heterogeneity). Cancer evolution can be described as two phases of evolution (punctuated and stepwise), which correspond with macro- and micro-cellular evolution. Genome level changes contribute to macro-cellular evolution by creating new systems, while gene level and epigene level changes contribute to micro-cellular evolution by modifying existing systems. The power of cancer evolution is the combination of the two phases. Genome chaos is independent of the stepwise accumulation of gene mutations, as macrocellular evolution is not equal to the accumulation of a series of microcellular evolutionary events over time.

(iv) The relationship among gene mutations, epigenetic changes and genome changes can be illustrated by the multiple level landscape model, where local landscape represents gene/epigene status and global landscape represents the status of genome replacement. Fundamentally, the impact of lower levels of alteration (which modify the system) needs to reach higher-level change in order to create new systems. In addition, there is no accumulative relationship between gene mutation and genome alterations. Since the genome represents the highest level of genetic organization, its alteration often has a much larger effect than individual gene alterations. Chromosomal changes can impact on hundreds or thousands of individual genes. Furthermore, rearrangement of the genome can change overall genomic information patterns without generating aberrations in specific cancer genes.

(v) The common mechanism of cancer is the unfortunate result of system response to various stresses (e.g. physiological, pathological including the aging process and infection, and direct DNA and chromosomal damage), the same mechanism for generating variation necessary for cellular adaptation. Under stress, multiple levels of genetic/non-genetic variation are needed for adaptation, but as a trade-off, some individuals also pay the price, as variations can also favor cancer evolution. Fortunately, the vast majority of variations will be eliminated, and cancer requires the perfect storm. In this sense, many cases of cancer represent bad luck. This makes it hard to predict cancer progression based on limited specific genetic and environmental factors, as the genetic information is fuzzy, and the environmental response is largely unpredictable. In particular, molecular specificity will not dominate in the long term, as due to stochastic genome instability (involving different pathways), cancer becomes an issue of evolutionary game theory, and each run of evolution is different. While some mutation signatures can be profiled for patients, the clinical prediction value is limited, as there is no fixed cancer genome. The same gene mutation holds a different role within different genomes, and the landscape of the cancer genome is highly dynamic and is constantly evolving, especially under harsh treatment. Thus, only the evolutionary mechanism of cancer can be used to explain the large number of molecular mechanisms and the limitation of each individual molecular mechanism in the patient population. For example, there are often no fixed drivers across the entire process of cancer evolution, as cancer evolution under crisis always depends on outliers to save the day. Most oncogenes, in contrast, are associated with the micro-evolutionary phase.

(vi) Due to the unique evolutionary features of cancer, the strategy for fighting cancer needs to be drastically changed. Future strategies need to focus on overall genome instability and evolutionary potential rather than specific pathways. To manage cancer, rather than directly targeting cancer cells with the intent of complete elimination, more approaches are needed to increase system constraint, reduce the speed of cancer, and maintain the specific phase of cancer evolution. For example, treatments with great killing power may have to be avoided in order to prevent the induction of genome chaos, which favors rapid cancer evolution and serves as the basis for drug resistance. As for the general population, more prevention methods are needed to promote healthy lifestyles and emphasize environmental impact on genome instability.

Chapter 6

Significance of the Insignificant: Why "Noise" is Essential for Cancer Evolution

6.1 Summary

Despite the fact that bio-noise (heterogeneity or fluctuations) overwhelmingly exists in all levels of biological systems, from the molecular components and sub-cellular processes to the dynamics of cells, tissues, organs, individual organisms and populations, the importance of bio-noise is largely ignored in the macroscopic world due to the law of large numbers. In addition, most current molecular methodologies are based on the "average profile" that makes noise/uncertainty less visible. Furthermore, when encountered, researchers have tried very hard to eliminate noise in the name of "separating signals from noise" for pattern identification. Such efforts are effectively helped by using more samples to attain statistical significance. This traditional strategy might have worked for some simpler processes where cell populations can be reliably measured through a dominant average behavior. However, it becomes highly problematic when dealing with bio-systems characterized by a large degree of heterogeneity that are outside the normal population equilibrium, such as cell populations under high stress conditions, and in particular, during the macro-evolutionary phase of cancer. Since most processes in nature are fundamentally stochastic, and multi-level heterogeneity is the key characteristic of life, the elimination of heterogeneity or noise can actually change the system itself, resulting in a knowledge gap between the "noise-less" artificial system under controlled conditions and the bio-reality where heterogeneity rules. In this chapter, following a brief review of the rationale as well as some misconceptions behind the elimination of bio-noise, the significance of the insignificant (noise) is discussed in the context of complexity and evolution. For many molecular analyses that focus on the characterization of bio-parts, the approach of eliminating noise is effective. In contrast, for the study of bio-dynamics under evolutionary selection such as cancer progression, noise becomes the key that drives evolution. Such analyses will not only reconcile the opposite view about the noise in biology, but also lead to the understanding of the mechanism of heterogeneity, and reveal a new type of inheritance: fuzzy inheritance. "Noise" is no longer the result of inefficiency, but the crucial component of the system evolvability that evolutionary selection favors. To support the above points, two independent essays will be presented that illustrate our own story to study genome level heterogeneity and its ultimate importance to understand cancer evolution.

6.2 What is "Bio-noise," and Why it is Even A Question?

To be perfectly clear, few like noise, scientists included. According to the common viewpoint, noise represents unwanted and/or insignificant sound, often associated with negative tone such as non-harmonious or discordant types of sounds. In physics, noise often refers to random error in observation or measurement, which varies between experimental duplicates. The way to eliminate noise is straightforward: by performing many measurements of the same subject. These measurements will fluctuate around some central value, and the more measurements obtained and averaged, the more precise this central value. Since measurements of some phenomena in nature fall into a Gaussian distribution, noise becomes relatively easy to deal with through statistical analysis.

It is known that more noise can be observed from biological systems. However, perhaps influenced by classical physics, the general attitude and approaches towards noise in biology is essentially the same in physics. In fact, it has long been realized that the major challenge for biological research is to eliminate noise (or randomness, even though not all noise is random) by searching for true patterns (deterministic and recurrent type of patterns), as the art of bioresearch seems to distinguish signal from noise.

We have casually asked bio-researchers about their immediate response to the word "bio-noise." Their responses varied; however most used the following phrases: observation error, operation error, no signal, inefficiency of specific reaction, genetic error, incapability of repeating, not significant, does not truly reflect the real mechanism, system is below optimal performance, and detrimental to reliable information transfer. When pushed, some will realize that noise may serve a secondary or tertiary/epistatic biological role, but is not that important in determining the primary relationship. A few have linked noise to stochastic regulation and heterogeneity. Within this minority, some have also concluded that noise could be good.

The majority's attitude explains the general approach in biology, as the magic tool to deal with bio-noise is to use a large number of observations and detect a pattern based on the qualitative summary. The underlying reasoning is: biological systems are highly effective due to their molecular specificity following millions of years of evolution, and noise might reflect the errors, accidents, or less effective statuses of the system (indicating that noise is generated when the system has problems). By collecting and averaging the large number of observations that filter out abnormal and insignificant noise, bio-specificity will surely be dominant. In fact, the beauty of science is to see through the noise to identify a pattern, period.

The supportive evidences are overwhelmingly documented by textbooks and a majority of scientific literature. Good biologists are very good at identifying patterns. Examples include Mendel's contribution that linked genetic elements to inherited traits, the identification of a number of hallmarks of cancer, and the characterization of various cancer specific pathways. In fact, a majority of these milestone experiments greatly enhanced our understanding of biological systems, in particular, the study of molecular mechanisms where pattern identification often plays an important role. In these studies, measurements of single-cell gene activity level seemed to fluctuate steadily around an average. The arithmetic mean was accepted as a representative measure because no large deviations existed between single cells of an isogenic population, therefore noise was fluctuations which could be considered background noise and were deemed negligible. Furthermore, noise could also result from technical or methodical errors that stem from variation in the experimental environment. The practice of eliminating outliers is generally accepted. Similarly, it is common practice for some researchers to disregard the results of an experiment if it was opposite or completely unlike experiments predicted by their hypotheses. Thus, noise represented an undesirable byproduct of biological systems and was ignored for the most part. So far, so good.

Far before the –omics era, researchers began to study bio-noise, the random variability in quantities that arise in biological processes, in microbial populations. It was observed early on that antibiotic treatment killed >99.99% of a bacterial population [Bigger, 1944a; Bigger, 1944b]. The surviving bacteria (one per million) were able to regrow and generate a population with the same genetic and phenotypic features as the parental population, a phenomenon known as bacterial persistence. What special properties (which are different from mutation mechanisms) do persistent colonies exhibit that enable them to survive antibiotic treatment? A key experiment demonstrated that bacterial persistence was due to phenotypic heterogeneity of a cell population — that is individual cells exist in different phenotypic states which contribute to the variability of an isogenic cell population [Wright, 1953a; Wright, 1953b]. Additional studies further confirmed that the frequency of persister cells can be linked to genetic elements [Moyed and Broderick, 1986]. This is an active area of research that focuses on the mechanism of population dynamics and stochastic phenotype switching [Balaban *et al.*, 2004; Patra and Klumpp, 2013; Rainey *et al.*, 2011]. As we later discuss in section 6.8.1 (fuzzy inheritance), we have linked this phenomenon to a new type of inheritance.

Interestingly, during the same time period, Waddington considered that developmental noise represents phenotypical diversity among individuals, and such diversity is not caused by gene mediated genetic factors [Waddington, 1957]. Epigenetic mechanisms have also been used to explain phenotype switching in bacterial populations.

Another active research area focuses on cellular noise in the context of gene expression, illustrated by Spudich and Koshland's initial work in the 1970s [Spudich and Koshland, 1976]. Currently, most molecular single cell analyses still focus on understanding the transcriptional and translational mechanisms of regulation that lead to phenotypic heterogeneity and other single-cell differences.

Furthermore, phenotypic heterogeneity of single cells in isogenic cell populations was later demonstrated in yeast and mammalian cells, indicating a universal and fundamental importance of phenotypic heterogeneity. Combined with studies of multiple levels of genetic and non-genetic noise [Heng *et al.*, 2009; Huang, 2009], it is clear that noise should be referred to as bio-heterogeneity. Thus, single cell variability is fundamental for survival and cellular evolution as we discussed in Chapter 5.

Up to this point, it seems obvious that noise research should become mainstream, as evolution is the most important principle in biology. This, unfortunately, has not been the case. A majority of researchers simply cannot appreciate this key concept. Some researchers do read these papers and occasionally talk about it, but few have linked their own research projects to this issue.

Clearly, there are two schools of thought regarding bio-noise. Both viewpoints contrast with each other, but which one is correct?

For a long time, this hardly was a question, as the general practice of biological study was to eliminate noise. While interesting, research related to noise study has failed to challenge the very core of many biological concepts, until recent cancer research, and especially in the sequencing and –omics era, where many concepts generated from reducing noise strategy no longer make sense.

6.3 The New Challenge for Cancer Research in the –Omics Era

Cancer research used to be much more straightforward: when a number of oncogenes were introduced into unstable cells in a petri dish, host cells gained the phenotype of transformation; when one gene is manipulated, the downstream genes can be clearly identified by studying their altered profiles; when a chemo-drug is used to eliminate cancer cells in *in vitro* and *in vivo* models, the results are impressive, illustrating obvious causation and prediction. These experiments have done well to build basic theoretical knowledge about cancer. However, it has been especially difficult to translate the theory into clinical practice. This reality should be sufficient to make noise a serious topic of conversation in cancer research.

As previously discussed in Chapter 4, the goal of the cancer genome-sequencing project was to sequence a large number of cancer samples for the identification of commonly shared gene mutations. The argument was simple: reduce noise to find the pattern. The same rationale was applied to many other large-scale –omics technologies. It was reasoned that the analysis of thousands of samples would surely eliminate "noise" and reveal the pattern of key disease genes or pathways.

Nature always has a sense of humor. The cancer genome sequencing project, against the initial rationale and expectations, only revealed more complexity and less certainty of a pattern, perhaps confusing researchers even further. The more samples that are sequenced, the more diverse gene mutations are discovered, paradoxically reducing the value of a pattern of key gene mutations we had hoped to identify.

Such experiences have profoundly impacted individual researchers, who have been focused on some favorite genes or pathway through their entire research career prior to the –omics era. In contrast to their expectation of confirming that the importance of their gene would be achieved, by and large, the –omics data suggested otherwise and in fact drastically reduced confidence in their own data generated through small scale analysis in a linear model. For instance, when studying the relationship between two genes, by dysfunction of gene A, the impact on gene B can be straightforwardly illustrated: expression pattern changes, which may change the amount of proteins. However, if global gene expression is monitored, hundreds of other genes also change their expression patterns, many of which are even more significant than their favorite genes. Furthermore, most of the "off target" genes with altered activities really do not make sense within the scope of their original hypotheses. What to do next? The easy answer is to "focus" on the few that do make sense and write up the papers. The "good news" is, there are so many genes that can be picked that can certainly be linked to some hallmarks of cancer [Horne *et al.*, 2015a]. Also, there are so many papers in PubMed, and often, their favorite gene can be linked to other genes or pathways associated with cancer that "justify" the reasoning behind their choice (a typical cherry picking approach, though generally accepted. But who says only these obvious changes are important?). Knowing that there is also a limited correlation between gene expression and protein amount further confuses researchers about what to measure and report. Furthermore, data from new single cell technologies have forcefully illustrated that the same pathway is not uniformly detected at the single cell level, and the population of cancer cells is highly heterogeneous in terms of its genetic profiles.

All of these findings are crucial for understanding the reason why there is such difficulty translating findings from the bench to the clinic. Perhaps the more astonishing finding is that many landmark experiments in cancer cannot be repeated, as the duplication rate is only approximately 11% [Begley and Ellis, 2012].

Part of facing reality involves asking why there is so much confusion and inconsistency. What makes cancer so special that it challenges much of our basic research? In addition to vigorously repeating experiments, what else needs to be changed? Why does more data paradoxically have a negative impact on our overall understanding? Some system biologists point out that the major issue is signal and noise. According to them, the key, is to separate signal from noise in order to find the pattern. We, on other the hand, prefer to ask the more fundamental question: what if the conceptual framework of molecular research that focuses on pattern identification through elimination of noise is the problem for current cancer research? Heterogeneity cannot be eliminated for understanding and fighting cancer. No heterogeneity (noise), no cancer, and no correct concept of cancer either.

6.4 Why Are Researchers Divided About Heterogeneity?

By quickly comparing the philosophical view of science, the bio-questions under investigation, the way scientists design experiments, and the approaches and model systems researchers have used, the reason why the attitude towards noise divides researchers becomes obvious:

6.4.1 *Reductionism or holism?*

As pointed by Thomas Kuhn, scientists who believe different paradigms see different evidence. For those who believe the reductionist view of science, they view a complex system as nothing but the sum of its parts, and that the understanding of any complex system can be reduced to its individual components (simpler or more fundamental parts) no matter how complex it is. Thus using high resolution methods that dissect the system is the way to do science. Following the understanding of each part, the system can be put back together, like the "bottom up" approach for knowledge. To clearly define each part or component, the reduction of noise represents an effective approach. Since reductionism is at the philosophical heart of the molecular biology revolution, such a strategy represents the mainstream of current cancer research.

In contrast, scholars who hold a holistic view of scientific understanding pay more attention to studying the system as a whole entity, as the emergent properties at this level may not be meaningfully dissected into the lower level of parts. They tend to study the holistic interaction, and in particular, focus on the dynamics of the system at the higher level that influences the system behavior the most. Therefore, they care less about reducing noise, and instead use the unavoidable noise to understand system behavior/dynamics.

The above difference seems obvious, but there is much confusion or misunderstanding in practice. First, a holistic understanding is not just a philosophical view, but an effective scientific approach. For example, the predictive power of quantifying the evolutionary potential based on measuring genome heterogeneity within cell population is more efficient than characterizing each gene mutation, as we discussed regarding the evolutionary mechanism of cancer. The new challenge for holistic researchers is to develop scientific methodologies to study emergent properties, and to demonstrate the usefulness of monitoring the system behavior without solely focusing on isolated parts [Heng, 2013c].

Second, there are calls to reconcile reductionism and holism. It is simple to say "why not use the combinational approach," but could be hard to do so, as their methods are very different at the core. For example, the strategies of dealing with noise are drastically different.

It is interesting to point out that current strategies employed by systems biologists that bring together these two ways of thinking seem promising, as the cellular and organismal structure and dynamics must be examined in intact cells and organisms rather than as isolated parts. However, as many system biologists were growing up in the molecular era, they are reductionists by training and still adopt many reductionist frameworks and approaches to study the system. For example, they want more data points that measure the parts and ignore the contributions from higher level of systems. One often mentioned strategy is to reduce the noise to identify the pattern. However, this strategy is doomed to failure, because noise is the essential component that defines the pattern. Furthermore, when pushed, systems biologists will acknowledge that noise may have biological significance, but it is unrelated to the research question at hand (e.g. causal relationship between gene A and gene B/phenotype, etc.), or plays a secondary or tertiary role. Clearly, current systems biologists need to incorporate methodologies that encompass the holistic framework. These strategies will help to understand which processes should be dissected and which should not. To truly understand how the system works in reality (with evolutionary selection), it is not enough to study the interaction of parts, as evolutionary selection is based on the emergent properties of the system rather than the feature of parts, despite the complexity of parts interaction.

6.4.2 *The research strategy also influences the scientist*

We then ask: why can the importance of noise be appreciated by a minority of investigators, given the fact that they also are trained in the reductionist school of science, especially in the molecular biology era? Is there anything special about their research experiences, including the models they use and the types of biological questions they are interested in?

These questions have revealed the quite surprising conclusion that the research strategy can fundamentally impact researchers on their attitude towards noise. Specifically, there is a strong linkage between using evolutionary models to study the system dynamics and appreciation of the noise. Similarly, there is a linkage between the study of specific molecular mechanisms and the efforts to eliminate noise.

In molecular biology, most researchers have focused their studies on isolated parts, such as: enzyme activities *in vitro*, a specific gene's structure, the regulation of a defined pathway, a given cellular structure/response, and molecular characterization of causality in a linear experimental system. To study these individual structures and mechanisms, focusing on the average is justified and even effective. At this level of understanding, reductionist approaches might work best, as noise becomes less visible and is less important for the understanding of parts.

The question is: what will be the problem for such a practice? If heterogeneity is a key feature of a bio-system, what will be the consequence of ignoring it? Just because some biologists ignore noise to simplify the understanding of the system, it does not mean the heterogeneities are non-essential for life forms and also unwanted by nature. But can we demonstrate this?

It turns out that our bio-knowledge can be classified into three types: (1) parts characterization: description of the parts and study how parts can potentially work. To achieve the goal, many parts are analyzed under the condition of isolation. Various large scale –omics projects can systematically characterize large numbers of parts; (2) parts assembly: study the conditions to put many parts together to characterize a specific pathway or bio-function, the assembly of ribosome, the stages of development, and interactions among parts. Current systems biology holds much promise to advance this type of understanding; and (3) how the system as a whole works under evolutionary selection in a laboratory setting or nature, and especially under various stress conditions where the physiological and pathological responses might differ drastically (**Table 6-1**).

Table 6-1. How the elimination of noise impacts research strategy and outcomes

Type of research	Features of model	Challenges	Examples
Parts characterization	• Requires average-based method • Isolation required	No real dynamic context	Characterization of specific gene or organelle function
Parts assembly	• Requires average-based methods • Ignore outliers	Unfit for unstable cells/pathological cases	Pathway characterizations; assembly of ribosome
Population dynamics	• Requires average-based methods for clonal cell populations • Requires single cell method at genome level for unstable cells to understand population dynamics under evolutionary selection	Difficult to study	• Cancer evolution; cancer selection under stress; response to drug • Population-level selection makes individual pathway less relevant

Table 6-1. Research strategies involving parts characterization are currently the most widely used. This includes a description of the parts (e.g. a particular gene, its function), followed by parts interaction (e.g. interaction of a gene in a pathway). However, in order to elucidate the mechanism or interactions of a particular gene, noise is eliminated through the use of highly homogenized lab models. The elimination of noise has some consequences: (1) the overall system and its emergent properties cannot be understood by simply studying the parts or the sum of the parts, therefore a higher level of understanding system dynamics is not achieved; (2) information gathered in linear systems about parts functions are extremely context-dependent, as the parts functions may change in different experimental settings; and (3) while noise is not a dominant feature in normal developmental processes, it is a dominant and key feature in conditions of the pathological state. Due to the extreme reduction in heterogeneity, lab models become artificial and do not reflect the natural situation of disease conditions where noise is a dominant feature. Different research strategies that incorporate noise as a feature of the

system seek to understand overall population dynamics. This allows for a more accurate characterization of what occurs in nature, and allows for the investigation of systems under stress-induced conditions.

To make the case even more complicated, there seems to be a knowledge gap or even conflict between these three layers of understanding. Traditionally, it is believed that knowing the mechanism of each individual part serves as a pre-condition for understanding the assembly. Now knowing the emergent properties and the fact that the part's function is context-dependent, it challenges the common approach that relies on parts characterization for higher-level understanding. For example, the same gene could have a different function under different conditions, or within a different genome-defined system. An even bigger gap exists between parts characterization/assembly and population dynamics. The different opinions on noise represent a good example.

For much of the research that focuses on isolated gene mutations and pathways, noiseless data and concepts can be generated. However, the problem arises in the natural situation, where selection is based on the package of the system rather than individual parts. At this level of evolutionary study, noise becomes more important as it defines the direction of evolution. Clearly, the importance of noise only can be appreciated by studying the evolutionary process of the system, as the parts view and evolutionary view are very different. In addition, under real life dynamic conditions, not all parts function at their optimal efficiency. When efficiency is sub-optimal, heterogeneity (from subcellular levels to the population-level) becomes the key feature to regulate the system. As the overall benefit of efficiency and robustness must be balanced to survive the constantly changing environments, and the real fitness of the bio-system depends on the genetic package (genome) that is capable of overall adaptation. It is convenient for researchers to eliminate noise, especially in model systems, but life forms cannot survive without it in the real world. Clearly, there is a significant difference between the biochemical and evolutionary views of life!

Unfortunately, most researchers cannot see how life works in this angle due to the lack of hands-on research experience in the natural setting where real evolution and complexity rules. The three types of knowledge also determine the knowledge gap between many basic concepts based on noise-free systems and the clinical reality. In contrast, for those who study systems (genomic) heterogeneity, they often have worked on system dynamics within evolutionary selection.

For evolution, variation or noise is the raw material. Since evolution is highly dynamic, the general mechanism becomes more about noise-defined potential than the fixed pattern of recurrent genetic change. For example, the level of transcriptional noise has been linked to positive evolutionary selection. A recent evolutionary experiment in yeast further demonstrated that the sequence variation in a gene promoter (which encodes an enzyme for central glucose metabolism) was affected more by selection for a low level of noise than by selection for a specific level of cis-regulatory activity [Metzger *et al.*, 2015]. In particular, the outliers are more important in punctuated macro-evolution than the average genotype, as the punctuated phase of evolution represents the phase that is outside of equilibrium. Thus, cancer macroevolution differs from stepwise gradual evolution in the above experiment in yeast, and favors outliers much more.

Obviously, the experience of directly performing evolutionary experiments on either bacterial cell populations or cancer model systems can force researchers to think about evolution in their daily research activities rather than just regard evolution as an abstract concept that is not applicable to their own studies. It is tempting to speculate: if most researchers were familiar with the population dynamics that occur during cancer evolution, would that change their practice of ignoring noise?

Our view changed as soon as we studied cancer evolution (see Chapter 5). As described, the stochastic non-clonal chromosome aberrations we observed were too overwhelming to ignore; we had to honestly change our view and go against the tide. Our realization that the evolutionary mechanism of cancer equals the collection of a large number of individual molecular mechanisms illustrates the relationship between the parts and the system under evolutionary selection. One perhaps can study some molecular parts by reducing the genomic heterogeneity that could contribute to cancer. However, he or she will not understand cancer evolution, as one cannot simply change the system in order to understand a system. Without heterogeneity, there would be no real cancer (for the vast majority of cases).

6.4.3 *Other non-scientific factors that promote the ignorance of heterogeneity*

Nevertheless, with the massive datasets of various –omics projects including cancer genome sequencing, some researchers "already know" the importance of noise. But why are they still resistant to acknowledging its true importance, and furthermore, why do they refuse to join the effort of searching for the truth?

There are also many non-scientific factors that discourage further research of heterogeneity. First is the art of proposing fundable hypotheses. Current popular hypothesis-driven research requires researchers to focus on a very specific hypothesis. Often, this type of hypothesis is designed for a specific model and within a limited phase of data collection without monitoring the whole system across the entire evolutionary process. Such approaches have proved many hypotheses work well in specific models without noise (or with significantly reduced noise levels), but might not apply to the clinical situation where noise is a general feature. One common response to this situation is: "well, we only tested the specific hypothesis. If it is not working in the clinic, we need to test other hypotheses. It is not our problem if our hypotheses that were proven in lab conditions cannot apply to the clinic. We just need to have more testable hypotheses, and hopefully, some of them will apply to the clinic." Few have asked whether a hypothesis for a noiseless system would work for a bio-system that is full of noise. Furthermore, statistical approaches can effectively help researchers to eliminate undesired noise. For most cancers, the elimination of noise changes the system of cancer. Under such situations, no matter how strong, the data become irrelevant. In a sense, many basic cancer researchers have played a game of testing countless hypotheses.

Second, many researchers carry out lab research without reflecting on clinical reality. The real world is much more complicated. They seem to understand this, except when they apply for funding and write their papers. They also do a good job convincing the public that basic research will usually take decades to determine its clinical implications, despite the fact that most pure research will never make it to the clinic due to theoretical disconnections with clinical reality. However, they failed to realize that what was analyzed in simplified systems might have very little to do with real cancer, as the real system that is dominated by noise is operated by different mechanisms and even different principles. In fact, it is easy to generate good data that demonstrate causation from simpler models with less noise; it is also easier to gain more visible publications as well as funding with simpler models. Why change these approaches when everything works fine? The problem is the knowledge generated from these studies matters little to real world cases. For example, targeting the average cancer population can often achieve a good, short-term response. However, short-term response is fundamentally different from long-term benefit, especially for an unstable system. This is part of the reason why most researchers only present short-term results to illustrate the success of their research, while imagining that the short-term success will translate into long-term success. Unfortunately, the short-term drug response often conflicts with long-term survival in cancer cell populations with unstable genomes (see Chapters 7 and 8). Clearly, the prediction power of systems behavior and its dynamic relationship between short- and long-term responses depends on which evolutionary phase (punctuated or stepwise) we are investigating.

6.5 It Is Not Noise, Stupid! It Is Bio-Complexity

According to the genome theory, multiple levels of heterogeneity are the key features of a biological system. To further illustrate this point, we need to consider heterogeneity as a new feature of bio-system complexity.

Comparatively speaking, nonliving systems are noise-less and thus less complex. "Noise"-mediated adaptation is a requirement for living systems. For living systems, the requirement of adaption to a dynamic environment demands the increased complexity based on increased heterogeneity. In fact, heterogeneity is not a by-product of the system;

rather it represents a new mechanism of increased complexity. It is not difficult to propose that evolution favors the system with increased heterogeneity-mediated complexity/dynamics.

By increasing heterogeneity from all levels (including the parts variation [gene variation], the interaction variations among parts [genetic networks]), both the bio-specificity and efficiency of a testing system for a fixed environment or given experimental setting become reduced. This might support the importance of reducing noise for bio-efficiency. However, increased heterogeneity can increase fitness in a changing environment of the real world. Unfortunately, this important point has often been ignored when we study bio-systems in fixed experimental conditions, as the effects and importance of "noise" and noise-mediated evolutionary potential become invisible. In contrast, when a dynamic system is used to study the specificity and efficiency of a bio-system, heterogeneity-defined potential becomes obvious.

For example, when transcriptional master regulators were rewired in a panel of 598 gene networks in *E. coli*, most newly formed genetic networks were either functionally unchanged or superior in terms of growth [Isalan *et al.*, 2008]. This surprising result demonstrated that there is a high level of heterogeneity in network structure and regulation, and the natural configuration does not operate at its maximal efficiency. Similarly, a large portion of yeast genes can be deleted experimentally without drastically altering their functions, further supporting this point, as many genetic alterations are part of system noise and are only essential in dynamic environments.

Recently, large scale genomic studies have revealed even more surprises: non-protein-coding RNAs (ncRNAs) represent the majority of human transcripts. The noncoding transcripts have been considered by many as noise. But the overwhelming level of noise has raised the question: why so much noise? Is it noise or choice? [Louro *et al.*, 2009]. Despite interesting discussions that link ncRNAs to evolutionary constraints, biogenesis, responsiveness to physiological stimuli, and the RNA-based layer of regulatory processes in eukaryotic cells, there is less discussion regarding the possible new mechanism of noise-mediated complexity.

Similarly, there is a high level of noise during protein synthesis and post-translation modifications. For example, up to 30% of newly synthesized proteins in human cells are rapidly degraded by the ubiquitin-proteasome system [Wang *et al.*, 2013]. In terms of bio-specificity and efficiency, the rapid degradation of newly synthesized proteins does not make sense at all if the molecular specificity is high, as a well-regulated system should not allow for so much waste generated by system non-specificity. This observation, however, is favored by the genome theory, where genomic topology and timing are important for specificity. According to the genome theory, the higher layer of specific interaction is achieved by genomic topology and time interaction, plus the specificity at individual molecular level. In our view, the lack of specificity at the parts level is essential for higher level regulation based on genomic topology and timing. Therefore, such higher level regulation ensures the stress response for an effective system that is based on the noise-mediated low efficiency at the parts level, but higher level specificity that is adaptive to environmental dynamics. Clearly, the balance between specificity and non-specificity is not just quality control that marks proteins for destruction while they are being synthesized, but it is essential for a robust system.

This point can be further illustrated by a study investigating the protein ISG15 [Durfee *et al.*, 2010]. ISG15 is an interferon-induced and antiviral ubiquitin-like protein (Ubl). ISG15's major E3 ubiquitin ligase (that recognizes the target protein that is to be ubiquinated) can target over 1,500 substrates, including many exogenously expressed foreign proteins but with low efficiency. Interestingly, ISG15 conjugation is restricted to newly synthesized pools of proteins, and its E3 ubiquitin ligase is physically associated with polyribosomes. In the context of an interferon-stimulated cell, newly translated viral proteins may be the primary target of ISG15.

We realized that the case of ISG15 might have illustrated a new mechanism of how a system can have low efficiency between specific molecules and a large number of substrates to achieve needed specificity simply through topology (such as the physical association with polyribosomes) and time (the window of newly synthesized protein). Such a mechanism represents a nice example of how a system uses noise to build complexity (regulate many substrates with less specificity) and achieve environment-defined specificity (timing and physical location to increase specificity).

We suggest that there might be different ways noise can contribute to complexity and even the survival of a bio-system. First, the "buffer effect." Among many specific pathways, a system needs "noise" to slow down the bio-reaction at all times. Otherwise, the system can be easily burned out (only under certain conditions high specificity is needed). That is the reason why a stable system has to balance its specificity (efficiency) with stability and robustness, as the advantage of specificity at certain conditions could become disadvantageous in altered conditions. There are plenty of examples in nature where species that are too specialized do not survive. Clearly, the balance between specificity and noise-buffered stability represent the ability of bio-complexity to deal with changing environments. It is worth mentioning that specificity is not an absolute term and is defined by less and no specificity. In addition, specificity and the lack thereof play different roles in biology. That is why noise is so important for the homeostasis of a bio-system.

Second, noise can promote stochasticity by providing many available options. These options are not drastically different from each other, and functional switching among them only has a limited impact at the current population status (so that it can stochastically occur without drastic changes in the phenotype), but has potential advantages or disadvantages in future environments. Therefore noise represents evolutionary potential. This point has been illustrated in our linkage between karyotype heterogeneity and transcriptome dynamics, for example (see Chapter 8). Also, as discussed in a later segment, this is why unstable cancer cell populations pass down NCCAs, which have been considered "noise" for decades until our studies suggested otherwise.

Third, while increasing the noise of the parts interaction can reduce the specificity at the parts level, it can achieve specificity of systems behavior at a higher level of regulation (such as stress response, regardless of the type of the stress source). The case of ISG15 represents such an example. The low interaction efficiency of E3 ubiquitin ligase and the large number of substrates makes the ligase less specific for each substrate, which can actually increase specificity for the overall stress response using the same mechanism. The low efficiency of the parts interaction level plus time and topological control open the probability for the ligase to interact with a large number of substrates in a time-specific fashion (not just viral, but other factors). Even though this means that energy is wasted, it is essential for the system to pay this price for survival. Imagine if each different type of stress needed a very specific mechanism, the bio-response system would be too complicated to operate, as so many specific stresses with defined mechanisms exist. Furthermore, the environments are constantly changing, and focusing on a certain type of specificity is not the best strategy. In daily life, we use a combination of tools to solve a particular problem even though the tools may not be highly specific to the task. We do not need to reinvent the tools every time; some less specialized tools can do the job just as well.

Fourth, small portions of "wasteful noise" are only important under extreme conditions. This type of noise can be considered as insurance that is expensive, but necessary. This case is particularly important for punctuated cancer evolution and especially during drug resistance. Of course, as we have discussed, for macro-cancer evolution, noise is also reflected at the genome level as NCCAs.

Finally, it is likely that cellular heterogeneity itself can serve as a strong selection force to favor heterogeneity. For example, the highly heterogeneous cellular status can function as a stressor or even the platform to maintain the heterogeneity of future cellular populations.

In summary, evolution mainly favors the package that balances effectiveness and robustness, where noise and complexity play an important role. It is the overall efficiency and robustness of the system that matters most, not just the parts efficiency, as we previously believed. Since the genome identity (species identity) is preserved by sexual reproduction (see Chapters 5 and 7), multiple levels of genetic and non-genetic heterogeneity become key for improving system complexity. Interestingly, recent large scale genomic studies also revealed high levels of heterogeneity at the somatic genome level for many normal and disease conditions. We further hypothesized that genome-level noise is essential for somatic cell adaptation, but is accompanied with increased potential for developing diseases [Heng *et al.*, 2015].

6.6 Flawed or Untested Yet Generally Accepted Concepts

The consequence of ignoring noise in cancer research is very serious: it promotes researchers to accept many concepts without vigorously testing them, as they seem to make sense in noise-less experimental systems. We discuss some assumptions below.

First, just a few years back, many molecular researchers have assumed that cancer cell populations are mainly isogenic. That is, one cancer cell can divide with seemingly high fidelity into two near-identical daughter cells. Cell division with high fidelity ensures that every cell that originated from that first common cell is nearly identical at the genome level, thus earning the term isogenic. Even though there is a certain degree of heterogeneity (occasionally, there are some genetic variations like gene mutations, but overall, parent and daughter cells should be nearly genetically identical, or at least very closely related), tumor cells that originate from one cancer cell should be clonal or contain multiple directly related subclones. Thus, clonal expansion of one or few deregulated cells dominates tumor growth. In fact, clonal expansion is the foundation for the most widely used cancer models, and has thus dominated the conceptual thought process and underlying assumptions of cancer research. With this conceptual mindset, cancer evolution can easily be traced by comparing the clones, primary tumors and their metastases, and should be (more) genetically similar (than analysis shows them to be). This idea was supported by clonal evolutionary studies of a number of blood cancer systems where clonal evolution is more common than most solid tumors. Unfortunately, it has been challenging to trace clonality at both gene and genome levels. Limited experiments have been done to experimentally validate the genetic/epigenetic clonality of "isogenic" cancer cell populations. By tracing a few well-known cancer genes, some reports have claimed that clonal expansion represents a major pattern of evolution in solid tumors. As we pointed out in previous chapters, tracing the clonality of DNA "parts" differs from tracing the clonality of karyotype that defines the genome system. In addition, single cell sequencing projects have revealed that even at the DNA level, evolution is punctuated. If both inter- and intra-tumor heterogeneity is a key feature of cancer, and punctuated cancer evolution is the general rule rather than the exception, it is time to change our belief on how widespread clonal expansion actually is.

Another common assumption harbored by most cancer researchers is that a recurrent pattern can be identified by analyzing a large number of samples to reduce the noise. This assumption is closely related to the first assumption. If clonal evolution is dominant, and if evolutionary convergence is strong, we should expect to identify the common genetic alterations during stepwise cancer evolution. Such efforts have dominated cancer research since the successful identification of the Philadelphia chromosome and the BCR-ABL fusion gene for CML. However, there is limited luck in solid tumor research. As illustrated in Chapter 4, even the large-scale cancer genome sequencing project that sequenced thousands of samples failed to identify highly penetrant common gene mutations. It is interesting to note that, for these two key assumptions, the concept of clonal evolution and the recurrent genetic pattern have been nicely illustrated in the case of CML. Unfortunately, the CML story clearly represents an exception to most cancer types. Even for the case of CML, there is the issue of noise, as many patients also display a high frequency of other types of chromosomal alterations, especially during the more punctuated phase observed in the blast crisis stage, which might play a key role in drug resistance as well as different patient responses to treatment [Horne *et al.*, 2013a].

A third assumption is that, since cancer populations are isogenic, gene mutations and epigenetic dysfunction are the key drivers of clonal expansion while chromosomal change is less important. This assumption has already been analyzed in Chapters 3 and 5. Such an assumption is very misleading, given that chromosomal aberrations are overwhelmingly present in most cancers.

6.7 Portrait of the Heterogeneity of Cancer

The key feature of cancer is its heterogeneity [Foulds, 1975; Heppner, 1984]. Despite the obvious pathological observations, researchers have tried very hard to establish a clear pattern of common and recurrent genetic alterations by ignoring heterogeneity. Such a practice has generated much of our molecular understanding through "idealized cancer models" without including the heterogeneity-mediated reality of cancer. It is no wonder that many scientific breakthroughs in basic research fail in clinical trials. It is now time to address the key feature of cancer, noise, in order to establish its real characteristics.

The following are some important yet less appreciated facts that are caused by cancer heterogeneity. Knowing these facts will help researchers to understand cancer.

6.7.1 *Evolutionary potential increased through "noise"/genome heterogeneity*

For cancer initiation and progression within a karyotypically homogenous normal tissue environment, the favored winning ticket is genome system heterogeneity. Without genome heterogeneity, there is no success for cancer, as it is very difficult for a new cancer system to become dominant within well-controlled normal tissue. As illustrated in Chapters 4 and 5, genome-level heterogeneity of cancer cell populations has been linked with survival and evolution of cell populations. The diverse molecular mechanisms that are associated with cancer have been linked to transcriptome dynamics. The degree of cancer genome heterogeneity has also been linked to potential diagnosis and monitoring of the treatment. Therefore, the key to studying cancer is to study its heterogeneity.

6.7.2 *Multi-level heterogeneity makes persistent clonal expansion a rare event for most cancer cell populations with unstable genomes*

It is well established that multi-level heterogeneity is a defining feature of cancer cell populations. However, the presence of such heterogeneity grossly contradicts clonal expansion models. While clonal expansion is the accepted model of cancer evolution, it was only illustrated in some cancer types within a defined window of observation [Horne *et al.*, 2013]. In addition to the large number of karyotype analyses for a majority of solid cancers, monitoring evolution in action experimentally has illustrated that NCCA/CCA dynamics make a recurrent CCA less likely for most cancers (see later section). Furthermore, recent single cell sequencing data also show that punctuated cancer evolution is generally detected from tumor samples. Together, analysis of the clinical data indicates that recurrent clonal expansion does not occur in the majority of tumors, and clones come and go during cancer evolution.

6.7.3 *Genome-level heterogeneity is ultimately important for cancer evolution*

The multiple levels of heterogeneity include genetic (from gene to genome) and non-genetic (such as epigenetic) components. Since the genome defines macrocellular evolution, it is obvious that heterogeneity at this level is more profound and decisive. A few lines of evidence support this view. The evolutionary mechanism of cancer involves genome level selection rather than gene level selection (during the macro-evolutionary phase); the NCCA/CCA-mediated transition represents the transition of evolutionary potential (reflected as different individual NCCAs or "noise") to reality (the winner that becomes a specific CCA). Even though gene mutations and aberrant epigenetic variants can contribute to cancer progression, they do so mainly through micro-evolution, unless they lead to genome alterations. Recently, an experiment investigating how genome alterations define the overall transcriptome has illustrated this case [Stevens *et al.*, 2013a].

As illustrated in Chapter 8, a time-course analysis was completed that investigates the relationship between genome heterogeneity and the transcriptome. In passages with high genomic instability, transcriptome profiles were highly heterogeneous, where entire gene expression profiles would dramatically change from one passage to the next. In contrast, for cell populations that were more stable/clonal, transcriptome dynamics were also more stable, as gene expression profiles largely remained constant and unchanged from one passage to the next. This study demonstrates that: (1) increased genome heterogeneity resulted in elevated transcriptome dynamics; and (2) when the genome changes, the transcriptome changes. Thus the genome defines the evolutionary potential through transcriptome dynamics. Such a relationship has been confirmed in two independent studies using spontaneously transformed mouse ovarian epithelial cells [Creekmore *et al.*, 2011; Lawrenson, 2010]. Gene expression was analyzed at different passages that exhibited elevated genome heterogeneity. In both studies, gene expression profiles changed at each passage, and there is no significant overlap of expression profiles between the two studies, including the final transformed cells. Interestingly, transcriptional heterogeneity resulting from genome change has also been demonstrated in other studies [Harewood *et al.*, 2010; Pavelka *et al.*, 2010]. Moreover, elevated transcriptional heterogeneity is mirrored in clinical samples, as patients with the same cancer subtype exhibit widely different transcriptomes. While some efforts, including identifying prognostic gene signatures, may have some clinical value, comparison of two or more gene signatures significantly reduces any overlap to 3–8% [Abdallah *et al.*, 2014]. This indicates that, not only does genome heterogeneity lead to unstable transcriptome dynamics, but each different experiment represents one evolutionary run, where the genome-defined transcriptome is unique.

6.7.4 *"Noise" or outliers define the trajectory of the cancer cell population*

Different from normal tissue where average cells display normal karyotypes and average cell profiles represent a relatively consistent and unchanging population state, the population state of cancer cell populations that are highly heterogeneous is constantly changing, as it is driven by few cells that are outliers. As soon as an outlier dominates a cancer cell population, a new outlier is ready to emerge because the interaction between fuzzy inheritance and environmental changes are unavoidable.

In such a dynamic process, genome heterogeneity impacts the emergent properties of a cell population through the continual replacement of outlier cells that dominate a cell population. In genomically heterogeneous cell populations, individual cells with different genomes will have different behaviors (e.g. individualized transcriptomes, growth rates). In contrast, cells of genomically homogeneous cell populations will exhibit similar behaviors. The relationship between genome heterogeneity and systems heterogeneity was demonstrated in a study that characterized both genome heterogeneity and growth heterogeneity. In this study, the average cancer cell concept is questioned, and the data strongly suggested that so-called noise, or outliers, are the key factor for cancer cell evolution (more details, see section 6.8.2).

6.7.5 *Heterogeneity provides cancer cell populations with complexity and robustness*

Since a high level of heterogeneity is the key feature of cancer, there is no fixed genome for most cancer types (Chapter 4). There is also no concept of the fixed individual cancer cell. Therefore, in a sense, the cancer cell population as a whole becomes a unit of evolution, and each cell with a different genome becomes every other cell's noise. Since the genome defines the system, heterogeneity at the genome level becomes the most powerful, and the complexity and robustness of cancer population entirely depends on the collective noise. Cells with different genomes can cooperate or compete with one another; all genomes can evolve independently, or evolve in convergent or divergent ways. The combinations are endless.

In such a heterogeneous and noisy cell population, a single cell may not be sufficient enough to survive on its own, as these cells have serious genetic variations (many of them could be considered as defects in a given condition) and a high likelihood of cell death (making it difficult to predict the winning surviving clone that will pass the identical winning genome to future cell populations). However, if we consider the whole highly heterogeneous cell population as a single entity, the cell population with different individuals wins the game of survival by having the winning ticket of being noisy and complex at the population level. As we will discuss in section 6.8.1, heterogeneity of cancer cell populations is maintained through the mechanism of fuzzy inheritance.

That is the reason why it is so hard to fight cancer. The concept of tumor societies has been suggested to acknowledge the importance of tumor heterogeneity and the emergent properties of the cell population [Heppner, 1989] (see Chapter 3).

6.7.6 *Average-based methods result in improper targeting and drug selection*

Noise gives a new meaning to understanding drug resistance in cancer. Cancer cells are believed to acquire resistance through the (in)activation of a certain gene, target site, or pathway. While this may be true for tumor cells within the stepwise clonal phase, or normal cell populations under normal physiological conditions, it is not applicable to most cases of cancer resistance where genome replacement-mediated macroevolution dominates. In the vast majority of cancer populations, high level noise exists, reflected by a high level of NCCAs. Furthermore, drug treatment (a form of selection pressure) of a heterogeneous cell population will further destabilize the cell population resulting in increased noise, leading treated cells to become resistant through genome replacement-mediated resistance, among highly unstable cell populations (for more information, see Chapter 5, Genome Chaos section).

Obviously, because drug treatments can change population dynamics (by changing the individual cell's genome), the use of average-based technical and analytical methods in unstable cancer cell populations can result in improper drug targeting. Identified targets represent only more dominant subpopulations and do not account for smaller subpopulations or single cells that can be more important for driving cancer evolution. Therefore, single cell analysis is required in order to understand and identify characteristics of an unstable cell population that can be used for therapeutic purposes.

6.7.7 *Statistical averages are not suitable for assessing unstable cell populations*

Heightened heterogeneity in cancer cell populations requires theoretical and technical strategies to understand the manner by which cell populations behave. Unstable cell populations characterized by elevated heterogeneity cannot be reliably assessed through measures like statistical means. Average-based methods such as gene expression profiling and western blotting disrupt cellular membranes and use whole-cell lysates to generate an average measurement across an entire heterogeneous cell population. For example, expression of a particular gene will vary among cells, as a small number of cells will exhibit high expression levels, while others will exhibit low levels or none at all. In contrast, average-based measures can be used in clonal cell populations, as cells are (relatively) genomically identical and expression should be similar [Abdallah *et al.*, 2013]. This same problem presents itself in the analytical assessment of data, as data generated from unstable cell populations that are subjected to average analytical measures do not result in representative measures (please refer to section 6.8.2 for more details).

Again, in stable cell populations with homogeneous karyotypes, noise can be eliminated by averaging techniques. However, in unstable cell population with heterogeneous karyotypes, outliers are dominant during evolutionary selection, even though the initial number is very small. The similar idea can be applied to solve the paradox of why it is hard to validate most cancer genes among patients when most patients have their own mutation profiles (see Chapter 4).

6.7.8 *Modeling without true heterogeneity is not correct for cancer*

The discussion regarding noise in cancer once again helps us to understand why most of our current cancer models (including most animal models, see Chapter 8 for more details) do not accurately portray tumors found in patients, as they have fatally ignored the noise issue and therefore no longer reflect tumors that grow in patients. In most of our molecular models, we have pushed one gene mutation, one pathway, or a specific molecular mechanism without considering the competition of genomic noise under evolutionary selection. We actually have only tested how a specific mechanism could be linked to cancer under a specific condition but did not address the question how cancer occurs within a natural patient population. Many specific molecular mechanisms that have been detected are present in cancer patients, but more often occur with very low probability, reflecting a much more complicated situation than *in vitro* and *in vivo* models provide. Due to their low predictability, these models offer little for our understanding of cancer evolution in the clinical setting. Knowing the evolutionary mechanism of cancer and its relationship with diverse molecular mechanisms (see Chapters 4–5), the future of cancer research should not continuously focus on specific pathways that are interesting, but have low penetrance in patient populations. The challenge is to put all types of noise back into the system of cancer, and study the pattern of heterogeneity and its impact on the pattern of cancer evolution.

6.7.9 *The insignificance of the significant*

Understanding the importance of heterogeneity (noise) in cancer, researchers now should appreciate the significance of the insignificant. Equally important, researchers should also realize another paradox: the obvious insignificance of the significant in current cancer research. Specifically, there is no clinical significance for many well-accepted basic concepts of cancer, well-known key cancer genes and pathways, and the strategies that exist to fight cancer. The noise-less knowledge (one example of artificial knowledge) is highly insignificant when applied to the clinic, where heterogeneity is the key issue for cancer.

Many such examples are discussed throughout this book. Below we discuss a few points to explain why this is the case.

A simple PubMed search will illustrate that there are many well-studied cancer genes. Each of them has a few dozen publications and authors have claimed that these genes are highly significant. They have convinced NIH study sections to fund them for years, and have convinced themselves to study these genes for decades. However, the insignificance of these genes has become increasingly obvious.

First, over the last few decades, many of the "favorite genes" have been linked to several categories: oncogene, tumor suppressor gene, cell death gene, tissue organization gene, etc. If a favorite gene is involved in everything, then what is its real function? Second, following the cancer genome sequencing project, many favorite gene mutations are only detected in a minority of cancer patients. Third, most well-characterized gene mutations have limited translational value in the clinic. While targeting specific genes often results in a good initial response, targeting these mutations often fail to achieve long term success, even for the best cases. For example, most melanoma patients treated with Vemurafenib (targeting the BRAF V600E mutation) experience a drug resistant relapse within 1–2 years.

Now, the answer becomes clear. In basic research, one can link a favorite gene to many biological significant pathways by eliminating noise. But in the clinical reality where heterogeneity mediated evolution is key, the highly significant information gained by only focusing on the favorite gene loses its utility and prediction power. This has been illustrated by the relationship between the evolutionary mechanism of cancer and diverse molecular mechanisms of cancers (Chapter 5). Among the almost countless combinations of gene mutations and other environmental factors, any significant gene mutation in fact is insignificant during the cancer evolutionary process. While difficult to accept, this statement is now confirmed by clinical reality. Even targeting gene mutations of the famous p53 gene has failed to deliver; so why, all of a sudden, do researchers believe their less penetrant gene mutation will be the magic one? We understand that this concept is difficult to accept for researchers who have invested much time and passion in their favorite gene mutations, but on the other hand, we have to face reality and search for new directions/strategies.

6.8 Short Essays: The Mechanism and Significance of Heterogeneity in Cancer

To further illustrate the mechanism of heterogeneity and its significance in cancer research, the following two essays will focus on the genetic mechanism that regulates and maintains heterogeneity, as well as how outliers are the key for cancer evolution. Through our own search for the meaning of noise at the genome level, we have discovered that the evolutionary patterns of the normal homeostatic state and the pathological state differ drastically, which challenges traditional approaches that are based on the knowledge of normal processes (with high predictability) to understand the abnormal pathological process (with much higher stochasticity), or vice versa.

6.8.1 *Fuzzy inheritance*

Throughout this book, we have introduced a new emerging theme that redefines the positive contribution of heterogeneity to biological functions and its evolutionary trade-off. Heterogeneity provides complex systems with needed stochasticity to survive dynamic environments. Knowing its biological significance, an ultimate challenge has been how to illustrate the molecular mechanism of the creation and maintenance of heterogeneity.

6.8.1.1 *Question: How do cell populations pass heterogeneity?*

The phenomenon of inheritance in a cellular population seems well known: a stable cell population faithfully passes the same stable genome to future populations until it becomes unstable (usually after many generations through the faithful passage of the same genome in each generation, combined with the accumulation of mutations at a very low rate). An unstable cell population, in contrast, will faithfully pass instability to the next generation. Interestingly, it is thought that both the normal and "abnormal" (altered) genetic information are passed to their own "daughter" cells with certainty. At the end of one population doubling, the degree of heterogeneity in the parent cell population (reflected as the proportion of the abnormal cells within the population) should be the same as the newly generated cell population. In other words, the daughter cell population inherits the same degree of heterogeneity present in the parent cell population.

There are a few assumptions to explain how the overall heterogeneity of a cell population changes, according to conventional thinking. First, there are gradual changes of cell population heterogeneity over time. When the mutation rate is extremely low under the current concept of inheritance, a long period of time is needed for the population to accumulate a set of mutations. Second, the mechanism of changing the overall degree of heterogeneity for a given cell population is by natural selection or genetic drift or draft, as it fits well with the stepwise evolutionary progression of cancer and that the degree of heterogeneity is accumulated. In agreement with this idea, it has been suggested that clonal expansion is the key to passing heterogeneity (terminology includes divergent clonal evolution and clonal heterogeneity). Third, cells with a mutator phenotype can generate a much higher rate of gene mutation mediated heterogeneity. Under stress conditions, many gene mutations can be linked to elevated population heterogeneity.

As we discussed in Chapter 5, this is not the case for most cancers, unfortunately. Many examples have been extensively discussed in this book: (1) For normal cells, there is always a certain degree of genetic/genomic heterogeneity. Recent studies of copy number variations, epigenetic profiling and the personal genome project have forcefully demonstrated this point, which supports our claim to focus on NCCAs [Conrad *et al.*, 2011; Iourov *et al.*, 2008; Landan *et al.*, 2012]. (2) A high level of NCCAs is often observed in unstable cell populations. (3) During the punctuated phase of cellular evolution, inheritance at the karyotype level is often interrupted as new karyotypes are formed. Yet, life goes on through the newly formed karyotypes. And finally, (4) Genome chaos can produce novel karyotypes without inheriting the genome from the mother cell. Since the various NCCAs represent different genomic topology-defined systems, each with different system inheritance, it has made us wonder: If the parent cell cannot pass the same karyotype, how is the genetic information passed to the daughter cell? It is clear that there is a limited and precise system inheritance among these cells; could it be the mechanism of maintaining genome level heterogeneity?

Following years of data collection and synthesis, a number of concepts have been developed. First, the heterogeneity of cancer cell populations is not typically maintained by passing various sub-clones, as during the punctuated phase, discontinuous NCCAs dominate. Second, there must be a new type of inheritance responsible for maintenance of the high degree of heterogeneity, if heterogeneity is caused by the constant generation of different NCCAs rather than the preservation of CCAs. In other words, in contrast to passing the same genetic information among cells within each sub-clone (which requires a precise mechanism), there is uncertainty in inheritance between the mother cell and daughter cell. We call this new type of inheritance fuzzy inheritance. This inheritance obviously differs from classical inheritance where the mechanism of passing the same system is precise.

6.8.1.2 *How do we illustrate that there is a new type of inheritance?*

Bio-inheritance refers to bio-processes of transmission between parents and offspring that account for heredity and the mechanisms responsible for them [Lamm, 2014]. In other words, it refers to the genetic process by which parents provide their children with specific traits. Multiple inheritance processes may contribute in the heredity of a particular phenotypic trait both at the individual and cellular level.

Despite the fact that there are many types of inheritance that are discussed and accepted (including epigenetic inheritance, mitochondrial mediated extra-genetic inheritance, cultural and behavioral inheritance in humans and various animals, and ecological inheritance), the most basic genetic inheritance system in biology is surprisingly not well understood and in fact largely unrealistic. For example, trait inheritance and the mechanisms of molecular inheritance of genes have been a primary principle of genetics for nearly a century. However, they do not address the real blueprint of inheritance, the plan for gene-gene interaction and their response to developmental and environmental changes. Even though genetic structure, function, variation and distribution are studied at the cellular, organismal and population levels, the main focus has been gene-centric. However, neither the gene-mediated parts inheritance nor epigenetic inheritance serve as the blueprint; the blueprint for inheritance lies in the genome context-mediated system inheritance (Chapter 5); therefore the basic genetic inheritance system should focus on the genome structure, function and variation.

Few types of genetic inheritance have been illustrated experimentally by the mechanism of passing genetic materials. Parts inheritance is achieved by passing genes from parents to offspring, extra-genetic inheritance is shown by passing mitochondrial DNA to offspring and the system inheritance is ensured by passing the same karyotype from parents to offspring. As for epigenetic inheritance, experiments have illustrated that the epigenetic status can be passed. For example, there is stochastic epigenetic variation (highly variably DNA-methylated regions) associated with development, and the methylation status (the loss or gain of CpG dinucleotides) is heritable [Feinberg and Irizarry, 2010]. Yet the mechanism for passing altered karyotypes has not been identified nor systematically analyzed. An interesting question is: through what inheritance mechanism are altered karyotypes inherited from parent cells during the punctuated phase of cancer evolution?

On one hand, it must be inheritance, as there is a clear reproductive relationship between mother cell and daughter cell, and passing inheritance is the key biological feature. All genetic materials (e.g. genes) of daughter cells are inherited from the mother cell. On the other hand, the karyotype is different between mother cell and daughter cell, illustrating altered system inheritance (and likely includes parts inheritance as well as epigenetic inheritance). Together, such inheritance is altered with limited precision for system inheritance, which differs from gene/genome mediated genetic/genomic inheritance (due to the variation), and epigenetic inheritance (due to actually altering the genetic materials).

Therefore, we need to compare mother and daughter cells to investigate whether: (1) heterogeneity can be inherited from mother cell to daughter cell by illustrating the mother cell population shares the same traits with the daughter cell population; (2) some important traits are linked to the degree of karyotype heterogeneity; (3) a single cell can pass heterogeneity (both genome and growth heterogeneity) for a cellular population; (4) the degree of altered inheritance is directly related to genome instability; and (5) while the disadvantage of imprecise or less faithful inheritance is obvious (leading to a heterogeneous cellular population with slower growth coupled with increased cell death due to genetic fuzziness, it also interrupts the accumulated selective advantage favored by previous cellular evolution), the advantage is also huge (even though it has been underappreciated). It represents the most powerful and costly strategy for cellular survival under crisis.

6.8.1.2.1 *Fuzzy inheritance: Passing the degree of change, but not specific changes*

In any cell population, there is a given degree of heterogeneity. It has been demonstrated in many experimental systems (bacteria, stem cells, various cancer cell lines) that single, and often small numbers of isolated cells can reconstitute the genetic and phenotypic heterogeneity of the original cell population. However, the mechanism behind this phenomenon remains elusive. Specifically, how is heterogeneity restored, and why does the degree of heterogeneity remain constant in daughter cell populations? To address this, we compared the karyotypic inheritance patterns of cells at both single cell and population levels using both genomically stable and unstable cell populations. We find that in cell populations that are genomically stable, as expected, a mother cell will pass the same karyotype to its daughter cell, a process that happens in most cells, and a small minority of cells will generate NCCAs. When cells with NCCAs divide, they either die off, or produce new NCCAs or generate new CCAs. In unstable cell populations (most cancer cell populations), when a single unstable mother cell undergoes cell division, rather than passing an identical karyotype, it passes an altered karyotype to its daughter cell. The parent cell may have specific structural or numerical NCCAs; however, these specific NCCAs are not found in the daughter cell, as different karyotypes are passed. In other words, fixed genomes or specific NCCAs are not inherited in unstable cell populations, and the only constant heritable feature is the proportion of cells with NCCAs. The mechanism of passing the degree of heterogeneity rather than specific genetic changes demonstrates a new form of inheritance, where rather than a fixed state with fixed genetic information is inherited, a stochastic state is inherited. We name such inheritance fuzzy inheritance. Fuzzy inheritance is low in stable cell populations and high in unstable cell populations.

6.8.1.2.2 *Karyotype heterogeneity is associated to many other traits*

Karyotype heterogeneity has a significant impact on the cellular dynamics and behaviors of a cell population. It is a known fact that karyotype heterogeneity is directly related to cancer evolution. Karyotype heterogeneity is also related to other traits such as growth heterogeneity, cell death heterogeneity, and drug resistance heterogeneity. In these cases, when karyotype heterogeneity is high, the heterogeneity of cellular dynamics (cell growth, cell death) is also high; however these associations can only be determined at single cell resolution, as the average profile washes away the heterogeneous dynamics (as previously discussed). For example, in cell populations that are unstable, measurements of single cell growth demonstrate a large degree of growth heterogeneity. Because nearly all cells have different karyotypes, the rate of growth difference is high within unstable cell populations. In contrast, in genomically stable cell populations, the rate of growth difference is lower because all cells have identical or nearly identical genomes, and genome heterogeneity does not contribute to growth heterogeneity. Cell death is also related to karyotype heterogeneity. In stable cell populations, cell death occurs using the same or similar mechanisms. However in unstable cells, there is a considerably larger degree of cell death heterogeneity where many different modes and mechanisms of cell death are active, leading cells to die in different ways. Finally, drug resistance is also directly related to karyotype heterogeneity. Drug resistance is established through multiple cycles of genome chaos, leading to cell population of cells that are genomically unstable. While the phenotype of resistance is the same between individual cells, the cells all have different karyotypes, indicating that karyotype heterogeneity is the main cause for resistance. Fuzzy inheritance links the heterogeneity of these traits, as there is little precision in the inheritance of specific karyotypes, growth rates, or cell death mechanisms in unstable cell populations.

6.8.1.2.3 *A single cell can pass heterogeneity for an entire population*

Single cell analysis has ultimately proven how heterogeneity can be passed to the next generation by one randomly picked cell. By isolating a panel of single cells from cell populations with different degrees of genome instability and following their single cell karyotype and growth profiles, we demonstrated that a single cell can pass a given amount of heterogeneity to its daughter cell population. We found that stable cells were able to inherit identical karyotypes and similar growth rates from the single isolated cell. In contrast, a single unstable cell could not pass down the same karyotype or growth rate. Rather, it generated a cell population with altered genomes. Furthermore, the level of heterogeneity in the daughter cell population mirrored that of the parent cell population, thus a single cell can restore the same degree of heterogeneity found in the parent population.

To confirm, a comparative analysis was completed in both stable and unstable cell populations where fast- and slow-growing cells were selected over multiple generations to determine whether specific features (karyotype, growth) could be faithfully inherited and selected at the population level. We found that unstable cells isolated from both fast-growing colonies always produced cells that proliferated at very fast and very slow rates. Similarly, unstable slow cells also produced fast- and slow-growing cells. The vast heterogeneity of growth in both fast- and slow-growing cells was similar to the growth heterogeneity of the parent population. The growth feature cannot be passed even through a single cell. In contrast, cells with stable genomes produce more homogenous growth patterns that are similar to the parent population. Over multiple generations, selection of stable fast- and stable slow-growing cells did result in the selection of a specific karyotype and growth rate.

This indicated that a single cell will pass its potential to generate population-level heterogeneity and evolvability to the next generation of cells. In other words, the degree of heterogeneity (not specific alterations) is inherited though a single cell. Such fuzzy inheritance represents the very mechanism of passing heterogeneity. Even though it is a well-known phenomenon that unstable cell populations can pass instability to their daughter cell populations, prior to this experiment, however, it was not clear whether most individual cells had the same capability. One common assumption has been that most cells can faithfully pass their "normal" genome, while only a small number of "abnormal" cells pass altered genomes. Our experiment has proven this assumption to be incorrect (Abdallah *et al.*, submitted).

6.8.1.2.4 *Fuzzy inheritance is a key feature for somatic cells and especially for cancer evolution*

All somatic cell populations exhibit some level of fuzziness in their inheritance patterns. The degree of fuzziness, however, differs drastically for cell populations that are karyotypically stable and karyotypically unstable. In unstable cell populations, as in most cancer cell populations, the degree of fuzziness is high. Fuzzy inheritance thus plays a fundamental role for somatic cell and especially cancer cell evolution by providing the essential evolutionary potential. If classical precise genetics were dominant, and the same genomes were inherited (regardless of how heterogeneous those genomes are), there would be limited potential for change (because evolution is gradual and accumulated among cells with the same genome), and the cell population as a whole would quickly die off when exposed to high levels of stress. In order to survive highly dynamic environments, cancer cells require a high degree of evolvability, of which fuzzy inheritance plays the most crucial role. In a sense, a highly dynamic population has a higher probability to win. When cancer is finally formed, the degree of fuzziness likely will be reduced due to micro-cellular selection, which is often reflected by increased CCAs. In contrast, drastic environmental changes can quickly elevate the degree of fuzziness, such as following harsh drug treatment.

6.8.1.3 *The biological significance of fuzzy inheritance*

There is an important distinction between inheritance in germ line cells and inheritance in somatic cells. Under the constraint of system inheritance, most normal germ line cells have identical karyotypes when passed to the next generation. Individual germ line cells can display different genotypes achieved by the meiotic mixing of gene combinations during the reproduction cycle. For a given individual, all of his or her somatic cells are derived from a single fertilized egg with fixed system inheritance. While precise inheritance is important for maintaining a given species, it could become disadvantageous at the somatic cell level, as the environments are constantly changing. For the highly dynamic needs of the somatic cell level, inheritance must be less precise to some degree, which ensures increased heterogeneity for somatic cell adaptation. To solve this conflict, a new type of less precise inheritance for somatic cells is needed which separates inheritance from germ line and somatic cells. In addition, many cellular processes also represent somatic cell evolution where variation is an essential element. Accordingly, there must be some mechanisms to generate high levels of variation. Fuzzy inheritance fits these needs well.

There are a number of key differences between classical inheritance and fuzzy inheritance. First, the degree of variability during the transmission of genetic information is different. The former inheritance occurs with high certainty: the precise mechanism of passing genes is achieved by semi-conservative replication, and the error rate is very low, while the precise mechanism of passing the genome is achieved by sexual reproduction, even though genome variation is high in gametes (Chapters 5 and 7). Classical inheritance cannot be used to explain the high level of variation observed from normal somatic cells as the frequencies of gene mutation is low; fuzzy inheritance, however, explains this well.

Second, the classical concept of inheritance is more suitable for the passing of inheritance over generations, while fuzzy inheritance is more about passing inheritance among somatic cells during an individual's life time, or among cancer cell populations in this specific case.

Third, classical inheritance might be more relevant for average cellular populations under normal homeostatic conditions, while fuzzy inheritance explains inheritance in cell populations under high levels of stress where there is a need for more variation, especially for outliers observed in somatic cell populations, such as some subpopulations of cells commonly detected in cancer.

It seems that fuzzy inheritance plays an ultimate role for cancer evolution as it serves as the mechanism for the creation and maintenance of a high degree of heterogeneity. In the case of cancer, due to high selective pressures, outliers become dominant and rely on such a mechanism. Different from germ line cells and even normal somatic cells, the key for success in cancer cells lies in their deviation from normal cells. As we have discussed, genome level stochasticity represents a driving force. And the beauty of fuzzy inheritance is passing the state of stochasticity rather than specific genome changes.

This variable inheritance is particularly visible during the genome chaos process, where drastically genome alterations dominate, leading to common genome alterations among a majority of cells. Fuzzy inheritance has a great advantage for a cancer cell's survival and rapid macrocellular evolution. We have initially named such inheritance as inherited heterogeneity to illustrate the mechanism of how cancer cells pass heterogeneity. Recently, by further comparing the inheritance types between germ line and somatic cells, and in particular, between stable somatic cells and unstable somatic cell populations, we realized that it is more proper to name it fuzzy inheritance in order to distinguish it from the classical precise inheritance that is mainly applicable to germ line and stable somatic cells.

6.8.1.4 *Potential mechanism of fuzzy inheritance*

While we only illustrated fuzzy inheritance using genome level alterations among *in vitro* cellular populations, the high degree of genome heterogeneity among different tissues and in particular within various diseases/illness conditions have been well documented, even though it has been largely ignored [Biesterfeld *et al.*, 1994; Heng *et al.*, 2013]. Equally important, knowing that the main function of sex is to eliminate major (chromosomal) genetic alterations and preserve species identity, it becomes important to separate germ line inheritance from somatic cell inheritance in order to provide both maximal species preservation (by maintaining system inheritance) and the needed plasticity for cellular adaptation (by increasing heterogeneity on multiple genetic levels). Stress-induced fuzzy inheritance therefore holds the key to understanding many types of common and complex diseases/illnesses [Heng *et al.*, 2015].

Furthermore, fuzzy inheritance functions at gene, epigenetic and even sub-nuclear levels as the epigenetic variations and gene level variations including copy number variations, and mtDNA dynamics are overwhelming in somatic cells [Pujadas and Feinberg, 2012]. Similarly, the function of the sexual filter can also be observed at the gene and epigenetic levels, as some nuclear DNA damages can be repaired. Cytosine methylation and chromatin marks can be reset to levels that worked for both parents [Gorelick and Heng, 2010], and mitochondrial defects can be eliminated by sexual reproduction [Silliker *et al.*, 1997].

Recent personal genome sequencing and especially single cell sequencing have revealed high levels of *de novo* gene mutations [Conrad *et al.*, 2011; Veltman and Brunner, 2012]. Similarly, a high level of epigenetic stochasticity has been demonstrated [Landan *et al.*, 2012]. These types of variations can reach to even higher levels in cancer and stressful conditions, as illustrated by the mutator type [Loeb, 2010].

Interestingly, under highly stressful conditions, a bio-system often can switch to survival mode with a new mechanism of reduced molecular specificity. Despite the high level of heterogeneity, such as increased errors during alternative repair systems, survival mode is pushed to a hyper-survival state with elevated survival potential. Examples include the elevated mutation frequency of bacteria under stress, genome chaos as response to high stress for eukaryotic cells, and mitochondrial DNA re-organization in disease conditions. Note that under more moderate stress conditions (within the physiological variation range, for example), some other mechanisms such as pathway switching or chromatin changes should be sufficient.

Increased stochasticity is shared by all survival modes. It is known that adaptive mutations can be induced by stress [Ponder *et al.*, 2005; Torkelson *et al.*, 1997]. Growth-limiting stress can lead to mutations that allow growth in an otherwise growth-limiting environment. Interestingly, the normally high fidelity DNA repair mechanism is switched for a fuzzier version using a special error-prone DNA polymerase as a result of the cellular stress response. Such fuzzy inheritance that is associated with error-prone mechanisms and increased mutagenesis is particularly useful for survival under altered environments. As soon as a new system is suitable for the environment after selection, precise inheritance returns, as it is the better fit in homeostatic environments.

Genome chaos represents a good example of fuzzy inheritance for cancer cells in survival mode [Liu *et al.*, 2014]. Genome chaos results from high stress (such as drug treatment) that triggers chromosome fragmentation. In the very few cells that survive, chromosomes rapidly reorganize into totally different genomes through stochastic rejoining, the mechanism of fuzzy inheritance, rather than orderly mechanisms for genome level repair. Since the key is to create cells with new genomes that can survive, more drastic changes are better. Furthermore, despite the fact that most cells die off because massively altered genomes are not survivable due to a high level alterations, with the large number of cells involved, survival becomes a certainty [Horne, 2015, submitted].

Fuzzy inheritance seems contrary to concepts like directional mutations or directional selection. Many experiments have shown at the surface that there is a direction of mutation or genome re-organization based on the type of inducer. Our data show that there is no directional selection, as the vast majority of the chaotic genomes are not survivable, illustrated by our *in vitro* model [Liu *et al.*, 2014]. In survival mode, error-prone enzymes takeover (through fuzzy inheritance) and elevated mutations are random and not specific [McKenzie *et al.*, 2001]. Therefore, there is no directional selection of a specific cell harboring a specific genome rearrangement or mutation with high frequency. However, the specific inducer represents the environment for selection, which gives the impression that the selected mutation favors the inducer.

Fuzzy inheritance also answers the puzzling question regarding the biological significance of increased mutation or genome re-organization rates under stress. Many researchers still doubt that higher rates of change exist under stress because they are maladaptive, and most random mutations will be harmful to the organism. Therefore, more frequent mutations would likely generate a less-fit population. As a result, evolution would not select such a mechanism. The confusion lies in distinguishing between macro- and microevolution, specifically, the confusion between the adaptive landscape with the survival landscape. For somatic cells in survival mode under high stress, cells employ fuzzy inheritance to maximize survival. As soon as a stable population is established, cells leave survival mode and enter a normal homeostatic environment, precise inheritance will be used in the replication process, or the micro-cellular phase of evolution. The fuzzy inheritance fits the survival and adaptive landscapes well.

Further research is needed to incorporate many previous studies into the framework of fuzzy inheritance. For example, the switching of fuzzy inheritance and more precise inheritance can occur within a cell (using different enzyme systems, pathways), between cells (when some individual cells are more unstable than others), between germ line and somatic cells, and among different tissue types of the same individual.

6.8.2 *Cancer evolution: The game of outliers*

Biological noise is a well-observed feature of cancer cell populations. However, its significance and contribution to somatic cell cellular evolution is vastly underrated. Traditional biological concepts that employ average-based techniques generate the "average cell" profile. Such a strategy is in particularly popular in cancer research, as clonal evolution has been the key focus in cancer research, despite the fact that cancer samples are often highly "noisy." However, the notion of the "average cell" is becoming increasingly incompatible with the high degree of heterogeneity that defines cancer, as it is becoming increasingly evident that heterogeneity is integral to cancer evolution including progression and drug resistance.

Single cell heterogeneity has been previously researched mainly to understand the mechanism of microbial resistance, transcriptional noise, and population dynamics [Cohen *et al.*, 2008; Elowitz *et al.*, 2002; Lidstrom and Konopka, 2010]. It has been accepted that isogenic cell populations exhibit a large degree of single cell phenotypic variability. These variations largely stem from stochastic fluctuations that result in the existence of distinct cell states [Balaban *et al.*, 2004; Elowitz *et al.*, 2002]. Individual cell states serve to diversify cell populations and thus become crucial for cellular evolution [Acar *et al.*, 2008; Cohen *et al.*, 2008; Kussell and Leibler, 2005]. While studying gene level and non-genetic level heterogeneity including epigenetics can explain a certain degree of phenotypic variation or novel regulatory mechanisms, the assumed key notion that cell populations are isogenic is becoming increasingly at odds with many recent studies.

Multi-level heterogeneity is a characteristic feature of cancer cell populations. While cancer cells have unique gene mutations, copy number variations and other genomic alterations, karyotypic heterogeneity is the most prominent type [Heng *et al.*, 2013b]. Furthermore, karyotype heterogeneity is also found in normal tissue and is linked to other somatic cell diseases/illnesses, including a phenomenon called somatic chimerism [Heng, 2013; Iourov *et al.*, 2008; Ye *et al.*, 2007]. Finally, according to Genome Theory, the macro-evolutionary unit of selection is the genome. Genome replacement (macrocellular evolution) drives cancer evolution, which can be observed by monitoring NCCA/CCA cycles. Structural or numerical changes to the karyotype alter the genome context, leading to cells with newly created systems.

Despite this impressive progress, non-clonal karyotype heterogeneity is still largely ignored in favor of finding clonal determinants of cancer cells with the hope of identifying common targets, even though NCCAs are overwhelming and ubiquitously present in most cancer cell populations. According to our own experience, common responses to the significance of genome heterogeneity are mixed and often conflicted. Someone might comment: "chromosome heterogeneity is well known, what is new?" Others may state: "studying chromosome level heterogeneity cannot really help our understanding of cancer, and especially cancer treatment, unless we can identify a specific pathway." Any one of these strong opinions can jeopardize research investigating genome level heterogeneity, not to mention their effectiveness of rejecting manuscripts or grant proposals. The main reason for the research community to continually ignore karyotype heterogeneity is due to the lack of understanding of the crucial biological significance of NCCAs in cancer evolution. According to traditional concepts of somatic cell evolution, most "noise" is insignificant (either due to being neutral or nearly neutral, or being eliminated due to the fact that most alterations are deleterious). As we previously discussed, this viewpoint is correct during microevolution, where changes in gene mutation frequency within an isogenic cell population accumulate over many generations (average profiles are often dominant), and drastically differs from macro-evolution where survival by new system formation (cells with altered genomes) is the key (outliers have more chance for survival).

To change the status quo, it is necessary to illustrate how stochastic genome alterations (which can be seen as evolutionary potential) can be beneficial in cancer evolution. In particular, tracing the specific NCCA/CCA transition and demonstrating that NCCAs, or "noise", under normal conditions can determine the direction of a population, especially in macrocellular evolution under crisis.

We designed some very straightforward, yet powerful experiments to directly examine the clonal nature of cancer cells by monitoring the degree/rate of genomic change in culture, a panel of single cells were isolated from cancer cell populations with different degrees of genome instability; genomic analysis was performed on each single-cell derived subpopulation.

After 3–5 weeks of growth, SKY was used to assess the degree of karyotype heterogeneity. Among genomically unstable cancer cells (Brca1/p53 DKO and HCT116/E6 human cancer cells), we found that each subpopulation exhibited the same degree of karyotype heterogeneity as the parent cell population, prior to single cell isolation. Each cell had a unique karyotype and the comparison of subpopulations revealed no karyotypic intermediates between the parent cell population and daughter subclones. Conversely, the subcloning of karyotypically stable cells (HCT116 human colorectal cancer cells without E6 transfection) showed that parent and daughter subpopulations share the same karyotype, with minimal karyotypic change even after continuous culture.

It is generally assumed that cell populations originating from a single cell are isogenic, and that heterogeneity develops over an extended time period. In other words, within limited generations, daughter cells should share karyotypes that are identical or very similar to those of the mother cells. However, this experiment demonstrates karyotypic heterogeneity returns in the same degree within weeks of cloning. This indicates that genomically unstable cells cannot be karyotypically cloned even with a short time period of continuous culture. Since all cells examined displayed different karyotypes, the only explanation is that, their karyotypes were constantly changing during cell division. In contrast, stable cells with stable genomes can be karyotypically cloned under the same culture conditions. As we discussed in Chapter 5, stable clonal karyotypes are often observed in the stable micro-evolutionary phase, while dynamic karyotypes are associated with the punctuated phase.

The non-clonable nature of unstable cells questions fundamental concepts at the heart of cancer evolution. The word clonal infers that cells originating from a single ancestor should be identical, or at least very similar. However, we observed that karyotypes are continuously changing, suggesting that traceable clonal evolution usually does not occur for cell lines that are genomically unstable. Since genome instability is a characteristic feature of most tumors, stable clonal evolution should occur rarely and within a limited timeframe (but is quickly followed by the unstable punctuated phase). This has been confirmed by genome sequencing of patient samples and other cancer cell lines [Baca *et al.*, 2013; Navin *et al.*, 2011; Wang *et al.*, 2014]. The genome context (gene contents + genomic topology, simply referred to as karyotype) changes once the karyotype is altered. Individual genes may play a different role in cells that exhibit unique karyotypes. Therefore, even if a gene is mutated, it may not necessarily exhibit the same properties across different samples of the same cancer type. Furthermore, commonly mutated genes are not found in a majority of cancer patients, and single cell sequencing has demonstrated that single cells display unique mutational profiles that are similar to patient mutation profiles. This point has been illustrated in Chapters 3 and 4.

Next, we assessed the effect of bio-noise on population behavior. Since karyotype heterogeneity has been previously linked with transcriptome, growth, and survival heterogeneity [Stevens *et al.*, 2013a–b], we compared the relationship between karyotype heterogeneity and the behavior of cell growth. Using the same karyotypically homogeneous and heterogeneous cancer cell lines, population-level growth assays were performed and compared to single-cell growth profiles. Specifically, population doubling was used to measure population-level growth due to its frequent use of measuring tumor growth in the clinical setting. Population doubling rates of karyotypically heterogeneous cell subpopulations exhibited growth rates unique to each subpopulation. Regression analysis of subpopulation growth showed no correlation. Furthermore, the doubling rate fluctuated greatly between passages, suggesting that growth is heterogeneous at the single-cell level. In contrast, karyotypically homogeneous cell populations exhibited less fluctuation in cell growth. Heterogeneous growth of karyotypically heterogeneous cells suggests that each unstable cell population is independently evolving.

Genome heterogeneity-linked growth diversity raises interesting questions into how the genome can impact cell growth. First, by what mechanism does this occur? And second, since all of the cells are genomically different, how do single cells contribute to population-level growth? According to the cancer hallmarks, cancer cells exhibit fast proliferation. Does each cancer cell contribute the same, average amount? Or do some cells exhibit some special features that enable them to dominate population growth?

Single cell/colony growth assays were performed for the same karyotypically homogeneous and heterogeneous cancer cell lines using *in situ* growth assays. Since all single-cell colonies were in the same culture flask, the culture conditions were the same. Therefore this comparison can be done for individual cells with less culture variations. This method also allows us to compare cell populations with different degrees of genomic heterogeneity.

Growth was monitored daily for up to one week, or when single cell-derived colonies began to merge (which can interfere with measuring colony growth). Single cells in genomically homogeneous cell populations exhibited limited variation in their growth rates, as each single cell-derived colony exhibited similar proliferation rates. In contrast, single cell-derived colonies in karyotypically unstable cell populations were significantly more variable by almost three-fold. Most cells exhibited growth at a slow to moderate rate, while only a few cells grew at an extremely high rate.

This stark contrast in growth indicates that the statistical average may be a misleading measure of population-level growth in unstable cell populations. Most individual cells in the karyotypically heterogeneous cell population did not contribute significantly to overall population growth. It was the outlier cells with significantly high growth that dominated population dynamics, demonstrating that the average is an unsuitable measure for unstable cell populations. In contrast, the average can be reliably used in relatively stable cell populations, as each cell made the similar, average contribution to population growth.

Genome heterogeneity has significant implications related to biological research. Genome heterogeneity-mediated transcriptome heterogeneity and growth heterogeneity allows for the enhanced evolutionary potential of unstable cancer cell populations. However, profiling unstable cell populations is increasingly difficult, given the overwhelming degree of heterogeneity present. Average-based technical and analytical methods are commonly misused when characterizing genomically unstable cancer cell populations.

To demonstrate the inadequacy of the average to accurately describe unstable cell populations, single colony growth of homogenous and heterogeneous cell populations was compared to their averages. A sample dataset demonstrated the broad range in growth between genomically homogeneous and heterogeneous cell populations. Single colony growth of karyotypically homogenous cells was relatively uniform, exhibiting a normal distribution. Genomically unstable cells, however, exhibited much more diverse growth, as cell growth manifested in a bimodal distribution. A majority of cells contributed little to overall population growth, leaving a few highly proliferative outlier cells to drive population growth. A comparison of single colony growth to population averages reveals wide discrepancies in unstable cell populations. According to the statistical mean, average colony size is measured at 73 cells, when proliferation ranged between 1–593 cells. Variation in actual colony size was high, as measured by the coefficient of variation (CV), 200%. In comparison, karyotypically stable cells made the same relative contribution to the total population sample. The mean was calculated at 41 cells, and proliferation ranged between 8–82 cells. Variation in colony size was low, as the CV measured at 44%.

Each cell in the unstable cell population has the potential to be an outlier that dominates population growth. The evolutionary potential, or "noise," is high in unstable cell populations, which is reflected by the presence of genomically unstable (and unique) cells. Under the right conditions, slow-growing cells can become the fast-growing outliers that dominate growth. For the one outlier cell with the right conditions for fast growth, the other cells become "noise." However, when the environmental conditions change that favor growth of a different cell, a new outlier dominates growth. Again, the other cells are "noise," including the original outlier cell.

These data have several important implications regarding chromosomal noise in cancer cell populations. First, unstable cancer cells cannot be cloned, leading to an overwhelming degree of karyotype heterogeneity. Since most cancers are characterized by genome instability, continued research on clonal evolution in cancer may only yield results of limited significance. Therefore, basic theoretical and technical research strategies that are based on averages (which are suitable for detecting clonal evolution) need reevaluation. Because the level of multi-level heterogeneity is overwhelming, finding a specific genetic change common to a majority of tumors of a specific cancer type will likely not occur. Rather, more effort should be directed towards studying the system and predicting the overall response. This is most effectively done at the genome level, and therefore an emphasis towards understanding genome dynamics is warranted.

Second, karyotype heterogeneity has significant effects on population dynamics, like cell growth and drug resistance. Genome heterogeneity leads to growth heterogeneity, where individual cells of unstable cell populations grow at their own rates; this is in contrast to the idea that all cells in a cell population share the same growth rate. Furthermore, outliers drive unstable cell populations, as most cells grow at slow to moderate rates and only few outliers are responsible for driving population level growth. The difficulty lies in predicting which outlier will become dominant next. Finally, the average represents a poor measure to characterize unstable cell populations.

Cell population heterogeneity is common, and its misrepresentation has significant implications. Cell populations traditionally characterized by statistical or even geometric means do not quantify heterogeneity, as means reduce heterogeneity, thus washing it away. To demonstrate this, a composite karyotype is generated by averaging the frequencies of all chromosomal structures. (**Fig. 6-1**) The composite karyotype reduces heterogeneity by excluding NCCAs and reducing the variation in aneuploidy. Therefore, genomic outliers and unique NCCAs that are fundamentally important for cancer evolution are excluded from analysis. Misrepresentation through the arithmetic mean is also observed in genome-mediated growth heterogeneity, which artificially makes the analysis less real. **Table 6-2** compares single cell growth data to both the statistical average and the geometric average. While the geometric average provides a better representation of single colony growth in unstable cell populations, it does not accurately portray growth of the entire population, as outliers are excluded from analysis.

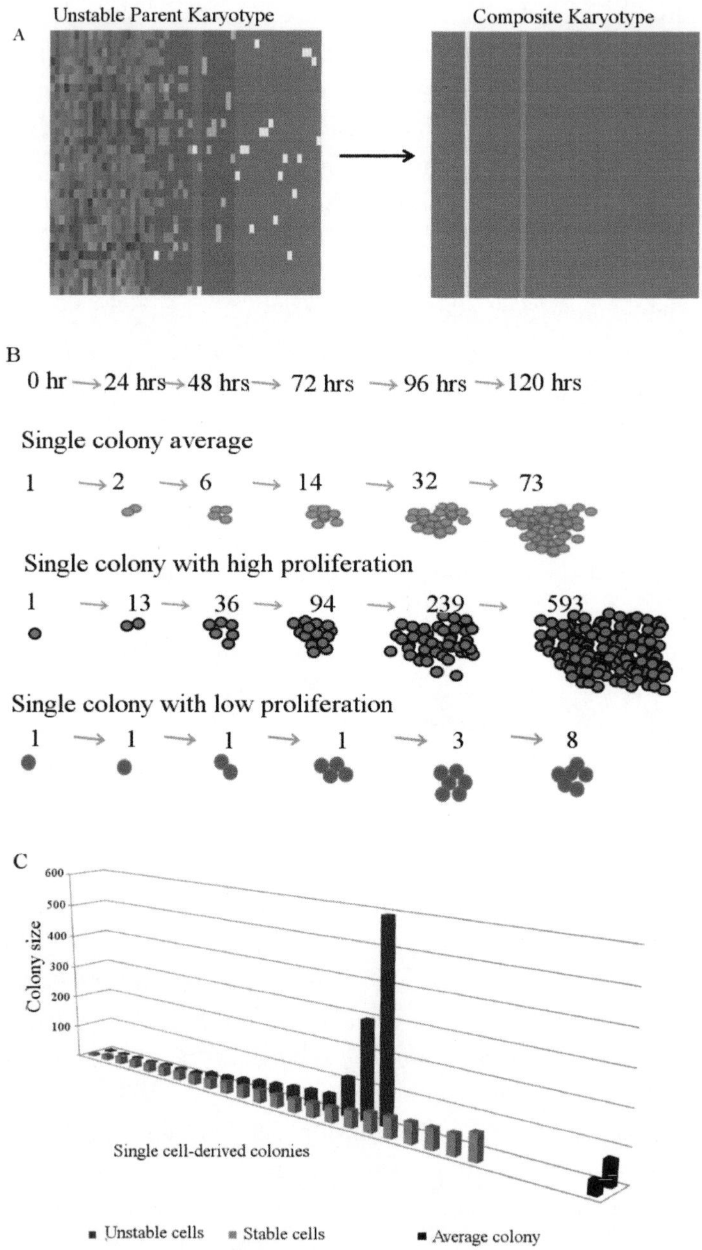

Fig. 6-1. Unstable cell populations are not reliably characterized by the arithmetic mean.

(A) Heat map generated single-cell karyotypes compared to its composite karyotype in an unstable cell population. The y-axis represents individual cells; the x-axis represents chromosome frequencies (normal and derivative). There is considerable single cell heterogeneity in the heat map on the left. No two cells are identical, as each cell has different chromosome and NCCA frequencies. The averaging of these chromosomes into a composite karyotype reduces the heterogeneity observed. (B) Cartoon depicting single cell growth in unstable cells. Growth was traced for single-cell derived colonies over six days. The top panel demonstrates daily average colony growth followed by the daily growth of fast and slow single colonies. The vast range in total proliferation demonstrates growth heterogeneity observed in unstable cells. (C) Comparison of single-cell derived colony growth in stable and unstable cells. Tracing of single-cell derived colony growth in unstable cells demonstrates a large degree of growth heterogeneity between individual cells, and outlier cells dominate population growth. The average colony growth (black column) is not representative of actual cell growth. In contrast, stable cells (shown in blue) exhibit the same overall growth. Averaging the colony growth demonstrates that the average is a reliable measure for stable cells. (Reproduced from [Abdallah *et al.*, 2013] with permission).

In contrast, means are appropriate measures in stable cells. This highlights the individual contribution of outlier cells to overall population growth, and demonstrates that the arithmetic mean is unreliable in measuring genome-mediated growth heterogeneity in unstable cell populations.

Accurate representation of genome heterogeneity is fundamentally important in understanding how biological systems function. Often, the statistical mean is used to characterize cell populations with elevated genome heterogeneity for the purpose of developing cancer models and drug development. However, because means smooth heterogeneity and often exclude outliers that drive population growth, it becomes an inappropriate measure, and may lead to faulty modeling and inappropriate drug selection. For example, technical methods such as

Table 6-2. Statistical means are not reliable measures of cell growth in unstable cell populations

	Mean colony size	Est total population	Difference from actual population
Arithmetic mean, heterogeneous cell populaton	73 cells	1314 cells	8 cells
Arithmetic mean, stable cell population	41 cells	943 cells	7 cells
Geometric mean, heterogeneous cell populaton	19 cells	342 cells	980 cells
Geometric mean, stable cell populaton	37 cells	806 cells	99 cells

Table 6-2. A comparison between the most commonly used statistical mean and the geometric mean (which is more aptly used for datasets that contain outliers). In heterogeneous cell populations, the arithmetic mean does not accurately depict single colony growth. While the geometric mean may give a more accurate assessment of single colony growth, it fails to accurately incorporate outliers into its overall population size. The estimated population size of heterogeneous cell population using geometric mean is equal to 342 cells, while actual population size is 1322 cells, a difference of 980 cells. However the use of statistical means is accurately used in cell populations that are genomically stable. (Reproduced from [Abdallah *et al.*, 2013] with permission)

genome sequencing, RNA-seq, Western blotting and expression profiling are commonly used for the genomic characterization of cancer cells, biomarker selection, etc. These average-based methods use whole cell lysates and generate an average measurement for the entire cell population. Average-based profiling of many cancer specific pathways and their specific response to treatment may generate a mechanism/response that does accurately not reflect the cell population. Because the "average cancer cell" is nonexistent, the actual mechanism and response to treatment could be completely different, given that outliers are a dominant force in cancer evolution. Despite the initial clinical success enjoyed by these strategies, cancer patients often relapse because outliers that are more important for cancer evolution are not detected by these strategies that profile only more dominant subpopulations. Furthermore, because cancer cells are continuously evolving, profiling the current average profile will not predict future cell population dynamics. Single cell strategies that profile the outlier cell should be incorporated to understand the contribution of outliers in cancer evolution. Therefore, average profiling of unstable cell populations fails to depict the cancer landscape that is overwhelmingly noisy. Similarly, the noisy landscape can be used to explain the lack of overlap among transcriptome profiles in different studies of the same cancer type.

The incorporation of genome-level heterogeneity into basic science and clinical research may lead to significant advancement in our understanding how normal physiological cells are fundamentally different from cell populations of the pathological context. Normal physiological cell populations can be defined as a near isogenic cell population where most cells have the same genome context. Cells can faithfully pass their genome context and there is overall population stability. Traditionally, cells in the disease state are characterized by a deviation of the normal that can result from a single or series of nuclear or mitochondrial gene mutations, a deregulated pathway, or some type of genetic instability. Often, researchers try to study disease mechanisms by understanding and comparing to the normal state or vice versa. If this was a correct approach, then we should have solved the mystery of cancer, as the problem should be relatively simple to fix. Our data reveal that the many disease states of cell populations are highly unstable, especially cancer cell populations, which separates the disease state from normal state. Furthermore, the drastically different evolutionary pattern in cancer even challenges the rationale of using normal processes to understand cancer, as they are totally different systems! Cancer cells are actually in the macro-evolutionary phase where the characteristic feature is elevated genome heterogeneity, whereas normal cells are in the micro-evolutionary phase where gradual change dominates. The fundamental difference between normal physiological cell populations and pathological cell populations is increased genome heterogeneity-mediated stochasticity in disease state cells, which can become highly unpredictable, in particular for late stage cancers that can be ruthless to study. Currently, the theoretical and methodological approaches used to study both normal physiological cell populations and cell populations of the pathological context are the same: we dissect the activity of a particular gene and take the average behavior. While this approach may be suitable for normal cell populations that have low or minimal genome heterogeneity, they are inappropriate for genomically heterogeneous cell populations where the average cell clearly does not exist. Therefore, we must analyze genome heterogeneity as an essential first step in our research practices.

One recent example of how genome heterogeneity can play the ultimate important role in cancer cell survival is illustrated by genome chaos after drug treatment. As discussed in Chapter 5, following drug treatment, the vast majority of treated cancer cells undergo rapid genome re-organization, and every cell exhibits a different genome. In this case, the majority of cells become "noise" and there is no average genome. By pushing the maximal potential, altered genomes represent an increased chance of survival. Within a three-week time window, these unstable genomes continue to change (keep trying their luck), until a few cells that are relatively stable and fit the environment to clonally expand [to become the winning cell(s)]. As soon as the winning cell(s) emerge, the specific noise starts to grow into an average. Clearly, without a high level of noise, these cancer cells are not able to survive. Quite different from adaptation at the normal growth conditions, noise or the outliers are the key factor for cancer cells becoming resistant under crisis. Now we understand for macroevolution, only an altered system (represented by altered genomes) can be successful, as gene-level modifications are no longer sufficient to ensure survival under crisis. By realizing this, we can highly appreciate the role of genome-level noise in somatic cell evolution.

As biologists, we should know better. Drastic evolution often occurs on the shoulders of outliers, while the average cell is usually left behind with no "surprise" story to tell. The average defines a given type, and maintaining the average cell profile is essential for normal physiological function. In contrast, outliers define the direction of macroevolution. Conversely, the formation of survivable outliers detected in the somatic cell macroevolution phase are not due to the accumulation of gene-level changes that are based on the average cell population, but rather large, drastic genome-based changes.

Averages reduce variation. Averages leave no hope for drastic revolution. Averages make a complicated process look relatively simple, but the average is not representative of cancer. Based on the continuous NCCA/CCA cycles detected in cancer evolution, the transition from outlier-dominant populations to average-dominant populations, and again new outliers and new averages, it is the outliers that define the direction of cancer evolution. In a sense, cancer evolution can be considered as a game of outliers.

Of course, it should be noted that the vast majority of outliers will be eliminated by the constraints above the cellular level. That might be the reason why over 90% of polyps in the colon are eliminated without any treatment (which makes our lives easier). Paradoxically, however, perhaps because of the power of system constraint, this same selection pressure that kills most polyps could also increase the level of fuzzy inheritance in those that survive, making them harder to kill in the future when they become cancerous (another example of trade-offs).

6.8.3 *New challenge: Fuzzy inheritance vs. precision medicine*

Recently, "precision medicine" became a buzzword following a research initiative announced by President Obama to accelerate progress towards a new era of molecular medicine (Collins and Varmus, 2015). Despite the rush to use this new term to replace "personalized medicine," there are increased concerns whether the goal of precision medicine can fit the reality of medicine [Joyner, 2015], as the large scale omics studies have pointed in the opposite direction. Now, knowing that the genetic information is fuzzy by nature, and the subsequent somatic cell evolution is highly dynamic especially for disease conditions with high heterogeneity, a series of debates is needed: Is precision medicine wishful thinking for most common diseases, especially for the majority of cancers? Is the precision movement yet another misstep for predicting complexity based on fixed average molecular profiling?

Chapter 7

Do Different Cancers Represent Different Species?

7.1 Summary

To address the question why cancer cells are so different from normal cells and how they can destroy hosts despite being derived from the host's cells, it was suggested that selfish cancer represents a new species of its own. While highly controversial, it is very beneficial to discuss this topic, as it is relevant to many concepts essential for understanding cancer. These include the species concept, the key difference between sexual and asexual reproduction including somatic cell reproduction, as well as the distinction of gene-centric and genome-based evolutionary views of cancer. We put this "cancer as new species" topic up for debate in this chapter following a brief introduction of the historical perspective of this subject and some main arguments for and against considering cancer as a new species. Specifically, we determine whether cancer fits into current species concepts and why the newly discovered types of inheritance mediated two phases of cancer evolution holds the answer, which favors the idea of using genome similarity to define species. Furthermore, the concept that cancer serves as a trade-off at both the cellular level and in natural animal populations as well as its genetic basis are discussed, which addresses an important question of why natural evolutionary processes will not eliminate cancer. Finally, we discuss how the cancer species concept could shine new light on cancer research and explore cancer as an ideal model for studying evolutionary processes beyond the cellular level.

7.2 Brief History of Considering Cancer as a New Species

Despite the fact that many cancer researchers are strongly against this seemingly drastic idea, considering cancer as a new species has its historical roots. In 1897, David Hansemann proposed that during cancer transformation, "the cell change(s) [its] character in every regard morphologically and physiologically to a new species" [Hansemann, 1897]. Hansemann's idea regarding the origin and nature of cancer was later echoed by Theodor Boveri in 1914. Boveri described that the starting point of malignant transformation was the result of aneuploidy, a major genome alteration. He insightfully described that aneuploidy changes the cell's character, crosses species borders and makes cancer its own species [Hansemann, 1897; Boveri, 1914; Knauss and Klein, 2012].

From a historical point of view, such idea was rather ordinary for Boveri. He was the father of chromosome theory, and at that time the chromosome represented the only known genetic material (this was prior to the suggestion of the conceptual linkage between cancer and gene). Unfortunately, as described in Chapter 1, the popularization of the gene concept left many chromosome-based concepts unpopular and disregarded, aside from a few scattered literatures that studied the karyotype of infectious cancers (such as canine transmissible venereal tumor or CTVT) and the effort to illustrate the parallelism between chromosomal changes in speciation and neoplastic transformation [Oshimura *et al.*, 1973; Pathak, 1990]. In contrast, there are more papers/books that linked karyotype changes to speciation during the same period, due to the various molecular cytogenetic advances that can compare karyotypic evolution.

Interestingly, two well-known evolutionary thinkers have pushed this idea. Julian Huxley clearly supported this concept in 1956 with apparent emphasis on autonomy. He suggested that this transformational event is indeed a speciation event, as "Once the neoplastic process has crossed the threshold of autonomy, the resultant tumour can be logically regarded as a new biological species" [Huxley, 1956]. Even though Huxley was a major player in evolutionary synthesis during the mid-twentieth century, his viewpoint failed to generate excitement in the field. Similarly, Leigh Van Valen, who proposed the "red queen hypothesis" to explain the main function of sexual reproduction, suggested that HeLa cells be defined as a new species, which he named Helacyton gartleri [Van Valen, 1991]. Van Valen defined the ecological species concept in 1976. According to his concept, populations form the discrete clusters with similar phenotype that we recognize as species. The ecological and evolutionary processes that control how resources are divided up tend to produce those clusters. Unfortunately, these insightful ideas did not lead to further research, perhaps due to the fact that cancer is not a popular topic in evolutionary studies.

In the early 2000s, Peter Duesberg and David Rasnick pushed the concept of aneuploidy and cancer speciation stating that aneuploidy, as somatic mutations, makes cancer a species of its own [Duesberg and Rasnick, 2000]. Based on the facts that many solid tumors are comprised of aneuploid cells, aneuploidy is highly associated with some key features of cancer, and variation of the numbers of chromosomes is the basis of speciation, they concluded that aneuploidy falls within the definition of speciation. Since Boveri's time, our understanding on genetic material has expanded beyond the chromosome. However, Duesberg understood cancer genes and their limitations in cancer. He was among the first who cloned the oncogene and has done extensive research to study the oncogene and cellular transformation. His transition from focusing on the cancer gene to aneuploidy is thus highly significant. During these years, Duesberg's group has provided more evidence to advance the importance of aneuploidy in most features of cancer such as immortality, metastasis, and drug resistance based on the theory that carcinogenesis is a form of speciation. It should be noted that, in recent years, Duesberg's group has focused more attention on structural re-organization rather than strictly aneuploidy, as these two types of genome changes are different in evolution [Ye *et al.*, 2007; 2009]. Their current viewpoint is that cancers are generated from normal cells by random karyotypic rearrangements and selection for cancer-specific reproductive autonomy [Duesberg *et al.*, 2011].

By analyzing the striking parallel between the heritable genomic variation in cancer and natural selection in organisms, Mark Vincent recently proposed that carcinogenesis and the clonal evolution of cancer cells within animals are speciation events in a strict sense. Based on the similarities between tumor formation and bacterial populations, Vincent argued that "the tumor genome may be swept clean and homogeneity reestablished by a form of genetic cohesion, as with bacteria" and cancer would thus fit the cohesion species concept (CSC) criteria for a new species [Templeton, 1989; Vincent, 2010; Knauss and Klein, 2012]. It is interesting to point out that Vincent's argument was published in a major evolutionary journal and attracted the attention of evolutionary biologists.

Perhaps not realized by many, in recent years, cancer evolutionary studies have quietly generated some very important information that can really help connect the dots of whether or not cancer should be considered as a new species. These studies have redefined the types of inheritance and linked karyotype changes to formation of new systems, and this lays out the framework of using genome similarity to define species [Ye *et al.*, 2007; Chapter 5].

Due to previous work involving the classification of frog species using chromosomal banding patterns, we have high appreciation for using karyotypic features to separate closely related species [Heng *et al.*, 1985; Heng *et al.*, 1987]. In addition, we have studied the chromatin topology both for mitotic and meiotic chromosomes and valued the ultimate importance of the overall chromatin topology within the nucleus both for gene expression and organization [Heng *et al.*, 1994; 1996; 2000; 2004b]. Even though we believe that genome-reorganization drives speciation based on obvious karyotype relationship among different species, it was hard to argue with researchers who believe that novel, specific genes are the key for speciation, while chromosomal changes are incidental. Furthermore, it was hard to imagine how chromosome-mediated genome reorganization happens, especially during an extremely slow and long process. Surprisingly, the watching evolution-in-action experiment suddenly provided novel understandings for genome reorganization and speciation, both for cancer and organismal evolution (Chapter 5).

As we described in Chapter 5, the decade long molecular investigation and synthesis using *in vitro* immortalization and genome chaos models have provided the following conclusions:

(1) The karyotypic differences between host cell and cancer cell can be drastically different, and these changes often are much more extensive than the differences observed when comparing humans with other animals. However, all of these different altered karyotypes are clearly generated from host cells through somatic cell evolution.

(2) The evolutionary process of most cancers is not simply driven by clonal expansion but by punctuated genome replacement, even though the clonal expansion phase is involved and in particular can be detected during the later stages when cancer is dominant. In fact, clonal populations often co-exist with heterogeneous cell populations as a "society" [Heppner, 1989], where heterogeneity is necessary for cancer evolution as different cells together can emerge new properties which cannot be achieved by independent clonal populations.

(3) The cancer cell population can maintain its heterogeneity through the mechanism of fuzzy-inheritance (see Chapter 6). As the cancer cells are highly unstable with less constraint, the fuzzy-inheritance overshadows the original parts and system inheritance of the host cells in terms of generating high level of variations and extraordinary outliers when under extreme conditions. The alteration range, or the fuzziness, becomes the maxima for cells with unstable genomes.

(4) The evolutionary pattern of cancer cells is similar to other asexual species, including bacterial species. Due to the lack of constraint, they display typical punctuated evolution. In contrast, for sexual species, the genome is preserved, and these species display a stepwise pattern of evolution similar to the cancer population of stable genomes [Heng, 2007]. The degree of genetic variation among cancer cells is more diverse than bacterial populations due to multiple levels of heterogeneity and, in particular, the effective mechanism of genome reorganization. For example, cancer cells can achieve macro-cellular evolution rapidly through the mechanism of genome chaos.

(5) Genome context rather than individual genes defines the system through system inheritance, and the specific karyotype is linked to the transcriptome [Heng *et al.*, 2008; Stevens *et al.*, 2013; 2014]. Thus the cell line with the same karyotype should be considered as the same overall system, despite differences or modifications at the gene and/or epigene levels.

(6) Outliers with specific genomes drive the direction of cancer evolution [Abdallah *et al.*, 2013]. This illustration of the mechanism of new population formation represents cohesion of cell populations with similar core genomes. When a core genome emerges, a new "species" may be formed. The stability of the core genome is an important feature of the emergent cell population.

(7) Perhaps the most important relevant synthesis about the species concept of cancer is the illustration of the relationship between micro- and macro-cellular evolution to organismal evolution. The key difference between micro- and macrocellular evolution is not the time scale where micro-accumulation leads to macro-evolution, but that micro- and macrocellular evolution represent different mechanisms, defined by different levels of genetic organization.

In fact, the above message was promoted in our publication in 2007 [Ye *et al.*, 2007]. Quoting Ye *et al.*:

> *"Based on the concept of genome context and its essential role in cancer and organismal evolution, we agree with the viewpoint that various cancer cells represent a 'new species' when compared to normal cells from where cancer cells are derived from. Strong support comes from the fact that most of the cancer cells display altered karyotypes coupled with many variable abnormal phenotypes. As it is known that cancer progression can be monitored by population replacement as illustrated by our recent experiments, it is logical to suggest that the process of cancer progression in fact represents the generation of a series of new species. Interestingly, the stochastic process can generate a variety of different cell populations or 'species' with similar 'out of control growth' phenotypes. This viewpoint has gained experimental support as all stages of cancer progression tested so far including immortalization, transformation and drug resistance are always associated with novel karyotypes [Heng et al., 2006b; Heng 2007b; Heng et al., in preparation].*
>
> *Following this lead, cancer should be considered as a genome-based disease rather than a gene-based disease, as in the majority*

of cancers it is the altered karyotype that is the universal feature where no universal mutations have been detected so far [Heng, 2007b]. It is also possible that the progression of cancers mainly depends on karyotype evolution. In this sense, gene mutation alone will not be able to generate new 'species' of cancer...

The concept of genome context is important to cancer evolution and gains support from similar evidence in natural evolution. By comparing the karyotypic patterns of a large number of species, distinctive karyotypes are the most important features among various species [King, 1993]. Ample evidence shows that the karyotypic changes are crucial for species evolution. The mammalian karyotypic evolution is one good example [Navarro and Barton, 2003; Murphy et al., 2005; Kohn et al., 2006]. Even though traditionally, the molecular evolution studies have focused on DNA or protein sequences rather than karyotypes, increasing evidence points to the importance of karyotypic evolution and its role in evolution, especially in light of the genome sequencing projects where there is a shortage of species-specific genes among various species and genome stochastic reorganization could play an important function. On the other hand, our novel observation that cancer evolution is mainly driven by karyotypic evolution will provide a new perspective on natural evolution. The whole set of chromosomes defines the genome context and determines the speciation [Heng et al., 2006c], while gene mutations modify the genome context to make it fit better [Heng et al., in preparation]..."

7.3 The Main Arguments For and Against Considering Cancer as a New Species

One of the common arguments against the "cancer is a new species" concept is that all human cancers are derived from cells of a host. Cancer shares the host's genes and mutations, so it is unacceptable to consider cancer as a totally different system. By the way, if the cancer is derived from clonal expansion, it is hard to tell where to draw the line when it is no longer human, as all changes from host to cancer cell are gradual, representing a process of accumulation of a number of small changes. In practice, all human cancer cell lines are treated as human materials. In other words, a patient may argue that since their cancer originated from their cells, it belongs to and is still a part of them. This argument may also be extended to established cancer cell lines that were originally extracted from patient samples. For example, the HeLa cancer cell line is one of the most used human control cell lines in current molecular research. As we mentioned in Chapter 4, the statement that the HeLa cell line is no longer human due to its drastically altered genome has generated heated debate. To further discuss this issue, we need to address the species concept and determine the relationship between normal host tissue and derived cancer.

7.3.1 *Traditional species concepts exclude cancer as a separate species*

To demonstrate cancer as a species separate from host, we need to appropriately fit cancer into one of the traditional species concepts. However, it is highly challenging to achieve this, due to the fact that most of the available species concepts have nothing to do with key characteristics of cellular evolution. For example, application of cancer to Ernst Mayr's "Biological Species Concept," one of the most accepted species definitions, is hindered by its mode of reproduction. Quoting Mayr: "Species are groups of actual or potentially interbreeding natural populations, which are reproductively isolated from other groups" [Mayr, 1942]. As sexual reproduction serves as a species-specific barrier in distinguishing species under the Biological Species Concept, cancer is not a species under this definition.

Since cancer cells reproduce exclusively through asexual reproduction, perhaps we could support cancer as a separate species using concepts that apply to asexual organisms. One such concept that applies to bacteria is the notion of defining "ecotypes" to define species. With respect to bacteria, an ecotype is a set of strains that use the same or similar ecological resources, or in other words serve a particular niche in the environment [Cohan, 2002]. This ecotype then serves as a separate evolutionary lineage. However, cancer does not display unique features from its host, as normal cells can display overgrowth and invasion under certain physical conditions. In addition, cancer persists in a wide variety of tissue environments and can utilize different metabolic pathways. The bacterial ecotype concept therefore does not extend to cancer as a separate species.

Bacterial species systematic approaches have been employed to distinguish species at the molecular level. For decades, the gold standard of distinguishing strains of different species has involved whole DNA sequence hybridization [Johnson *et al.*, 1973; Wayne *et al.*, 1987]. With this approach, genetic similarity of two strains is determined by measuring the fraction of total sequence that is homologous. A 70% level of homology was determined as the upper limit for determining whether two strains should be considered different species [Wayne *et al.*, 1987]. Application of DNA-DNA hybridization between cancer and healthy somatic cells results in high degrees of sequence homology, refuting the notion that cancer is a separate species. However, given the drastic differences in genome (karyotype) complexity and heterogeneity observed in cancers, it is difficult to argue that cancer is not different from healthy somatic cells. Further, considering the simplicity of the bacterial genome compared to multicellular organisms (most bacterial genomes consists of a single DNA molecule rather than many pairs of chromosomes and include a very limited number of genes and regulatory systems), we need a new set of criteria to account for this karyotypic diversity to make a better definition, especially when it is now obvious that the genome context rather than individual gene defines the species.

7.3.2 *Cancer fits the concept of using genome similarity to define both asexual and sexual species*

In order to effectively use genome similarity to distinguish species effectively, we need to apply a genome-based system level measurement, which accounts for multiple genetic levels and can extend to both sexual and asexual species. We propose comparing the genome topology when making species designations, as the species-specific genome context may be the key distinction when defining a species. To fully appreciate this concept, we need to understand what the genome is. The genome is not just a collection of all genes or other sequences, arranged like pearls on a string. While individual genes produce functional parts or proteins, the genome structure has a function defined by the karyotype that provides the blueprint that determines how these parts function together. In other words, the genome defines the network structure by providing the "system inheritance" (Chapter 5). It is the karyotype that defines the identity of the cell line as different derived cell lines could have similar DNA sequences but different genome contexts.

As discussed, distinctive karyotypes were previously noted as the most important features among various species [King, 1993; Kohn *et al.*, 2006; Murphy *et al.*, 2005; Navarro and Barton, 2003]. Recall that the genome represents the highest level of the genetic system. Within a given genome, genes and their interactions form a genetic network. It is the given karyotype and genome topology that govern the form and boundaries of the network, which has been termed the genome context [Heng, 2009; Ye *et al.*, 2007]. Thus, even cells with similar gene contents could have drastically different genomic contexts, which could significantly change the behavior of the overall system.

While sexual reproduction has been described as a distinguishing barrier between species, the newly established paradigm for the main function of sexual reproduction emphasizes that species identity is retained in the genome topology [Heng, 2007c; Gorelick and Carpinone, 2009; Wilkins and Holliday, 2009; Gorelick and Heng, 2011]. In contrast to the traditional reasoning that sexual reproduction (in this context, we will define sexual reproduction as meiosis and syngamy) functions to increase genetic variation, under this new paradigm, sexual reproduction mainly acts to reduce genome alteration despite its secondary function of mixing genes. This new concept is supported by genetic, genomic, and epigenetic status restoration that occurs during sex. Further, some DNA damage is repaired, and many mutations are avoided. Cytosine methylation and chromatin marks are reset to levels that worked for both parents. Most importantly, large chromosomal abnormalities are purged, including number changes and structural rearrangements. Through this mechanism, the karyotype is preserved as well as the identity of the species. Thus, sexual reproduction serves as a conservation force for biology, and under normal circumstances, this species will be maintained despite variance in gene mutations and environments over time, as the genomic framework is species-specific rather than the presence or absence of specific genes. Only very rarely do new karyotypes pass through this constraint, and even so, require a partner of compatible genome to mate successfully in order for a new species to emerge [Ying *et al.*, manuscript in submission].

The similarity of karyotype-defined genome topology is thus the key to identifying and distinguishing sexual species, as the integrity of the genome also ensures gene content, resulting in individuals of the same species sharing the same or very similar genomes (including gene loci) within the population. In the case of humans, despite that some minor variations can be detected, such as individuals with sex chromosome abnormalities, certain rare balanced translocations, as well as Down's syndrome with an extra copy of chromosome 21, for example, the core genome of 46XX or 46XY is maintained for the vast majority of individuals. These mentioned large-scale genome level alterations usually do not pass to future generations, as altered genomes often reduce successful reproduction and likely would be eliminated (even those viable balanced chromosomal translocations are often associated with infertility). Under some specific circumstances, some balanced Robertsonian translocations can be passed [Martinez-Castro *et al.*, 1984]. However, these will be diluted by the majority of the population. The majority of genome alterations detected within a population occur through *de novo* mechanisms, as most babies with unbalanced chromosome translocations and aneuploidy have parents with normal karyotypes.

The similarity of the genome can also be used to distinguish asexual species. Without complicated multiple chromosomes, the genome topology becomes much simpler. However, the orientation of genes within their genome still matter. Despite the traditional, yet inaccurate impression that asexual species produce identical clones, new evidence has amply illustrated that asexual reproduction is associated with a high level of diversity [Heng, 2007c; Gorelick and Heng, 2011]. Our recent computer simulation data also showed that asexual reproduction could result in the gradual accumulation of alterations [Ying *et al.*, manuscript in preparation]. In contrast, sexual reproduction can rapidly form distinct clusters, mimicking different species. While bacteria lack typical eukaryotic chromosomes, and they can be appropriately classified by the similarity of select DNA sequences, cancer cells display complicated reorganization at the genome level. Thus, cancer cells cannot simply be classified by using sequence similarity; therefore, using genome similarity to distinguish different cancer cells is the most appropriate approach.

Together, the similarity of the genome provides a better species definition for both sexual and asexual organisms, and cancer cells fit such definition quite well. In fact, tracing karyotypic change in evolution has been an effective way to study mammalian evolution. This similar strategy can now also be applied to cancer cell evolution, mimicking new species formation.

7.3.3 *Individual constraint and contagious cancer*

Another often used argument against the cancer species concept is that cancer cells must grow within a host's environment, and they ultimately die with the host without migrating to additional hosts. It seems that it does not make sense to consider cancer as a new species, as the success of cancer is also the means to its demise. In contrast, for organismal species, they can pass on from generation to generation by migrating to different environments.

This argument comes with its own limitation, however. Certainly, there are many constraints for cancer cells, but there are always constraints for any species, including bacterial and viral. Viruses depend on host cells for replication, and many viruses kill host cells. However, they are considered different species. In addition, there are many bacteria that also die with their hosts, and many bacteria only live within specific regions of the host's body. They are not shared between different individuals, thus belonging to an individual host-specific population or even species. Since these bacteria are classified as different species, such definition could also be applicable to cancer cells.

The last issue is the ability of many micro-organisms to be successfully transmitted between different hosts by infection. Does this feature exist in cancer cells? Indeed, contagious cancer exists in the wild, and examples include canine transmissible venereal tumor (CTVT) and Tasmanian devil facial tumor disease (TDFT). CTVT is sexually transmitted between dogs, wolves, coyotes and jackals, whereas Tasmanian devils transmit TDFT by biting [Belov, 2012]. In both cases, the contagious cancer is basically a somatic cell line that has become a transmissible parasite [Welsh, 2011]. These two systems display different prognoses, illustrating varying degrees of constraint. TDFT is often fatal within six months, while the majority of CTVT instances are eventually rejected by the impacted individual and coupled with lifelong immunity. More drastically, there were reports that the contagious reticulum cell sarcoma of the Syrian hamster can be transmitted to different individuals by mosquito [Banfield *et al.*, 1965].

It is worth pointing out that even though the reports of contagious cancer in humans are scattered, these cases seem real. In addition to the most frequent cases that involve organ or hematopoietic stem cell transplants and cancers occurring during pregnancy that are transmitted to the fetus, there is also an instance of a surgeon contracting a malignant fibrous histiocytoma from a patient after cutting his hand during an operation [Gärtner *et al.*, 1996].

To explain why contagious cancer has not evolved in humans, one idea is that human population genetic diversity is the key. From an evolutionary point of view, however, a potentially scary idea is that since overall cancer incidence is drastically increasing with prolonged human life span, there is increased opportunity for cancer cells to gain communicable features through increased rounds of evolution. Recall that the result of the punctuated phase of evolution is the generation of new systems and increased evolutionary potential, and this can lead to the emergence of new features. It is known that contagious cancer in canines arose over six thousand years ago [Belov, 2012]. Thus, considering continued increases in the human population, life expectancy and cancer incidence, it would not be a surprise if select human cancers eventually gain the capability of transferring to different individuals.

In a broad sense, most species have their own environmental constraints. Examples include fish restricted to a water habitat, most trees anchored to soil, and animals to specific climates. We should not use environmental constraint to disqualify a species.

7.3.4 *Many cancer cell lines represent artificial species*

The above-mentioned constraint could be removed by using an artificial culture system. For example, the HeLa cell should be considered at least as an artificial species, as it has been passed for many generations and distributed all over the world, each sample having a separate lineage and evolutionary fate. *In vitro* cell culture has allowed many cancer cells to survive outside of its initial host and beyond a host's lifespan. Due to its highly dynamic genome, however, only the core genomes of cell lines are preserved, and there are high levels of karyotype change among different sublines. Therefore, they should be considered as a collection of many artificial species, which are in fact still evolving.

When discussing the artificial "breakdown of natural constraints" in order for certain systems to proliferate, the preparation and preservation of cancer cell lines (tumor cell extraction and maintenance in a culture system) has a key parallel to the application of assisted reproductive technologies to help infertile individuals. In both cases, the genome constraint that would be applied under normal physiological conditions is lifted, allowing these genomes to pass beyond the "restricted" individual or host. Tumor cells are freed of tissue and immunological constraint and grown in a nutrient-rich environment, which are no longer bound by the constraint of an individual's life span. Depending on the technology employed, gametes may bypass either the selective obstacle course of the reproductive tract (e.g. *in vitro* fertilization, intrauterine insemination) or be compensated for a lack of otherwise necessary capabilities, like effectively penetrating the oocyte zona pellucida (e.g. intracytoplasmic sperm injection) [Horne *et al.*, 2013b]. By avoiding the filter of normal sexual reproduction, reproductive technologies could preserve certain chromosomal variations such as some rare balanced chromosomal translocations from parental germ line to artificially fertilized embryo. This prediction has been illustrated by a real clinical case [Moreira *et al.*, 2013]. These altered karyotypes can often be eliminated by natural processes associated with infertility such as spontaneous abortion.

By considering cancer cell lines as different species or artificial species, it is much easier to accept that proliferation is only one key selection factor of cancer formation in patients, among many other factors, especially at the initial stage where proliferation might not be that important. When more stable cancer cell lines were established, they were already the product of multiple rounds of evolution, meaning that the established cancer system is at the micro-evolutionary phase. This can be linked to rapid proliferation, which is often favored by artificial association with many oncogenes (when proliferation becomes the key feature for evolutionary selection *in vitro*). As for highly unstable cell lines, they are still entrenched in the dynamics of NCCA/CCA cycles, and their genetic profiles are highly diverse. This diversity can be observed when comparing the "same" cell line types from different laboratories and at different time periods. Thus, careful attention is needed by researchers to characterize their experimental cell lines to ensure they are working with the same system rather than different artificial species. The key is to consistently karyotype cell lines prior to any molecular analyses [Ye *et al.*, 2007; Heng *et al.*, 2013a].

7.3.5 *The principles of macro- and microevolution support considering cancer as a new species*

The realization that cancer evolution needs to be divided into punctuated and stepwise phases along with its linkage to macro- and microcellular evolution further support the concept of considering cancer with altered genome after the punctuated phase as a new species. The separation of these two phases rejects the concept of speciation by gradual accumulation of gene level changes and favors the concept that an altered genome defines a new species. This also fits well with the concept of genome defined system inheritance. Since the system is changed, it must not belong to the old system it was derived from, and genome change is the key feature of most cancers.

Considering cancer as new species also addresses the time scale difference of speciation between cancer and other forms of asexual and sexual species. One should not use millions of years of the population life span of natural organisms to disqualify the cancer species concept as cancers only have a years-to-decades cell population life span within hosts. However, we do need to appreciate the time difference required for speciation of cancer cells and other types of sexual and asexual organismal species. The long historical time scale for species with sexual reproduction is easier understood due to the sexual filter that prevents speciation by stabilizing the existing species. For common asexual species such as bacterial, the situation is a bit more complicated. On one hand, speciation could be much quicker if genome similarity was used for classification, and there should be more realizable species or sub-species. On the other hand, the process of bacterial speciation could be slower than cancer cells, as cancer cells have high numbers of typical chromosomes, which can be easily reorganized to form new species through the mechanism of genome chaos. In addition, both bacterial and cancer populations often exist as a society of species rather than a species of their own, or fixed independent species, and they are connected by the higher level fuzzy inheritance rather than fixed system inheritance (Chapter 6).

Due to these differences, even among researchers who favor the cancer species concept, there are different viewpoints. For example, Duesberg thought that "Cancer is comparable to a bacterial level of complexity, but still autonomous, that is, it doesn't depend on other cells for survival; it doesn't follow orders like other cells in the body, and it can grow where, when, and how it likes...That's what species are all about." [Duesberg *et al.*, 2011; Sanders, 2011]. However for us, we see many differences and constraints of cancer.

Another different point relates to aneuploidy and speciation. Most of the speciation in mammals is associated with genome or chromosomal reorganization but not simply aneuploidy. In plants, B chromosomes are common. It would be interesting to investigate whether B chromosomes represent one form of fuzzy inheritance. The core karyotypes otherwise are the same for a given species, and the B chromosomes are not present in every individual and tissue type. Also, the B chromosomes present some adaptive advantages.

7.4 Conclusion: Different Cancers Represent Different Species

From this synthesis, we can take this discussion a step further. Rather than cancer representing a new species, we argue that different cancers represent different species (or cellular species). Each cancer has its own evolutionary fate, as each cancer is the result of many rounds of the NCCA/CCA cycle, which has been shown to stochastically generate different products with every run. Under certain conditions, they can be cultured for decades or likely much longer (as artificial species). Importantly from this debate, cancer is still a biological system and warrants consideration in the field of speciation.

In order to appreciate the idea that "different cancers represent different species", an appropriate species classification framework and evolutionary theory is needed that aptly accounts for both sexual and asexual organisms, including distinguished behaviors of cancer. This requires shifting emphasis on species-specific genome context as the key distinguishing feature. Placing this feature in the framework of genome-mediated evolution is necessary, as this framework demonstrates the importance of species-specific genome identity. Interestingly, only a few decades ago, many "animal centric" researchers would not consider classifying microorganisms as species due to the sexual reproduction barrier. It is now time to include cancer cells as well, as they still represent biological systems despite any bizarre features.

Like any other new concepts, the cancer-somatic species idea warrants some further clarification. First, while it is relatively easier to call a tumor that is mostly clonal a population of a single cancer species, how should a highly heterogeneous cancer where there are a large number of diverse genomes be named? In such a real somatic ecosystem, one cancer mass is made up by large number of new species. It is clear that the research community needs to develop some quantitative index to deal with diverse cases.

Second, during the punctuated macro-evolutionary phase, new cancer species are being created with each generation, creating many transitional species that exist for very limited time periods. Do transient, unstable cells qualify as a distinct species, even though vast majority of them will be extinct? And third, what degree of karyotypic change would qualify a cancer cell as a species or subspecies? Karyotypic differences between cells can be very large (involving multiple chromosomes) or very small (involving a fragment of a single chromosome). Speciation between two cancer cells with significant karyotypic differences would be apparent. However it is not clear how smaller changes to the genome topology would affect overall systems dynamics. Therefore it becomes necessary to characterize the functional differences between karyotypically distinct cells to understand what degree of karyotypic change is used to designate a species or subspecies, and the differential contribution from numerical or structural chromosomal changes including copy number variations as well as transposable elements. Understanding these differences will further the understanding of evolutionary dynamics of the cancer ecosystem. Similarly, the same argument can be applied to somatic mosaicism found in normal tissue. How parts heterogeneity (individual genome) can impact the emergent properties at the level of system behavior (ecosystem) represents a key challenge.

This scientific and philosophical debate we have presented so far in this chapter can pose as a platform for further discussion, which may be very helpful not only for cancer researchers to better understand cancer evolution, but also for evolution scholars to better understand natural evolution. The following sections serve as further discussion and illustrate our points.

7.5 The Potential Impact of Considering Cancers as New Species

A drastic research mindset change requires drastic concepts as well as new terminologies. Clearly, the discussion of whether or not cancers should be considered as new species is not just a philosophical debate, as such concept offers greater influence on the cancer research community. It can enrich our understanding of cancer (knowing the enemy by its true colors), offer new avenues for research and improve current strategies to fighting cancer.

7.5.1 *Help to shift the research focus: From gene to genome*

With the acceptance of this new genome-based cancer species concept, the research emphasis must change. As new species, cancer cells differ from normal cells at the genome-defined species level, and in order to better diagnose and treat this disease, a departure from the traditional gene mutation-based approach is necessary. Further, as an evolving species, tracing and targeting specific genes is an approach with limited, short-term success for most cancer types. This is because genome change leads to species wide genetic framework change, which can re-define an individual gene's function.

7.5.2 *The cancer species concept demands new strategies for reducing cancer drug resistance*

Currently, the primary standard of care in cancer therapy typically consists of systemic administration of cytotoxic drugs. Using a wide variety of treatment regimens and agents, the shared goal in the clinic is to eliminate the highest possible number of tumor cells [Frei and Canellos, 1980; Frei *et al.*, 1998]. Thus, most research efforts regarding chemotherapy (along with other approaches) are focused on conceiving and developing therapeutic strategies that maximize tumor cell death. The basis for such rationale is that a cancer cell population of smaller size will have a lower probability for a new gene mutation to occur that drives drug resistance and would take a longer time to come back by generating a high tumor cell count. Since the primary factor that limits cytotoxic drug dosage in most clinical therapies is patient tolerance, cancer patients are typically treated at or near the maximum tolerated dose with the intention of eradicating the tumor. Unfortunately, tumor response even to the harshest regimens is often transient, and therapy frequently fails due to the emergence of drug resistant populations. As an alternative to general cytotoxic agents, research has increased focus on developing and administering cancer-specific targeting therapies. This, in part, has been inspired by the overwhelming success of imatinib application against chronic phase chronic myeloid leukemia. However, with rare exceptions, specific target approaches have offered little in terms of long-term patient prognosis, as drug resistance remains the inevitable consequence of cancer treatment [Baldo and Pham, 2013; Horne *et al.*, 2013a].

Interestingly, mathematical and evolutionary modeling have previously predicted that therapeutic intervention can provide selective pressure for the expansion of resistant variants [Maley *et* al., 2004; Pepper, 2012], and initial cancer drug response has been noted as "not a strong predictor of reduced mortality" [Pepper, 2011]. From an evolutionary standpoint, these models explain a paradox of current cancer therapeutic strategies, where initially effective treatment both eliminates a large number of tumor cells and favors the formation of resistant clones. However, the exact mechanism is hard to illustrate based on the gene mutation theory. One popular explanation is that drug treatment selects on pre-existing gene mutations and favors its domination. This is, however, an over simplified explanation, which failed to account for treatment-induced resistance especially at the genome level (Horne *et al.*, submitted).

By accepting the cancer species concept, it is no longer crucial to identify specific gene-based targets, as these moving targets are highly dynamic under treatments and only are responsible for some switchable features. Since the emergence of new species requires new genome formation with many features of plasticity, the key mechanism of drug resistance in cancer is to increase the opportunities for genome alteration mediated macrocellular evolution. As a survival response, treatment-induced genome chaos is the real driver of cancer drug resistance [Heng *et al.*, 2006a; Heng, 2007c; Horne *et al.*, submitted; Chapter 5]. In contrast, targeting specific gene mutation-defined features within the stepwise microcellular phase will not change the existence of a given "species", but likely will reduce the population of these species. Meanwhile, the same treatment will perhaps allow for new species to occupy the niches of the "old" species. Thus, a high-dose drug treatment or even specific target-mediated killing initiates a speciation event in cancer, where surviving cells represent different species as they display new, altered karyotypes. As a result, these new species could display new phenotypes, including highly aggressive phenotypes. This also provides explanation for the paradox of cancer therapy, where initially effective treatment (i.e. associated with an early high cell death count) consistently results in rapid drug resistance.

Extending this mechanism to the cancer species concept opens two doors of discussion. The first is that cancer is not an end product or destination. Rather, it continues to evolve and remain subject to selection as the disease progresses and the environment changes. The second is that high dose treatment-related stress (or from other sources of stress, for that matter) can promote the formation of new variants or species, meaning that multiple rounds of maximum tolerated dose treatment would initiate additional speciation events beyond initial carcinogenesis.

Fortunately, alternative therapeutic strategies that do not incorporate the administration of maximum tolerated doses have had early success. These include adaptive therapy, where the goal is to maintain a stable tumor burden by adjusting treatments and dose concentrations in response to tumor cell growth/death [Gatenby *et al.*, 2009; Gillies *et al.*, 2012; Silva *et al.*, 2012]. Metronomic therapy aims to maximize tumor cell death, but this strategy involves a rhythmic regimen consisting of more frequent treatments of lower concentrations than the maximum tolerated dose strategy [Kerbel and Kamen, 2004]. Lower concentrations of therapeutics may reduce both the onset of genome chaos and formation of aggressive outliers; however, this needs to be explored. We recently have studied the transition between the initial effective killing of cancer cells and paradoxical long-term cancer cell success (see Chapter 8 for more information). Obviously, appreciation of the fact that cancer represents a heterogeneous collection of different systems or species can lead to the systematic and standardized examination of karyotypes of cancer samples, which is the most important information regarding the cancer landscape. Understanding the heterogeneity and genome stability/instability of a specific tumor will provide necessary information regarding the level of constraint that can be applied to that tumor in a particular regimen. Furthermore, the understanding that different cancers represent different species can promote shifts towards a "personalized" or "precise" approach in cancer medicine.

7.5.3 *The cancer species concept will attract more evolutionary biologists to study cancer*

The current –omics study, including cancer genome sequencing, has forcefully demonstrated that the cancer evolution holds the key for cancer research. Unfortunately, the vast majority of cancer researchers lack evolutionary training. In fact, very few investigators have a holistic appreciation for the evolution of systems. It is thus urgent to recruit more evolutionary biologists to change the landscape of cancer research. The linkage between cancer and speciation is an important factor to attract evolutionary biologists' attention, as speciation is a key question in evolutionary biology. If cancer could provide insight on this issue, then many scientists will join this exciting journey (more will be discussed in the next section).

7.6 Will the Evolutionary Process ever Eliminate Cancer?

With increased acknowledgment that cancer represents an evolutionary process, related questions can be asked, such as why has evolution so far failed to eliminate cancer occurrence in humans? Many believe that evolution cares less about life after reproductive age. This is based on the fact that many cancer cases occur later in life, typically past reproductive age. This idea, however, cannot explain why there are still cases of cancers prior to puberty. It is also known that for human societies, elders do contribute to the success of the young within the family structure. Alternatively, some suggest that, as human is a relatively young species, time is still needed for human evolution to get rid of cancer genes. But this view point is not solid as these so called "cancer genes" are presented in organisms even prior to the evolution and emergence of multiple cellular organisms, and these genes often are essential for normal human development and function. Others suggest that human genes fail to adapt to a fast-changing environment, and some old way of living or life style might be helpful. For example, increased hygiene and delayed reproduction age have been suggested as the causative factors to increased diseases/illnesses. While these suggestions seem to have been made with evolution in mind, some of them in fact are at odds with evolutionary principles.

7.6.1 *Both humans and cancers are products of evolution*

The wide variety of life that currently exists is the result of billions of years of evolution. Even though most runs of somatic cell evolution that result in cancer formation only takes years to decades, the evolutionary principle is the same. We are who we are because of the evolutionary process, and perhaps it is arguable that the high level of evolutionary phenotypic plasticity is a winning ticket for humans. Unfortunately, we cannot separate the package of evolution nor stop evolution entirely because of the risk of cancer. As we discuss later, adaptation needs variation, and variation is also the basis of cancer. Thus, we as humans cannot fully separate ourselves from the possibility of cancer. Changes that result from evolution come with both advantages and disadvantages, and in terms of cancer and humans, this is a package deal.

7.6.2 *The evolutionary process selects evolutionary potential rather than perfect products like fixed "good" genes*

One persistent misconception (or wishful thinking) held by the general public regarding evolution is that it ultimately acts to improve or better an organismal lineage through stepwise, Darwinian selection focused on increasing good genes and reducing the frequency of bad genes. As a result of this constant tinkering and renovation, we see the emergence of beneficial and removal of harmful traits, improving the organism's fitness and reproductive success. Based on this trend, the products of evolution should get better and approach perfection (the rationale of hope that humans will someday cure and eliminate cancer and other diseases).

This perception, however, does not make sense with reality, as no species is perfect despite the long history of evolution (the extinction of a given species is not simply governed by the time factor either). Humans will be the same and the disease-free world will not be on the horizon. Only when the environment is fixed, and the future is highly predictable, the so called perfection of selected products can be reached. In reality, the selection criteria are constantly changing, as reflected by the environment, and selection acts upon the total genome package rather than isolated genes. This concept directly applies to cancer, where during the punctuated phase, mother and daughter cells display different genomes. Since future environments are hard to predict, for the success of a given species (or cancer cell population), increased evolutionary potential is the key rather than fixed "good" genes, as good or bad is context-dependent. For example, for cancer, the best winning strategy is to increase the genome heterogeneity-mediated robustness rather than specific driver gene mutations, as no given driver can finish the job of cancer evolution. All drivers are constantly changing during cancer evolution.

Further, a newly realized form of inheritance known as fuzzy inheritance can effectively contribute to the evolutionary potential as it delivers the inherited degree of change rather than fixed specific changes (Chapter 6). This mechanism ensures change (by providing the potential pool of different candidates) and allows the evolutionary process to select the winner based on the future context. So while evolution may favor a particular subgroup during a specific selection event, fuzzy inheritance is still passed along with karyotype-defined system inheritance, and that surviving subgroup can recreate the parental population's evolutionary potential when needed.

Interestingly, environmental stress that may occur during evolution can result in the increase of evolutionary potential of the disease by creating additional fuzzy inheritance. This was demonstrated in the genome chaos study, as multiple sources of high stress can induce genome chaos with the highest degree of fuzzy inheritance. When surviving cells become dominant, displaying stable altered genomes, the fuzzy inheritance becomes low.

Since variation is a key component of evolution, and variation exists and persists even after the long process of evolutionary selection (e.g. through fuzzy inheritance and stress-induced variance), the evolutionary process will continue and will not aim to put an end to evolution.

7.6.3 *Cancer can have a positive impact on organisms*

Due to its horrible burden on human health, cancer is often viewed in a negative light, especially when discussed in terms of being caused by mutations or good genes gone bad. Such negative view also prevents the acceptance of considering cancer as new species. Paradoxically, in the wild, cancer may also play a role towards the survival and persistence of the overall host species. The impact cancer holds on ecosystem functioning, stability, and diversity was recently proposed [Vittecoq *et al.*, 2013]. For instance, older individuals with cancer may be weaker, more vulnerable to predation and have higher pathogen susceptibility.

These individuals still require a share of resources to survive, thus they represent a "cost" to the overall success of the species. Elimination from the population frees up resources to the healthy and/or younger members, increasing reproductive success and ultimately increasing chances for continued persistence of the species. Another example is selection of would-be melanoma formation sites in Xiphophorus cortesi fish [Fernandez and Morris, 2008]. This species of fish displays a spotted caudal melanin pattern, which serves as the site of later melanoma formation. It was shown that sexual selection appears to be maintaining the oncogene due to a mating preference, serving as an example of a given gene that has opposite effects on fitness at different ages (beneficial in early life but harmful at a later stage). There are also some discussions about the potential benefits or conflicting functions of cancer in humans. A possible trade-off involving sexual selection has been implicated. Similar to the Xiphophorus cortesi fish story, some people use sun tanning beds to increase their attractiveness, which also increases the risk of melanoma. A shorter CAG repeat region within the human androgen receptor gene can be linked to both increased fertility and the increased risk of prostate cancer formation [Summers and Crespi, 2008]. Some essential genes for fetus implantation and placentation appear to play roles in invasion and metastasis in cancer [Crespi and Summers, 2006].

Interestingly, some tumor types have been shown to play a role in artificial selection. For example, members of the goldfish var. oranda can display a large benign tumor on their heads, which can be described as resembling a "hood" [Kozlov, 2010]. This feature provides the goldfish a unique appearance that is favored by people interested in keeping goldfish as pets. In this case, this type of benign tumor is key in the selection of these goldfish by humans and thus the continued growth of the goldfish population that possesses this trait.

7.6.4 *Cancer may also represent a trade-off for cellular stress adaptation*

The concept of fuzzy inheritance, supported by drastically increased observations of the multiple genetic and non-genetic variations from normal and disease human tissues, has promoted us to consider the benefit of the high degree of variations in physiological and pathological conditions. Our synthesis has led a trade-off model that unifies the stress-induced genomic/non-genetic alterations and disease (including cancer), which also provides reasoning for the presence of genome alterations in normal, healthy tissues [Heng *et al.*, 2013b; Horne *et al.*, 2014]. In humans, these include the polyploidzation of many tissues such as liver cells, skeletal muscle, ovary, placenta, thyroid gland, blood, urothelium and Purkinje neurons, and blastocyst mosaicism and trisomy 21 mosaicism in the general population, as well as detected stochastic karyotypic changes caused by physiological and environmental challenges [Biesterfeld *et al.*, 1994; Celton-Morizur and Desdouets, 2010; Davoli and de Lange, 2011; Fragouli and Wells, 2011; Heng *et al.*, 2004; Hultén *et al.*, 2013]. In addition, an increase of genome-level alterations has been revealed by whole-genome sequencing of healthy individuals [Abecasis *et al.*, 2012]. Chromosomal aneuploidy, chromosome non-disjunction, and micronuclei formation in peripheral lymphocytes are associated with age [Ohshima and Seyama, 2010], and polyploidy increases with age in hepatocytes [Gentric *et al.*, 2012]. Somatic mosaicism as a result of aneuploidy and chromosome instability has been proposed to play a role in brain aging [Faggioli *et al.*, 2011; Iourov *et al.*, 2012].

At any given time, under either normal physiological or pathological conditions, cells are subject to a wide variety of internal and external stress. Stress, in general, results in many infrequent genome alterations [Heng *et al.*, 2004; 2006a]. These types of seemingly random chromosome aberrations, or non-clonal chromosome aberrations (NCCAs), have been ignored in the field and thought of as genetic noise. Following years of research, we have concluded that the elevated frequency of NCCAs represents increased genome level system dynamics [Heng *et al.*, 2013b]. Many forms of stress can be linked to elevated NCCAs, including virus infection, temperature change, drug treatment, as well as many types of environmental stress. Interestingly, certain levels of NCCAs can be detected from each individual's blood cells, and we have linked this index to many disease conditions.

As genome-level alterations are more effective at drastically changing the genetic system than gene mutation or epigenetic change, it can effectively provide adaptive advantages for cells against high levels of environmental stress. Further, genome diversity within normal, healthy tissues allows for complex organ function while providing the genome heterogeneity necessary to account for organ function-associated stress, such as liver-mediated blood detoxification. Interestingly, back in the 1970s and 80s, the degree of polyploidy of heart muscle has been suggested to be functionally linked. For example, normal heart muscles in the left ventricle have higher ploidy levels than those of the right ventricle, possibly due to the higher blood pressure [Biesterfeld *et al.*, 1994].

As the genome level variations have been largely ignored by many molecular biologists, the aforementioned examples are focusing on the genome level. The genetic and non-genetic variations are overwhelming [Heng *et al.*, 2009]. Together, the general mechanism of stress is to trigger the alteration of system dynamics at multiple levels, and increased dynamics serve bio-function.

Increased heterogeneity is necessary for successful adaptation to occur, with the trade-off being potential disease onset [Horne *et al.*, 2015a,b; see Chapter 8]. Taking the new function of sexual reproduction into account, we understand how system dynamics are promoted for short-term adaptation at the individual level while the accumulation of alterations that could harm at the species population level is prevented. This is due to the relationship between stochastic somatic genome dynamics (allowing for adaptation and function) and genome purification through sexual reproduction (providing long-term species-specific system persistence). In other words, the stochastic somatic genome-level aberrations that occur provide individuals with an evolutionary advantage against external and internal stress. However, somatic genomic aberrations could also lead to the onset and progression of common disease and cancer. In contrast to increasing evolutionary potential by stochastic somatic genomic aberrations, the constraint of germ line evolution through sexual reproduction preserves system integrity and the species. Through this separation of germ line and somatic cell genomes, somatic genome alteration ensures short-term adaptation, while the filtering process of the germ line genome ensures long-term genome system identity [Horne *et al.*, 2013a].

7.6.5 *Many factors that contribute to the dynamics of life cannot be eliminated*

Current cancer research also hopes to identify all the main contributing factors, genetic and environmental alike, and to eliminate them. Efforts range from asbestos removal to the development of the HPV vaccine. While these efforts and accomplishments are noteworthy, unfortunately after understanding the discussed trade-off, there are volumes of factors that contribute to the dynamics of life that can never be eliminated altogether. As all humans are linked with all microbiota and environmental factors, and all these factors are both good and bad, we cannot hope for selection of only the good ones and the elimination of the bad ones. Even in developed nations where certain infections are not the biggest threat for contributing to cancer, citizens may live a more sedentary lifestyle. Indeed overeating, poor diet and lack of exercise are risk factors for multiple diseases, including cancer. If we include the various environmental factors (e.g. UV, seasonal temperatures, fine particulates in the air) and internal factors (e.g. oxidative/ER stress) that we are exposed to on a daily basis, it becomes clear that stress in general cannot be completely removed or avoided, as they are essential for life as well. There will always be a need for adaptation to these stresses, and thus it is unlikely that humans will evolve away from cancer. Of course, the main causative factors could be different at different times of human history.

Even more interesting, diseases and illnesses will always remain challenges to us, although the disease spectrum will likely be different (we have witnessed shifts of the major disease spectrum from infectious diseases to metabolic diseases including cancer and likely to mental diseases). Furthermore, it is likely that future humans will face increased challenges and have to deal with different unfit variations. For example, the heightened mental health issue appears to be linked to changes in lifestyle. The incidence of autism spectrum disorders might be associated with increasing father reproductive age, a by-product of extended education required in a fast-paced, high tech world. More importantly, deficits in social communication and interaction might not be as crucial in the past, when most people lived in a less technology driven and more isolated society, where social deficits were not a problem. In addition, many labor works were highly repetitive, and individuals with autism features would not display obvious disadvantages. Based on the fact that the drastically changed human life style will create less fit individuals, and the so-called diseases/illness fundamentally represents unfit variations, we have suggested the concept of future diseases/illnesses [Heng, 2009; 2011b; Horne *et al.*, 2014]. In agreement with this idea, it has been noticed that, tumors are more frequent in a more successful (in an evolutionary sense) group of organisms [Kozlov, 2010]. It would not be a surprise to us if cancer will be more often associated with future humans when more plasticity is needed for humans to deal with more challenges including increased life span, more environmental pressures, fast-paced work/lifestyle, and the overconsumption of food and medical drugs.

7.7 "Cancer-species" is an Excellent Model System for Studying Evolution

Directly studying speciation has been extremely challenging due to the unrealistic time window necessary for most researchers. Despite the fact that limited genes can be linked to reproductive isolation, how much gene mediated reproductive isolation contributes to the overall process of speciation is unknown. The most obvious mechanism of reproductive isolation is achieved by chromosomal or karyotype change, so any gene-based explanation has to incorporate this fact. More importantly, the hypothesis that speciation is a long process characterized by the accumulation of small gene mediated changes is unproven.

As stated by the genome theory of cancer and organismal evolution, speciation is a process of genome alteration and finding a compatible partner to form a sustainable population [Heng, 2007a; 2009; Ying *et al.*, in preparation]. This involves breaking the sexual filter-mediated genome constraint (through macroevolution), finding a match between altered genomes (by opportunities) and increasing the population size and fitness (through gene modification during microevolution). Cancer evolution includes the first and last stages. Importantly, new cancer species formation can mimic organismal speciation in a drastically reduced time scale due to the fact that there is no sexual filter for cancer cells to eliminate altered genomes.

7.7.1 *Unique advantages of using cancer to study evolution*

First, cancer evolution covers both macro- and microevolution in a short period of time, and this is especially true for *in vitro* models. As previously discussed, macroevolution refers to large genome level changes whereas microevolution pertains to lower level genetic/epigenetic alterations. Cancer-species models allow researchers to examine the contributions of different genetic levels in evolution. No alternative window exists for us to trace organismal evolution directly, especially within the macroevolution phase.

Second, multiple runs of evolution can be rapidly performed to watch somatic cell evolution in action [Heng *et al.*, 2006a]. This approach also allows for precise extraction of samples for analysis at a single-cell level of resolution, which can aptly profile the overall status of the disease. We have successfully demonstrated with this approach that both macro- and microcellular evolution can be observed by following the genome topology (karyotype) during each run of evolution.

Third, not only can we perform multiple runs of evolutionary experiments in parallel, but we can also compare the patterns of each run with similar or different initial genome systems and incorporate various environmental conditions to either slow down or speed up evolution. In both *in vitro* and *in vivo* experiments, experimental manipulations can be easily applied and different levels of genetic organization can be compared such as gene mutations and genome level alterations. Interestingly, we have demonstrated that each run of evolution results with its own, unique products at the genome level. Taking into account the vast amount of modeling and molecular efforts that have been performed in the field of cancer research, with each experiment representing a single "run," the abundance of heterogeneous data retrieved has demonstrated that individual genes and pathways hold very limited explanation power for the general cancer evolution. This is disconcerting, considering that the field of cancer research is firmly founded on molecular-based theories. Application of cancer models has shed insight on the evolutionary process, and we conclude this chapter by discussing some of these examples.

7.7.1.1 *The inconsistencies between the stepwise evolutionary concept and the clinical reality challenge the current thinking of cancer evolution*

As mentioned in Chapter 5, most evolutionary researchers have directly adapted Darwinian evolutionary concepts when studying cancer. Such practice has actually ignored the key difference between somatic cell evolution (the combination of macro- and microevolution) and micro-evolution for organisms based on stepwise Darwinian adaptation. That might be the reason why most clinical data do not fit the prediction of Darwinian evolution, as the current evolutionary framework is mainly based on and focused on microevolution, where genetic alterations at the gene and epigene levels are key factors. To understand the limitation of using current evolutionary concepts to explain cancer, we have to search for answers beyond microevolution and the assumption that the accumulation of micro-evolutionary events (e.g. gene/epigene mutations) over time leads to macro-evolutionary events (e.g. karyotypic alterations).

7.7.1.2 *Cancer speciation cannot be explained only by stepwise microevolution*

As discussed in Chapter 5, the evolution-in-action experiments resulted in the realization that genome-mediated cancer evolution occurs in cycles consisting of two phases, the NCCA-dominant punctuated phase and the CCA-dominated clonal expansion phase or stepwise Darwinian phase. Separating these two phases was important, as it suggested that cancer evolution differs from Darwinian evolution. Specifically, extending the two phases to organismal evolution provides a genome-based understanding of the relationship between micro- and macroevolution. Here, microevolution represents a tinkering and refinement of a particular lineage, whereas macroevolution can describe the formation of new lineages and speciation events through large genome-level changes. As soon as the concept that cancer represents new species is accepted, the generation of NCCAs can be considered as the formation of new species, and domination of CCAs can be thought of as population establishment following speciation. The relationship between parental cell population and NCCAs and CCAs can be considered as the entire process of speciation. Of course, there are some differences when compared with organismal speciation. For sexual species, they need to overcome a key challenge, which is to find a partner with the same genome (an extremely low probability event). Many asexual species do not have the typical chromosome-defined genome. Thus, the speciation process could also be slower for asexual species, as most of the changes are similar and will not separate them from the main domain of the parental species.

While it is not widespread, some have asked whether cancer evolution should be explained by traditional Darwinian selection. Since there are not significant accumulations of small evolutionary changes in cancer [Germain, 2012], and punctuated macroevolution cannot be explained by the accumulation of small changes over time [Heng *et al.*, 2006; 2013b], a new framework for the mechanism of organismal evolution is needed.

7.7.1.3 *Importance of non-clonal chromosome aberrations and outliers in evolution*

By considering each NCCA as a potential founder of a new cellular species, NCCAs, traditionally disregarded as noise, suddenly become the dark horses of cancer evolution. Increased NCCAs mediated genome heterogeneity yields higher macro-evolutionary potential at the species level, rather than the individual cell level within a given cellular species. This increases the robustness of the cellular species for continued persistence through drastic environmental changes (including drug treatment). Survival at the species level differs from the small and gradual adaptation that occurs within the species level, as the survival landscape differs from the fitness landscape (see Chapter 5).

Recent single-cell and population-based experiments have further demonstrated that, in a heterogeneous population, genome level outliers can strongly influence the growth and progression of the cancer cell population. This study also revealed that application of averaging methods to a heterogeneous disease has limitations. These include the disregard of disease instability and presence of influential outlier subgroups within the population. This study not only determined that an average profile is not representative of the entire population, but the average does not likely exist within the population. This study further supported that a single cell resolution-based understanding of the disease can aptly profile the overall status of the disease.

The above discussions extend to organismal evolution in terms of both speciation and extinction, which differs from micro-evolutionary adaptation, where variance within a population allows for increased environmental niche occupation while increasing the odds for survival against environmental change. When drastic changes are required for survival, gene/epigene level changes alone are not sufficient enough, and selection favors the formation of new systems (species) with new genomes and the extinction of old genomes.

7.7.1.4 *Cancer illustrates the importance of system inheritance and fuzzy inheritance in evolution*

Application of cancer models led to the discovery of new forms of inheritance. These came with the understanding of the importance of genome topology in terms of governing and defining the genetic network structure or system of the cell. As the genome organizes the interactive relationship among genes, including those of the same chromosome along with those of other chromosomes, and the overall karyotype is passed down the lineage, the inherited interactive network represents the "system inheritance." This is opposed to "parts inheritance," which describes the individual genes and sequences that are passed from parent to offspring or mother to daughter cell. This importance of this concept is further supported by the genome constraint function of sexual reproduction. In addition to genes and stretches of DNA sequence that are passed to offspring, what is crucial is that the genome topology is maintained and conserved as a result of sexual reproduction. Species-specific karyotype counts are consistent over generations and generations, as is species identity.

Furthermore, "fuzzy inheritance" was recently introduced as a new type of inheritance (see Chapter 6). Fuzzy inheritance, where a degree of heterogeneity is inherited but the specific genetic/non-genetic alterations vary, explains the mechanism of inherited heterogeneity. In cancer, this new understanding is highly important, as heterogeneity is the key feature of cancer especially during the punctuated phase of genome-mediated cancer macroevolution where karyotypes are untraceable. In this specific phase of cancer evolution, the unstable individual cell can produce a heterogeneous population, which makes it extremely challenging to avoid drug resistance. In organismal evolution, however, fuzzy inheritance is less obvious at the germline level, but apparent at the somatic cellular level. Due to the limitation of the fixed genome, the needed dynamics of somatic cells can be achieved by fuzzy inheritance during somatic cell division. Furthermore, fuzzy inheritance might also explain the individual variation mediated population dynamics.

7.7.1.5 *While there are multiple interactive genetic levels of heterogeneity, macroevolution is mainly driven by genome heterogeneity*

Studies of cancer have revealed vast degrees of heterogeneity, including heterogeneity at multiple genetic levels (e.g. karyotypic, genetic, epigenetic). For example, observed transcriptome elevation was associated with karyotypic alteration, suggesting that genome reorganization (including that observed during genome chaos) results in network restructuring and rewiring [Stevens *et al.*, 2014]. Evidence in yeast studies strongly supports that aneuploidy directly affects gene expression, resulting in phenotypic variation [Pavelka *et al.*, 2010]. Thus, karyotypic alteration influences gene expression profiles across the whole genome, meaning that genome-level alteration is more effective at drastically changing the genetic system than individual gene mutation or epigenetic change. Understanding this is crucial for evolutionary study, as focusing solely on lower genetic levels will only yield relevant information during some micro-evolutionary processes and will not be informative during macroevolution.

7.7.1.6 *Late stage cancer offers little information regarding disease progression*

Cancer progression is highly dynamic reflected by many NCCA/CCA cycles influenced by micro-environmental changes and tissue/immunological constraints. Yet the late stage products (such as the karyotypes or gene mutation profiles) only reflect limited changes, as the cancer evolutionary process is not just simply accumulating all small changes. The high level dynamic potential cannot be inclusively listed. Further, overall tumor cell growth could be driven by outlier subgroups (especially as a result of the punctuated phase), changing the overall profile of the entire population at a particular time of evolution. By the time tumor cells are analyzed in the later stages, they likely will display growth advantages achieved by microevolution. However, during the entire evolutionary process, many rounds of success might not link to growth at all, but rather other key features of cancer. For example, we have illustrated that an increased degree of genome heterogeneity (the most important feature of cancer) often reduces population growth. While the identification of overexpressed or "driving" genes within cancer samples is certainly feasible, it offers very limited clinical information as the roles of those genes vary within the tumor along with the genetic profiles. More challenging, these moving targets are not fixed under clinical treatment. Knowing the difference between the profiling of end products and the cancer process, cautions are needed to explain the history of organismal evolution where most of the process is not accessible.

7.7.1.7 *The phenomenon of genome chaos provides understanding for organismal evolution*

Recall that genome chaos is a common phenomenon in cancer that involves genome reorganization within a short period of time, often induced by high stress. The products of this process are altered karyotypes, representing different cellular species. Since genome-level reorganization is key in cancer speciation rather than a gradual accumulation of gene-level alterations, following the genome and population dynamics associated with genome chaos and later selection is an excellent system to employ in evolutionary study. A unique advantage of this model is that it allows for easy observation of macro-cellular evolution (Chapter 5).

In contrast, many speciation studies including field research are actually focused on microevolution rather than macroevolution. For speciation to occur in sexual organisms, new genome systems have to emerge, and individuals must find genomically-compatible mates in order for that new species to successfully persist. However, given the constraint applied to the genome in sexual organisms coupled with the low probability of finding an appropriate mate, speciation will not be easily observed, except for situations under extremely high stress that might push species to extinction.

Prior to directly studying cancer genome evolution, we had much less appreciation and understanding of how drastic genome alteration can happen during organismal evolution and how the vast majority of altered genomes are not viable and thus unknown to us. This is because we can only study extant species. For extinct species, the karyotype information is not available and only having DNA is not enough to illustrate the genome evolution, at least based on the current technologies. Despite the fact that cancer evolution, due to the lack of the sexual filter, can be much more dynamic than organismal evolution, the overall principles should be the same. For example, altered genomes can be detected frequently from both sperm and egg, but the chance of fertilization between sperm and egg with the same genome alteration is extremely low.

Interestingly, high stress conditions have been linked to an elevated gene mutation rate in bacteria, similar to the genome chaos detected in cancer cells. When the stress is too high to adapt, the survival strategy is to actively produce altered genomes. Despite the fact that the vast majority of massive chaotic genomes will be eliminated, some of these will survive. By creating new genomes followed by modifying the same genome with gene mutations, cancer cells are applying different evolutionary strategies through either macro- or microevolution. This information should apply to organismal evolution as well, as the adaptive and survival landscapes are different.

7.7.2 *Application of ecological approaches to study cancer evolution*

Studying cancer evolution within an ecological framework is becoming a popular approach [Basanta *et al.*, 2012; Merlo *et al.*, 2006; Yang *et al.*, 2014]. Ecology and evolution are intimately linked. As ecology encompasses the study of interactions among organisms and their environments, it provides the players and the scene for the evolutionary process. In typical ecological conditions, complex organismal interactions involve individuals from the same and different species. We thus cannot just consider cancer as a collection of mutated cells of the same type (or cellular species), but rather as an adaptive system consisting of interactions within and between species. The idea of considering cancer as an evolving species can provide the needed attention to correctly study the ecological evolution of cancer. In fact, many current modeling attempts fail to differentiate genome and gene variation, which represent an important limitation for understanding cancer evolution when discussing key features of the cancer ecosystem. This includes, but is not limited to, restricted resources, nutrients and space for growth, predation, parasitism, mutualism and commensalism.

Chapter 8

Facts vs. Myths

8.1 Summary

Distinguishing facts and myths is essential for cancer research. Since this field is highly diverse and most researchers only focus on select model systems or fixed molecular aspects within a very specific sub-field, making these distinctions largely depends on the results and claims from other investigators, especially when dealing with knowledge from different disciplines. As we illustrated in previous chapters, due to the limitations of the current theory of cancer as well as various linear models and molecular average profiling methodologies, many popular "facts" regarding cancer are actually myths. These myths have not only slowed down the understanding of cancer biology, but also misled the field by confusing exceptions with general rules, which has created many false hopes such as the search for a (non-existing) magic bullet. In addition, many cancer researchers have a pre-conceived gene-centric view of cancer, which leads them to only believe the facts that fit well to their own concepts, despite being at odds with the facts. By applying the new genome-based concepts we described in previous chapters, it is now easier to re-examine many important "facts" in the field, and to demonstrate that many of them are actually myths. Even though only seven topics have been selected as examples, the same evolutionary principles and genome-based analyses are repetitively used to achieve critical thinking.

8.2 The Usefulness and Limitation of the Oncogene Addiction Concept

Oncogene addiction is a hot topic of current cancer research. It has also been used by many as strong evidence to support the concept of the oncogene. Since its introduction, it has served as a key rationale for targeted anti-cancer therapy based on some oncogenes [Torti and Trusolino, 2011]. For example, it is believed that oncogenes not only initiate the oncogenic state, they also maintain it. Cancer cells die when there is no continued oncogene expression. Several tumor maintenance genes are often mentioned, including C-MYC in T-cell myeloid leukemia, H-RAS in melanomas, BCR-ABL in B cell tumors, MET in hepatomas, C-MYC, NEV, and WNT-1 in mammary tumors and K-RAS in lung adenocarcinomas [Koboldt, 2008]. In fact, targeting BCR-ABL in the chronic phase of CML (CML-CP) represents the most successful model of targeting oncogene addiction in the clinic. However, despite over a decade's worth of studies and high hopes, the real mechanism of oncogene addiction is not well understood and clinical results vary [Luo *et al.*, 2009; Weinstein, 2002].

The initial observation/idea was that despite extensive genetic changes in cancer cells, there are examples where the reversal of only one or a few oncogenes can strongly inhibit the growth of cancer cells and, in some exceptional cases, can improve patient survival. However, there are important facts that challenge the significance of this concept in particular for clinical implications.

8.2.1 *Challenges for oncogene addiction*

First, oncogene addiction is not universal. This phenomenon is associated with a few strong oncogenes within model systems where cellular proliferation often is a dominant feature. However, as we discussed, proliferation is not the only key factor. In contrast, genetic heterogeneity is (see Chapters 5 and 6). Despite the fact that oncogene addiction can often be illustrated in mice models, clinical reality is not usually that simple, and some models might be exceptions rather than the rule [Jonkers and Berns, 2004]. For the same cancer type where this phenomenon is applicable, it can only be detected within a subset of patients sharing a distinct etiology. For example, good responses of the EGFR inhibitor gefitinib against non-small cell lung cancer are only observed in 10–20 percent of patients. Even more interesting to mention is that for many publications that discuss the clinical implications of oncogene addiction, they always use the same success story of targeting BCR-ABL in CML-CP as the example. It is now well understood that the BCR-ABL story represents a real exception and only works for the slow growing phase prior to the blast crisis stage when the genome become extremely unstable [Horne *et al.*, 2013a]. Obviously, the clinical implication so far is extremely limited and cautions are needed when applying the CML story to infer the vast majority of solid tumors. On the surface, initially reduced cellular population size during drug treatment is the key marker of success and support for targeting the oncogene. However, the truth at the fundamental level is that the ultimate challenge of cancer drug treatment lies in the survival of the minority of cells that display newly formed genome-defined systems, which, as we discussed, is not determined by the behavior of a specific oncogene but the behavior of genome re-organization. In other words, there is life after death. When oncogene addiction is effectively targeted, many cells will be eliminated, and then the survival mechanism kicks in through genome chaos. This process produces new cell populations with newly formed genomes, making the original targeting now off-target.

Second, even in mouse models, unpredictability is high when studying oncogene addiction. For example, the suppression of MYC was well known to induce sustained tumor regression, strongly supporting the concept of oncogene addiction. However, for some models, after following prolonged inactivation of MYC in a conditional transgenic mouse model of T-cell acute lymphoblastic leukemia, a proportion of the tumors recurred. This is similar to the clinical situation where targeted therapies often fail. Interestingly, lymphomas that recur following MYC suppression continue to exhibit oncogene addiction [Choi *et al.*, 2011]. This observation can be explained by the importance of genome alterations and emergent new systems beyond the oncogene addiction mechanism, as genomic rearrangement seems to play the dominant role here, and gene expression profiles vary despite some similarity. Similar to the situation where it is much easier to link the contribution of specific oncogenes to cancer using linear experimental models, demonstrating the long term effect of targeting oncogene addiction is challenging even in animal models, as shorter term observation can ensure a more linear relationship with better prediction power. The fact that the recurred tumor continues to be addicted to MYC also suggests the possibility that while microevolution in these models often favors the oncogene with proliferation advantages, the recurrence event itself likely involved genome level alterations in the first place. It should also be pointed out that, despite the fact that this mouse model was related to the oncogene addiction phenomenon, it is very different from targeting oncogene addiction when genome chaos dominates (Chapter 5). Again, as illustrated in the example of using imatinib, targeting BCR-ABL oncogene addiction is only highly effective in CML-CP, but not the later, blast crisis phase where genome chaos dominates [Horne *et al.*, 2013a].

Third, so far, there is no common and convincing mechanism to explain oncogene addiction. It was proposed that this unique phenomenon is not simply a summary of the individual effects of multiple oncogenes and tumor suppressor genes, but some "bizarre" mechanisms which differ from those in normal cells (such as strange combinations of gene interactions within disordered cell circuitry). To explain why oncogene addiction is not observed from normal cells that also express the same oncogenes, some complicated ideas were suggested. One explanation is that there is a balance between proliferation-enhancing effects of a specific oncogene and another proliferation-inhibiting factor linked by negative feedback. By inactivating the oncogene, it might cause a relative excess of the latter inhibitory factors and thus lead to cell death. Another often cited idea is "synthetic lethality" (mutation of one of two genes is survivable but mutation of both genes results in cell death). Since normal cells differ from cancer cells in terms of different synthetic lethality relationships, and cancer cells are highly dependent on specific oncogenes due to the large number of gene mutations already existing that normally function in other pathways, cancer cells are more dependent on oncogenes than normal cells [Weinstein and Joe, 2006].

Even though their explanations seem reasonable under specific conditions, and there are some cases to support them, they are associated with many inferred and unknown details. By and large, they build upon the specific gene/pathway viewpoint, and mainly focus on the initial response of the cellular population following oncogene targeting rather than the long term effect following treatment where cellular evolution is the key. Furthermore, these ideas do not seem to represent general mechanisms, as there are many diverse examples that can be explained otherwise. If this gene-centric view of oncogene addiction is correct, we should have a large number of molecular targets for drug design. However, the very limited examples in the clinic have forced us to ask, why is this not true? Should we question the concept of oncogene addiction in the first place?

8.2.2 *Evolutionary selection defines oncogene addiction*

It is thus necessary to develop a general explanation for oncogene addiction and why it is only obvious in some cancer cells. From the viewpoint of cellular population growth and survival following a specific treatment (the perspective of cellular evolution), the mechanism becomes much less complicated, as the key answer is the change of the overall status of the cellular system under a given stress. This can be achieved either by changing the genetic/non-genetic components or their environmental constraints. All systems are more or less "addicted" to current stable status. Therefore, normal cells are addicted to the normal status that balances the contribution of individual genes. In contrast, for many cancer cell populations, when the relative stable status is achieved by specific oncogene generated growth, the breaking down of such homeostasis by targeting this oncogene could lead to initial cell death en route to establishing a new homeostasis. In other words, the so called addiction represents a cellular evolutionary mechanism: normal cell is addicted to normal status (when the environment is not normal, they stop dividing), and abnormal cell can be temporally addicted to an abnormal status, such as dominant oncogene generated tumor growth.

Since "status addiction" represents an evolutionary feature for cellular populations, it is expected that oncogenes and many non-oncogenes should display variable addiction. Surely, a new research direction is to study non-oncogene addiction. According to Elledge's group, non-oncogene addiction or NOA is also of importance, as (1) oncogenes closely rely on adaptations in cellular processes that are themselves not oncogenic; (2) cancer cells may display the addiction of genes that act in oncogenic pathways but are not themselves classical oncogenes; and (3) it is hard to distinguish many genes which are both oncogene or non-oncogene depending on the context. They have promoted the approaches to target both oncogene and non-oncogene addiction [Luo *et al.*, 2009; Solimini *et al.*, 2007]. They have classified non-oncogene addiction genes into two categories, tumor intrinsic and tumor extrinsic. "Whereas tumor-intrinsic NOA genes support the oncogenic state of the tumor cell in a cell-autonomous manner, tumor-extrinsic NOA genes function in stromal and vascular cells that provide heterotypic support for the tumor."

Clearly, many features they discussed are focused within a gene-centric view. If a broad evolutionary view is considered, both the oncogene and gene-based non-oncogene addiction would only account for a small portion of the "status addiction". Even though a large number of cancer genes can contribute to cancer, the dominant degree for each gene is low within the patient population. More importantly, the genetic landscape of each tumor is constantly changing especially under targeting, making predictions based on individual gene mutations difficult, if not impossible. In comparison, some non-genetic types of addictions are more consistent and powerful. Finally, phenotype selection can involve many genes in a stochastic fashion (among different individuals and even within the same tumors). Together, they result in the clinical fact that targeting gene-based addiction is very limited, despite having so many potential molecular targets.

This analysis can also explain why tumor suppressor genes are even harder to target compared with some oncogenes. It is well known that pharmacological applications of targeting tumor suppressor mutations have lagged behind efforts of targeting oncogenes. One common explanation is that it is often difficult to use a small molecule to either restore or mimic the function of a protein that is either mutated or absent. However, from an evolutionary perspective, another important point is that many tumor suppressors function as system constraints. Differing from oncogenes that can be directly linked to proliferation, reducing constraint often has less direct (and therefore more diverse) effects. Such changes, despite their importance, are difficult to predict or even study, as the patterns are less dominant.

In summary, by studying the evolutionary process of cancer and the relationship with various changeable statuses or "addictions", the limitation of targeting "oncogene addiction" is obvious for a highly dynamic system. In the long term, all of the patterns of addiction will change and new "addictions" will take over, which reduce the value of these statements: "Oncogene addiction represents an ideal target to fight cancer" and "the fact that there is oncogene addiction supports the importance of the oncogene concept". As we discussed in previous chapters, many facts have pointed out the limitation of targeting oncogene addiction. These facts include the highly diverse results of the current cancer genome sequencing project which shows the limited number of common targets; the two phases of cancer evolution where manipulating oncogene addiction is only suitable for the micro-cellular evolutionary phase; and the drastic change in addiction status observed in the clinic following targeting, which differs from animal models. There are some remarkable examples of initial success from targeting addictions in animal models, but this success has failed to translate to the clinic. A well-known case was the result of targeting the tumor cells' addiction to its nutrition supply from environments (e.g. angiogenesis therapy by targeting VEGF). The molecular mechanism of how angiogenesis works is well illustrated, and the resulting effect of tumor elimination is wonderfully illustrated in animal models. However, the clinical result is not nearly as good as expected. In fact, it also could harm some patients by paradoxically favoring tumors [Heng *et al.*, 2010b]. Recently, increased efforts have focused on targeting the metabolic pathway and tissue architecture, as there are some obvious differences between normal and cancer cells. However, similar to targeting angiogenesis, the vigorous cancer evolution occurring in patients is likely to win by increasing its heterogeneity and creating new statuses for the cellular population, despite any success that may be observed in various model systems with reduced heterogeneity.

The rationale of discussing the limitation of oncogene addiction is to promote the importance of applying the evolutionary mechanism of cancer "addiction" over individual mechanisms of gene-mediated addiction and the usefulness of monitoring or manipulating the population dynamic status rather than focusing on a few oncogenes. Despite the long list of individual molecular possibilities involved and their complicated "bizarre" mechanisms, it is the simple evolutionary mechanism that changes the older status and establishes new statuses. In agreement with the "out of the cancer gene box" thinking, we need to pay more attention to various non-genetic addictions. Interestingly, many features of non-genetic addiction in fact are linked to genome instability. Genome instability can generate cellular heterogeneity through fuzzy inheritance and especially favor cancer cell population survival through generation and survival of outliers. Thus, the new paradox is the following: Targeting the system status addiction works better for those populations with stable genomes, which often display increased addictive behavior and response due to the micro-evolution occurring in these populations. However, effective targeting on the average population can only work in a limited time period, which can also lead to new systems with more aggressive features. A new approach is to use the "status addiction" to impact the phase transition of evolution. While controlling the size of the cellular population is important, so is not triggering system instability, as this would lead to the macro-evolutionary phase. In other words, when applying different constraints, it is crucial to not trigger macro-cellular evolution by drastically targeting the population. More research focused on how to constrain moving targets in a longer time frame is needed.

8.2.3 *Needed experimental approaches to understand oncogene addiction*

For this purpose, some systematic analyses are very useful to (1) Demonstrate that oncogene addiction is dependent on system homeostasis, which is established through the process of evolutionary selection. The general mechanism of oncogene and non-oncogene addiction is accomplished through diverse individual molecular mechanisms. The working hypothesis is that "addiction" can be generated by both genetic and non-genetic factors through experimental or natural selection, and oncogene addiction might be more dominant in experimental model systems where other variable factors are eliminated or ignored. To test this hypothesis, specific oncogene addiction should be compared with other forms of non-oncogene addiction using various culture conditions or treatments. In particular, a given type of addiction should be systematically targeted in multiple ways to compare the system response. The system dynamics should be analyzed by comparing growth heterogeneity at the single cell and population levels. The following conclusions are expected: (a) Compared to normal cells, addictions should be easier to establish in cancer cells, and genome heterogeneity increases the probability of the acquisition of addiction in a system under evolutionary selection. In such evolutionary context, the targeting of oncogenes leading to cancer cell death is not a unique feature. (b) Non-genetic addictions such as those to environmental conditions, when targeted, can be more effective in killing cancer cells than normal cells. Due to the heterogeneous nature of cancer cells, alteration of many factors will trigger cell death. However, the emergence of a new genome system possibly through genome chaos (creating new genome systems under stress) or simple selection of previously existing genomes (in the form of non-clonal chromosome aberrations or NCCAs) will always be a consequence of massive cell death. (c) When multiple factors are involved in acquisition of addiction, it may not be worthwhile to illustrate the contributions of each single factor. Even though targeting oncogene addiction can effectively eliminate the dependent population, new addictions can quickly emerge for the whole population.

(2) Demonstrate that the degree of genome heterogeneity and the resultant evolutionary patterns impact the result of targeting oncogene addiction and environmental addictions. It is known that there is a correlation between oncogene, environmental addictions and the degree of genomic heterogeneity. Therefore, the working hypothesis is that targeting oncogene addiction and environmental addiction is most effective during the stepwise micro-evolutionary phase of cancer, and in contrast during the punctuated evolutionary phase, high levels of genome heterogeneity increase the probability of developing resistance to the targeting approach. Paradoxically, effective targeting applied during the

stepwise phase could trigger a transition into the punctuated phase, and this poses a major challenge for the targeting addiction approach. Unfortunately, in most cancer types, most treatments will cause genome chaos leading to the punctuated macro-evolutionary phase where genome replacement occurs rapidly and frequently (making targeting oncogene addiction less effective).

These demonstrations are urgently needed. First, they can unify the field by linking the known diverse molecular mechanisms of oncogene addiction to genome system status (stable or unstable under which evolutionary selection acts). Currently, most efforts focus on searching for individual molecular mechanisms. Second, they offer a methodology to monitor the system status using the degree of cell population heterogeneity at multiple genetic and epigenetic levels, and in particular at the genome level (determined from individual cells and cell populations). The comparative analysis of individual cells and population growth/death has and will further reveal significant information regarding growth and death heterogeneity. Furthermore, the concept of determining the system status through two distinct phases that involve gene and genome alterations respectively is of importance to understanding cancer evolution. Third, they explain the paradox between the initial effectiveness and subsequent resistance of the same treatment when targeting oncogene addiction. While inducing cell death represents a necessary means to weaken cancer cells, a byproduct is the induction of high levels of genetic alterations in the system that results in the establishment of a new, stable cellular system which renders the original targeting strategy ineffective. Finally, previous studies have shown that there is a big gap in the effectiveness of targeting oncogene addiction when *in vitro* models and clinical realities are compared. It was thought that such differences were caused by the disparity between *in vitro* and *in vivo* systems. Our recent studies have demonstrated that the main reason is that most current *in vitro* systems do not accurately reflect the genome heterogeneity that exists in natural or *in vivo* systems. By using well-controlled *in vitro* or animal systems it is much easier to achieve a more homogenous system than in real patients, which increases the differences in system behavior. Correctly analyzing *in vitro* systems without drastically eliminating their system heterogeneity will reduce the system behavior disparities between *in vitro* and *in vivo* approaches. It is anticipated that these analyses will prove this point.

8.3 Do Aggressive Growth and Invasion Make Cancer Cells the Toughest Cells?

Due to its aggressiveness and relentlessness, cancer is one of the most feared diseases. The general public considers cancer cells as ruthless monster cells. Also, the experimental success of many cancer cell lines especially the Hela cell line (can easily grow *in vitro*, contaminate other cell cultures and overgrow normal cells that hold a limited division number, and can be passed for decades worldwide) also gives researchers the impression that the cancer cell is the toughest cell of all. This impression has undoubtedly impacted our attitude and means towards fighting cancer. For example, for killing cancer cells, we chose to hit them using the hardest and deadliest weapon with the highest dosage. Since cancer cells are considered the toughest, any method that can kill more cancer cells than normal cells must be good. The overall killing power has become a standard benchmark in the screening of potential agents as therapeutic options.

As discussed in section 8.5, the approach to maximize initial killing has not been working so far in the clinic for most cancer types despite ample success using various *in vitro* and *in vivo* models in laboratory settings, as drug resistance seems more robust with effective killing. Interestingly, the limitation of current cancer treatments by aggressive agents is directly related to the popular myth of how tough an individual cancer cell is. In this section, we will examine this myth through the lens of cancer evolution.

8.3.1 *Facts supporting that cancer cells are often less powerful than we imagined*

There is an interesting paradox regarding cancer cells' robustness. On one hand, they are difficult to cure clinically when metastasis has been achieved. On the other hand, most cancer cells are easily destroyed under experimental conditions, and in fact, it is not easy to culture most cancer cells when isolated from patients. The following facts support this paradox:

First, in contrast to its aggressive reputation, it is still challenging to culture primary cancer samples and establish cancer cell lines that reflect the original cancer population. When the selection environment changes (e.g. being transferred from an internal organismal environment to culture conditions), the cancer's robustness could be lost for many cases and coupled with massive cell death. Similarly, despite the fact that cancer can be transplanted among animals, it is a very challenging process. That is the reason why immuno-deficient mice are often used *in vivo* studies with a high rate of success.

Second, despite their aggressive growth phenotype *in vitro*, many of the lab-created cancer cell lines through various molecular manipulations (from gene mutation to carcinogen treatment) fail to produce tumors in "normal" mice. Again, immuno-deficient mice are needed to increase success. The constraint of the immune system can block the aggressive growth of most cell lines. Even some environmental settings can reduce the penetration of cancer phenotypes [Cao *et al.*, 2010].

Third, together, these *in vitro* and animal data support the fact that it is not easy at all for cancer cells to successfully evolve into clinical full blown cancer. If the cancer cell was very tough, the situation would be the opposite. Considering that there are so many stress-induced genetic and non-genetic variations within so many somatic cells undergoing many runs of division, the overall cancer rate is remarkably low. In addition, many normal examined tissues already display multiple levels of genetic variations. For example, genome chaos can be observed from embryogenic developmental stages, and high levels of polyploidy can be observed from normal, healthy tissues, including liver. This also supports the limited power of cancer cells, as it takes the perfect storm to get cancer. It is well known that most cancers require many years or decades to fully finish formation. One generally accepted impression is that cells from earlier stages of cancer evolution are less aggressive than the late stage cells, and the later stage of cancer seems have more certainty in terms of out-of-control growth. However, it all depends on if we judge this situation under an evolutionary context based on the individual cells or the population.

8.3.2 *Power of cancer is shaped by somatic cell evolution*

Most altered cells should have disadvantages when compared with normal cells, so how do these disadvantaged cells become so aggressive? In other words, how do generally weak cells become monsters? The traditional idea was that advantages developed as a result of gene mutation accumulation through cycles of clonal expansion, which is now unlikely due to the realization of punctuated cancer evolution. The analyses of the two phases of cancer evolution revealed that the key of the punctuated phase is to pass altered genomes. Continuous genome alteration provides the heterogeneity that is essential for cancer evolution. Despite massive death, this will likely result in new genomes that could survive and become dominant at least for a short period of time, resulting in a population ready for the next wave of high heterogeneity. Clearly, genome-mediated evolution makes the cancer population powerful by preserving the potential of outliers and favoring their success in the changing environment.

During this process, in contrast to the popular viewpoint, most individual cancer cells are weak and are easily killed by many treatments. In fact, for those highly dynamic cancer cell lines, even without any treatment, large portions of the cell population are constantly dying. That is how cancer cells pass their fuzzy inheritance and continue evolving. Therefore, observed increased short-term cell death by treatment is not a big deal for these cancer cell populations in the long run. Highly dynamic cell populations are paradoxically harder to kill despite displaying high levels of death. The high level of heterogeneity would suggest that the population should be sensitive to any effective treatment coupled with the increased base line of individual cell death, however the population also gains robustness associated with rapid turnover. This is the power of a high level of heterogeneity, which achieves the success of the population by sacrificing individual cells. As long as selection is focused on the survival of the population, only high degrees of heterogeneity best serve the goal.

For each major step of cancer evolution, from immortalization, transformation, metastasis, and drug resistance, they use the same strategy of increasing heterogeneity (especially at the genome level) to achieve this goal. This strategy drastically differs from normal cell response, as the macrocellular dynamics observed under pathological conditions differ from normal physiological regulation.

8.3.3 *The mechanism of high level heterogeneity ensures the robustness of cancer*

Now here is the paradox: weak cancer cells actually promote or strengthen the mechanism of the evolutionary process where the generation of high level heterogeneity is the driving force [Heng *et al.*, 2013a–b]. The following are some specific mechanisms that contribute to this paradox:

8.3.3.1 *The importance of fuzzy inheritance*

As we discussed in Chapter 6, using the error accumulation theory to explain the high level heterogeneity in cancer does not fit, as the mistake rate is very low for a biological system. In contrast, fuzzy inheritance serves as the mechanism for generating and maintaining the degree of heterogeneity. Here, the high level of heterogeneity is not just caused by mistake, but rather by the specific mechanism that is essential and selected by evolution for cellular adaptation. When somatic cells are unable to deal with high stress using their available pathways, altered genes/epigenes/genomes become necessary in order to do the job. Fortunately, the cells' fuzzy inheritance functions to help. Since the fuzzy inheritance already produces some altered genes/genomes, these different cells can help the population's adaptation. Obviously, the stress conditions are constantly changing, and the capability of fuzzy inheritance fits this well by producing different genetic changes. That is the reason why fuzzy inheritance can ensure the maximum cancer evolutionary potential.

The result of fuzzy inheritance is the production of a certain level of altered cells in a dynamic fashion (allowing for constant introduction of newly altered genes/genomes). By constantly modifying the genetic information of a certain portion of the cell population, it prevents most of the altered genetic packages from becoming too dominant and competing with normal cells. However, as the environment is still changing, the best strategy is to fuzz up the system a bit. The question is then, how do some altered genes/genomes suddenly gain an advantage and switch to a low fuzzy inheritance model? It is reasonable to imagine that among the many altered genomes within a population (reflected by the large number of NCCAs), some of them will have better fitness than the surrounding cells, and then clonal expansion takes off. More general cases might involve the high levels of stress these cells face. The altered genomes are constantly facing challenges from the normal cellular system (cellular competition, immuno-attack, etc.). This high pressure likely serves as the evolutionary force to favor the success of some outliers. This idea fits the observation that, when without immune pressure, some altered cells will not evolve into cancer. When some altered cells form a significant population, the host will try everything to eliminate them. The battle is on. These cell populations then will switch to the high fuzzy inheritance model and push the population into new systems through this high level of heterogeneity.

8.3.3.2 *Heterogeneity-created cancer society*

One important feature of cancer evolution is the heterogeneity-created cancer society. Due to the high diversity of cells comprised of multiple genetic and non-genetic backgrounds, the cancer society is highly dynamic. They do not have fixed genomes, most of the cells are not equal in terms of their contributions to overall population growth and survival, and they can co-exist but will not separate from each other. We have collected some drug-resistant cell lines following drug treatment. Some clones can be grown continuously for years but never display pure clones, despite other clones displaying clonal features. Within the cancer society concept (see Chapter 3), it becomes clear that the whole (population) is not equal to the sum of all individual cells. Such cancer society has obvious advantages when compared to normal cells. Thus when the "matured" cancer society is formed, it is very hard to fight against. Cancer cells can always change to different cells to escape from treatment, but the normal cell cannot. When the normal cell changes, they will become bad cells, a no win situation. New data show that the response to drug therapy is less chaotic for these altered cells, representing an earlier stage of cancer evolution rather than a late stage. In general, aggressive cancer cells display much higher evolutionary potential. In addition, within the framework of cancer society, fragile individual cancer cells assemble robust populations by heterogeneity mediated complexity, a new layer of bio-complexity (see Chapter 6).

8.3.4 *Potential implications*

The correct understanding of the relationship between the individual cell and population, and especially the dynamics under evolutionary selection, will change many concepts and approaches for fighting cancer. When studying cancer evolution, we can no longer ignore heterogeneity and especially the outliers, as these outliers are the key reason for the success of cancer evolution as well as the failure of most drug therapies.

Dynamic evolution also explains the paradox of weak and strong features of cancer. Since evolutionary selection makes the weak cells strong, when the evolutionary selection is changed (by simply changing the environment such as removing cancer from patients and placing into culture dishes, or by reducing the competition from surrounding cells or immune system, or adding a new variable into the equation), the power of the current population will soon be lost, which displays weak features. Of course, there is potential to regain high fitness through fuzzy inheritance-mediated heterogeneity, but it is not a sure thing. In a sense, evolutionary selection makes weak cancer cells strong, and the change of the evolutionary condition also can make the strong population weak. That is the rationale behind changing the evolutionary dynamics to weaken cancer cells.

Knowing that the initial effective killing of cancer cells can induce genome chaos and promote more aggressive cancers to come back, we now need to reevaluate our strategy that focuses on the maximum elimination of cancer cells. In contrast, we might balance initial killing and the long-term benefits. In particular, the "hit hardest" approach is best at initial killing in terms of reducing tumor cell numbers. But massive cell death also triggers the mechanism of increasing heterogeneity and the production of survivable new genomes. This has been illustrated by the genome chaos experiments as well as drug resistance experiments (Horne *et al.*, submitted). To develop treatment strategies that account for surviving cancer cell populations long after treatment, we cannot focus solely on the number of cells we can eliminate initially, as we must also consider the induced level of aggression of the surviving cells.

The realization that outliers can easily dominate under high selection pressure also explains why targeting specific pathways is not very successful in the clinic. As outlier formation displaying different pathways can be easily induced through targeting approaches, fast domination of these induced outliers with new pathways can make the previous approach now off-target. Similarly, for using genetic profiling for diagnosis, the overall information of the genome dynamics (reflected by the degree of heterogeneity and the complexity of the chaotic genome) is more useful than the average profiling information of the cell population. One of the biggest complaints about cancer cells is that cancer cells do not follow the rules. These outliers, which represent different runs of evolution with different combinations of pathways, for sure are ruthless.

8.4 Are *In Vitro* Models Unreliable?

The current molecular understanding of cancer heavily relies on *in vitro* systems. For example, many biochemical characterization and genetic manipulations have been carried out using various cell lines. While it is more convenient to access cell lines than clinical samples (also in many cases, well-characterized cell lines represent the only accessible materials for many basic researchers), high degrees of variation have been observed from most *in vitro* models. Large scale –omics studies have further highlighted this issue.

The recent cancer genome sequencing project revealed a high level of heterogeneity from most clinical samples, which challenges the current theory of cancer (see Chapter 4). Rather than asking hard questions in order to search for the new conceptual framework to better explain these findings, some simply complain about the various cell culture systems, as if the problems will be solved if we use more *in vivo* mouse models to avoid cell culture "artifacts." According to them, the current confusion is caused by these cell lines. When "better" *in vivo* systems are used, the stepwise cancer progression as well as a small number of common targets can be finally identified. Along the same line of thinking, there also is a new trend of using 3D culture to replace 2D culture, as 3D should be closer to reality. While these "less artificial models" are helpful by incorporating more layers of complexity of cancer, one inconvenient truth is that the inconsistencies lie far beyond the differences between *in vitro* and *in vivo* models, as there is already a huge gap between mouse model results and real clinical responses. Furthermore, even for the same cancer type, cancers differ among different patients, and for the same patient, the genetic profile of cancer is constantly changing. In reality, the heterogeneity is much higher in real cancer than in established cell lines! The real artifacts of a cell line are not the result of too many changes but are actually from the artificial reduction of the true complexity of cancer.

So the key limitation of using an *in vitro* system is not artificially increasing the variation to an overwhelming degree, but the opposite: reduced heterogeneity. Therefore, the issue is not necessarily the widespread replacement of *in vitro* systems with *in vivo* systems, but using the correct concepts and approaches when applying these systems as well as taking full advantage of *in vitro* systems. Clearly, both *in vitro* and *in vivo* models can be reliable and unreliable, but the key is using the correct platform to conduct the research.

8.4.1 *Comparing in vitro and in vivo cells*

One general assumption is that *in vitro* culture conditions can generate many artifacts, increasing cellular diversity. A common concern may be: "You can observe so much more change (*in vitro*) that does not happen *in vivo*, as cells in tissues are well behaved." However, this comment would be an oversimplification and misunderstanding of tissue cell heterogeneity, which indeed exists and depends on many factors including the pathological status of the tissue, phase of cellular evolution, overall stability of the cell population and selective conditions. In fact, there is so much diversity in cancer samples and in a large number of normal tissues, but this has been traditionally ignored by many [Biesterfeld *et al.*, 1994; Ye *et al.*, 2007].

In general, *in vitro* cultured cells live in a relatively stable and simple environment, as culture conditions (e.g. types and concentrations of growth factors in media, temperature, and cell density) can be easily adjusted compared to the much more complicated interactions of multiple systems that occur *in vivo*. On the other hand, the tissue architecture and immune system can serve as strong constraints on cells and very dynamic evolutionary selection must be accounted for as well, as this could both reduce and enhance cellular diversity.

If researchers want to compare specific pathways of cancer, the differences between *in vitro* and *in vivo* systems need to be clearly studied, as culture conditions can impact these pathways. For example, O_2 concentrations, 2D culture and cellular interactions among multiple types of cells are all clearly different compared to *in vivo* systems. Not surprisingly, it was recently reported that different *in vitro* cell culture conditions could quickly lead to genetic and non-genetic changes in human embryonic stem cells, and these changes could alter the cells' clinical utility [Garitaonandia *et al.*, 2015]. It has also been known to cytogeneticists for decades that the karyotypes of established cell lines are often very different from the initial samples where the cell lines were derived from [Ye *et al.*, 2007].

Interestingly, if we take into account: the realization of high heterogeneity-mediated evolutionary complexity in individual cancers, the fact that there is generally low penetration for a given pathway among the patient population, and that most cancer cells have their own unique combinations of pathways, the *in vitro* and *in vivo* difference becomes less significant than previously thought, as the difference between cancer cases (the *in vivo* and *in vivo* difference) is already remarkably high! Even if we can add 1–2 key factors (like 3D culture or stroma cell interaction) back into the *in vitro* models to reduce the difference between *in vitro* and *in vivo* systems, the degree of complexity will not drastically alter, as the overall complexity is so high already in these *in vitro* systems. In other words, even if all systems we used to study cancer are *in vivo* models, we still cannot solve the problem of cancer variations in the clinic, as the gap between *in vivo* models and the clinic is much larger than the difference between *in vitro* and *in vivo* models.

In contrast, if we do not focus on the linear mechanisms such as a "fixed" pathway or specific gene mutation profile (which are often case specific), but rather study and compare the system behavior, and/or the degree of heterogeneity or evolutionary potential, by considering both *in vitro* and *in vivo* as different systems, the similarity then becomes much higher along with the relevance of the research. As illustrated by our research, by ignoring some of the molecular details such as individual pathways or gene mutations, the complexity can be reduced by comparing the level of dynamics or evolutionary potential. Without knowing which pathways will be involved, just by measuring the degree of genome heterogeneity of the population, tumorigenesis can be predicted. Since the pathways switch among different cells and within a given cell is highly dynamic, it is rather hard to predict the evolutionary outcomes based on pathway information only.

8.4.2 *Be aware of the power and limitation of in vitro systems*

8.4.2.1 *In vitro systems are ideal for "parts characterization" but with limitations*

Various *in vitro* systems have provided the foundation of our current molecular knowledge of biology through characterization of bio-components and defining their functions in relative, isolated experimental conditions. For example, how is DNA replicated, repaired, transcribed; how does genetic information pass from DNA, RNA and protein; what are the essential factors involved; what does the abnormal process look like and what is its linkage to diseases? This knowledge forms the conceptual basis for our cancer research and should not be simply denied. At the same time, however, we need to be aware of the limitations of this knowledge. First, as we discussed in Chapter 5, most biochemical studies are based on the average profile of the mixed cell population, and important outlier data have been ignored. Homogenous or "noiseless" data do not provide a true representation of the biological system. Second, many popular cell lines are cancer cells, which could be drastically different from normal cells. Comparative studies are thus needed even for some very basic concepts. Third, due to the unstable nature of cancer cells, many cultures of these well-established cell lines are no longer similar to the original line. As a result, some researchers actually work on very different cell lines than intended since most molecular researchers do not examine the karyotypes. Even some journals now request for cell line validation prior to publication, but they only focus on reporting the DNA markers but not the karyotypes. As illustrated, most cancer cell lines display karyotypic change despite sharing the same DNA markers. Since altered karyotypes represent different systems [Stepanenko *et al.*, 2015; Chapter 5], they are literally studying a given pathway or gene within different systems! No wonder there are so many conflicted data! In the future, for any molecular studies using cultured cells or even *in vivo* materials, the karyotype is needed to ensure the identity and pattern of the cell population [Ye *et al.*, 2007].

8.4.2.2 *In vitro systems are suitable to watch cancer evolution in action*

As discussed in Chapters 5 and 6, studying cancer evolution has become a new priority based on the high heterogeneity of cancer. To achieve this goal, various *in vitro* systems hold the key to search for the principles of somatic cell evolution.

Good *in vitro* models not only offer the window to watch evolution in action (with the capability of studying individual cells and cellular populations using multiple time points, which is hard to do by using *in vivo* tumors where excision is an endpoint), but also can compare different runs of independent somatic cell evolution with or without specific environmental factors. The speed of evolution can further be monitored by applying various constraints above the level of gene and genome.

We have used the *in vitro* immortalization model and drug resistance models to study cancer evolution and have revealed many significant features of cancer evolution, such as the punctuated stochastic pattern of cancer evolution and its links to system stresses. In addition, we have directly linked genome chaos dynamics to genome heterogeneity itself and the level of heterogeneity to transcriptome variation. Furthermore, since the degree of genome heterogeneity of each subline (including the frequencies of chaotic genomes) can be precisely measured, a quantitative relationship with transcriptome dynamics can be established, which is difficult to achieve in *in vivo* systems. The principles generated from these *in vitro* systems also explain and predict the clinical situation. For example, when the condition fits, a clinical cancer sample can generate a large number of cell lines, as each successful cell line represents a small portion of the real cancer cellular population defined by the evolution in *in vitro* culture conditions. Due to the drastic change of growth conditions, the chance of evolutionary success will also be different. It is known that multiple cell lines can be generated from the same tumor sample. The same cell lines can also generate different tumors *in vivo*. Therefore, there is a dynamic relationship between *in vitro* and *in vivo* models and as long as we focus on the system behavior during cancer evolution, the difference between these two systems become less significant. In addition, some disadvantages of the *in vitro* system become advantages over an *in vivo* system, as one *in vitro* model can represent a large number of *in vivo* models (as each outcome of one patient only represents one successful run of cancer evolution), which can be mimicked by parallel repeats of the same *in vitro* system.

More importantly, the field of cancer research urgently needs a new conceptual framework following the realization that there are major limitations of the gene mutation theory of cancer. The re-evaluation of various concepts of cancer using well-designed *in vitro* systems becomes essential. Systematic analyses to compare *in vitro* and *in vivo* systems are also critical, especially for determining when the correct methods are used, such as (1) monitoring both individual cell and population; (2) paying more attention to the overall level of heterogeneity and outliers; (3) performing parallel experiments and not cherry picking the patterns (such as favoring some gene mutations and ignoring others according to researchers' "intelligent guesses," or popularity of some genes or even convenience/agreement to what we already know); (4) treating evolutionary studies as a process rather than just the initial response (e.g. cell death studies where most researchers have focused on the mechanisms of triggering cell death without knowing that the death process can lead to genome re-organization mediated cancer evolution, see Chapter 5); (5) considering the non-specific stresses initiated by more specific molecular manipulations; and (6) establishing methods to monitor higher level system constraints (e.g. tissue organization, immune system, human behavior). Finally, more attention should be paid to principles and dynamic patterns of evolution rather than specific molecular mechanisms, as there are so many diverse and low penetrating molecular mechanisms that have already been identified and so many more will soon come. In particular, how fuzzy inheritance can be used to understand cancer evolution and its implication to the clinic. Judged by the highly diverse clinical samples and the stochastic nature of cancer evolution, *in vitro* systems should offer many general understandings to clinical cases.

It is worth pointing out that nearly all of our major discoveries based on *in vitro* systems have been recently confirmed by sequencing of *in vivo* or clinical samples. These include the two phases of cancer evolution and genome chaos. As mentioned in Chapter 4, when we first observed high levels of genome alterations in *in vitro* systems, many firmly concluded that they must be artifacts, as it is hard to imagine that these bizarre events could happen *in vivo*. Now, with confirmation from sequencing various cancer patients' samples, genome chaos has become a hot topic for study. Interestingly, in contrast, for studying these clinical cancer samples, the most we can do is to sequence them all and infer what has happened. For example, a large number of reports have appeared in various top journals illustrating genome chaos. Based on the sequencing data, many attractive models have been suggested to illustrate potential mechanisms. Unfortunately, without the capability to monitor the entire process, many proposed imagined steps may have never happened [Horne and Heng, 2014]. The key to studying the process of genome chaos is by using an *in vitro* system rather than focusing on specific chaotic events in specific patients, as the possibilities of coupling specific genome re-organization with certain gene mutations or pathways are endless. Moreover, not only can *in vitro* systems provide essential information for understanding *in vivo* systems, but the evolutionary patterns illustrated by *in vitro* somatic cell evolution have actually led to the realization of the main function of sex as well as organismal evolutionary theory (see Chapters 5 and 7). Clearly, it is not just a question of using *in vitro* or *in vivo*, it is how to perform less biased scientific research, data collection and synthesis.

8.5 Myths in Drug Resistance

Drug resistance in cancer is a key challenge. Despite extensive research to understand the molecular basis of drug resistance, which had led to the identification of a large number of different molecular mechanisms, the general mechanism of cancer drug resistance is still elusive. Recently, various large-scale approaches have been used to screen drugs that target various cancer genes, to improve the efficiency of drug delivery (e.g. molecular targets based on specific antibodies, receptor systems to design a "smart bomb," or application of nanotechnologies), to apply combinational targeting strategies with the aim of simultaneously and/or sequentially block multiple pathways in order to fight new mutations generated during treatment, to target some essential biological structures and pathways, to reduce multi-drug resistance, to identify and treat tumor stem cells, to develop new methods of immunotherapy, and to analyze network dynamics and modeling during treatment [Heng *et al.*, 2010b].

However, clinical success remains limited. Meanwhile, there are many myths in drug resistance research, which in fact have slowed down efforts of searching for the correct framework. A few examples will be briefly discussed using some recent experimental data. The clarification of these questions is important for understanding the mechanism of cancer drug resistance and to bridge the gap between basic research and its clinical implications.

8.5.1 *Is hitting cancer cells with maximum force the best therapeutic strategy?*

Cancer patients are typically treated at or near the maximum tolerated dose of chemotherapeutic drugs in the clinic. The rationale is that initial elimination of the maximal total number of cancer cells can effectively reduce the chances of cancer recurrence, as it would take a longer time for a small number of survivor cancer cells to grow back, which allows for the accumulation of new mutations responsible for drug resistance. Again, such assumption is based on the unproven stepwise theory of cancer progression. Despite the fact that the majority of basic research experiments seem to support this idea, and initial killing power has been used as an important criterion in the search and selection of new drugs, various therapies often fail in the clinic in the long term. Even more controversial, there seems to be a paradoxical link between the initial success of inducing cell death and the speed and aggression of tumor cells that survive treatment.

Both mathematical models and evolutionary studies have predicted that the effective elimination of cancer cells could paradoxically lead to increased cancer aggression, and the possible mechanism is that drug treatment can effectively eliminate cancer cell competitors. However, experimental evidence is not available. Other possibilities should also be considered, such as treatment potentially breaking down system constraint, which would allow for rapid cancer cell growth; another possibility could be that treatment itself can induce changes in cancer cells.

Taking into account the two phases of cancer evolution, high stress conditions such as drug treatment can induce genome chaos, and the fact that most drug resistant cells display altered genomes, we have hypothesized that induced chaotic genome formation may represent the key mechanism of rapid, adaptive drug resistance. This would result in the return of cancer cells post-treatment with increased aggressiveness.

To test this possibility, we need to directly demonstrate the following: (1) harsher treatment (i.e. higher dosage) can initially eliminate more cancer cells than less harsh treatment (i.e. lower dosage); (2) high-dose treatment also induces more chaotic genomes; (3) despite an initially lower number of cells that survive high-dose treatment, there is a transition where some outliers, whose formation was induced by the high-dose treatment, could outgrow populations influenced by lower dose treatments as well as no treatment control populations. As a result, the total number of cells will switch between harsh and less harsh groups; and (4) the fast growing cells will display altered genomes, indicating that these cells have gone through genome chaos.

For the first trial, we observed the perfect expected results. The transition was observed about 5 weeks following the induction of genome chaos. For the untreated control and lower dose groups, the overall number of cells was less than the high-dose group. The five weeks of continuous culture suggested that researchers might have missed the opportunity to previously observe this transition, as typical molecular experiments do not follow the recovery of treated populations for this long. For the majority of molecular studies, when massive cell death is induced, it is the end of the story.

To confirm the exciting observations, we quickly repeated another two trials of experiments. Surprisingly, we failed to observe the transition as the group of harshest treatment generated less cells compared with control groups.

These conflicting results led to the following questions: what was going on? How to explain the beginner's luck? Among the three experiments, which results are true?

Based on the behavior of the outlier and its unique contributions to cancer evolution under crisis (Chapters 5 and 6), we decided to systematically examine the odds of the game changing outliers. Specifically, we asked the question: under the same condition, what is the frequency of the winning outliers induced by harsher treatment that could generate more cells than the control group? Based on the stochasticity of cancer evolution, we realized that all three experimental trials were true, which accurately reflected the dynamics of cancer evolution.

To cut the long story short, the odds were around 2 in 63 treatments. In other words, among a total of 63 high-dose treatments, there was only two outliers that became a big winner in our experimental setting.

Two important messages were also obtained from this analysis: (1) Harsh treatment universally increased heterogeneity (measured by increased coefficients of variation) regardless of the final cell counts. This suggests that the potential for drug resistance was increased for all high-dose treated groups, even though many of them did not survive; (2) From the statistical average point of view, overall for all treated groups, the higher the drug dosage, the lower the surviving cell count. However, harsher treatment did generate some outliers, and those dominated an individual treated group. While formation of these outliers represents a low probability event, these outliers are powerful and should not be ignored. This explains why many researchers are convinced that harsher treatment is better for fighting cancer cells. One can imagine that, from time to time, researchers may have observed the opposite, where outliers overpowered untreated groups, but these were probably considered as bizarre events. These outliers are usually not reported, as they are not real in the majority of investigators' minds.

A series of T-75 flasks (each containing millions of cells) were examined following the induction of genome chaos under the identical drug treatment conditions. When the resistant cells became stable after 3–4 weeks, clones were isolated for comparison as they represented the end products of drug treatment. Different from previous experiments, each flask contained more cells before treatment, increasing the numbers of induced, aggressive outliers. Surely, almost all individual flasks contained survivors. Among the five flasks, one flask contained subpopulations displaying very high growth. Again, SKY analyses confirmed that these fast growing cells differed from the cell population prior to treatment and displayed new genomes.

At this point, four key components of our hypothesis have been examined. It should be pointed out that rapid growth is just one feature of the survivors we have investigated. Many other features, such as increased resistance to other drugs, the capability of creating a more cancer friendly environment, and hijacking normal cell function to push cancer progression, need to be investigated. The key is that harsher drug regimens result in increased heterogeneity and evolutionary potential. Heightened potential increases the disease's chances for evolutionary selection, which will favor rapid growth, especially at the late stage.

To further illustrate the importance of these outliers in drug resistance, we have examined many resistant clones that are selected from different drug treatments. These treatments are drastically different, from various chemo-drugs to small molecule inhibitor. Despite different treatment approaches, all survivors displayed altered genomes. This suggests that different drugs, including general chemotherapeutics or target-specific molecules, could induce genome chaos and resulting survivable new genomes.

Put together, hitting cancer with maximum force is not the best strategy. On the surface, harsher treatments can kill more tumor cells initially, but increased outlier formation from high-dose treatment can result in more aggressive resistant cells in the long term. Even though the chance is low (1 in 30), and the outliers have been traditionally ignored by most researchers (due to only monitoring an early, limited time window and applying averaging techniques in reports), when correct strategies are used, the low probability of drug-induced rapidly proliferating cancer cells becomes a sure thing. This conclusion is ultimately important, given the large number of cancer cells in patients, low probability events become certainties. That is why cancer cells come back in most cases following drug therapies.

In addition to the demonstration of the key transition (from initial drug effect to long-term resistance), the above experiments also offered new insight for the mechanism of drug resistance in cancer.

(1) Cancer drug resistance differs from the traditional concept of antibiotic resistance in bacteria. It was previously thought that resistance mainly came from pre-existing mutations rather than treatment induction. It is interesting to point out that stress can also induce an elevated rate of mutations in bacteria. As discussed in the genome chaos section, as genome replacement can be easily induced, the degree of inducible genome alteration is much higher in cancer cells than in bacterial cells.

Thus, drug induced resistance becomes more of an issue for cancer drug resistance. Similar to genome chaos, there are many other examples of stress induced genome changes through elevated fuzzy inheritance, including the loss of LIN1 retrotransposon repression, which results in increased dynamics.

(2) Aggressive therapy may work if it is successful at completely killing all cancer cells at once and not just initiating cell death for all cancer cells. As we have discussed, many instances of cell death cannot be completed, and instead, genome re-organization occurs as a survival strategy. Aggressive therapy may appear to work if a researcher only observes the cell death induction within a short time window of post-treatment and ignores the dynamic process of cancer evolution. This approach may work if the cancer cellular population is a homogenous mass, which can be accurately monitored by average profiles. This would suggest that the strategies must be different when dealing with homogenous cell populations (such as a normal cellular population) and heterogeneous cell population (such as a cancer cellular population). Unfortunately, these key "ifs" do not represent the clinical reality.

(3) Drug treatment can induce increased heterogeneity and promote outlier formation, which represents macro-evolutionary potential, the basis for any possible future phenotypes.

(4) Together, they explain why most experimentally proved strategies fail in the clinic. In addition to the large number of cancer cells in patients, the window of observation, linear models used for experiments, and data collection all contribute to this inconsistency. Finally, they might have provided the basis to support new approaches such as metronomic thermotherapy.

8.5.2 *The urgency to search for new concepts/strategies of cancer drug therapy*

The above explanations of the paradoxical effect of "hitting cancer with the hardest weapons" fit well with our analyses presented in previous chapters, and the implications are many and highly significant, since this paradox challenges the very foundation of most current cancer drug treatments.

First, the rationale of drug design needs to be seriously debated. The traditional strategy is based on the gene mutation theory of cancer and the stepwise evolutionary concept, as elimination of cancer cells and targeting cells with cancer-specific gene mutations are key concepts. Unfortunately, the old way of thinking is not working. In particular, if the principle of hitting the hardest is not working and, in fact, jeopardizes long-term benefit, new strategies need to be tested using the new genome-based cancer evolutionary framework.

Based on observed cellular population behavior under drug treatment, the new strategy should not continue to focus on the absolute elimination of cancer cells in order to cure cancer, but rather the new attitude of living with cancer by managing the cancer's long-term behavior and overall long-term benefit to the patient. To avoid inducing rapid cancer evolution while reducing the cancer mass, the key is not to trigger genome chaos with high unpredictability. New research is urgently needed to investigate some critical questions:

8.5.2.1 *Low dose or high dose?*

Is a low-dose, managing approach a better general strategy than high-dose cancer treatment, especially for patients displaying unstable cancer genomes (where genome chaos is easily induced)?

Low drug dosage is less effective for killing tumor cells; however, it will not drastically change the course of cancer evolution, which reduces the chance of triggering rapid resistance through rapid progression. It also introduces less stress to the overall system of the patient, improving quality of life and system self-recovery. Research is needed to study the lower ranges of drugs necessary to slow long-term cancer progression and balance the short- and long-term benefits. According to the prediction of genome theory, low-dose approaches should have better outcomes.

Interestingly, since 2000, metronomic chemotherapy, or administration of a chemotherapeutic drug at much lower dosages for a longer duration with no extended drug-free breaks, has become a potential alternative approach to maximum tolerated dose therapy [Pasquier *et al.*, 2010; Romiti *et al.*, 2013]. Furthermore, adaptive therapy as another alternative approach has demonstrated early success [Gatenby *et al.*, 2009]. The reduced chance of promoting genome chaos might be used to explain the success of these alternative approaches. It is also necessary to systematically compare low dose treatment for general chemotherapeutic and specific targeting treatments.

One immediate task is to study the question whether some patients are especially vulnerable to high-dose drug treatment due to their high personal level of genome instability prior to therapy. Based on our *in vitro* data, cancer cells with highly unstable genomes are most sensitive to genome chaos induction. In addition, fast relapse is also associated with a high level of genome instability [Chandrakasan *et al.*, 2011; Horne *et al.*, in preparation].

8.5.2.2 *Is the induction of genome chaos specific to chemotherapeutic agents?*

One of the key rationales behind the development of target-specific therapeutics is that general chemotherapeutics lack specificity, and drug resistance will be less of an issue with specific targeting of tumor cells. However, our experiments have illustrated that genome chaos can be induced by specific therapy as well. Thus, the key is whether or not the treatment can effectively induce cell death, regardless if the treatment is tumor cell specific or not. In a related study, we have linked mitotic cell death, which is often followed by genome chaos, to an array of different specific and non-specific treatments [Stevens *et al.*, 2011]. Based on these observations, we expect that rapid drug resistance will also apply to various target specific therapies. Sure enough, recent studies have demonstrated that drug resistance can be quickly developed.

8.5.2.3 *Is drug specificity crucial?*

The discussions of the negative consequences of aggressively treating cancer cells and the similar responses of specific targeting and general chemotherapeutics raise an interesting question: Is the specific targeting therapeutic approach really important in cancer therapy?

The answer seems to be yes to some highly exceptional cases such as the chronic phase of CML, where linear progression is dominant, as well as for some limited phase of clonal expansion where some gene mutations temporarily dominate. However, for most cancer types, high heterogeneity makes targeting specific genes or pathways less useful, especially for the longer term. This is because for most cases, there is no fixed specificity, and temporary specificity is often lost following target-specific therapy. These facts suggest that perhaps a lower dosage of chemotherapy might be a better option than specific targeting. In other words, it is possible that a low dosage of a less specific drug could have a better outcome than high dosage of a targeting agent for more diverse cases.

To illustrate this point, we have tested different drugs/reagents to represent general stresses that slow down cancer cell population growth. Interestingly, regardless of the different combinations of drug treatment, various drugs at low dosage can all slow down cancer cell progression. The similar phenomenon can be achieved even by non-drug mediated stress conditions. Specifically, we have examined the effect of a non-homologous end joining (NHEJ) inhibitor to reduce genome chaos. Surprisingly, many low dosage drugs can also reduce genome chaos. These experiments demonstrate the importance of using less specific drugs or conditions to manage cancer. It was reported that starvation prior to chemotherapy could help the effect of chemotherapy [Raffaghello *et al.*, 2008]. We suggest developing different methods to constrain cancer growth rather than direct targeting, as less specific constraint may have better coverage, especially when the cancer targets are moving.

Another example of using a general stress to increase the killing power of a specific drug is to change the epigenetic status. It is important to know that pre-treatment can increase system dynamics, which will increase the death caused by the following drug treatment. On the other hand, this will likely also induce genome chaos. Further studies are needed to determine the overall benefit of this approach.

8.5.2.4 *Suggestions for future strategy development*

In summary, the following considerations should be applied towards the development of future cancer therapeutic strategies:

(a) Separate patients into different subgroups, as their responses will vary to different treatment options. Criteria to distinguish these subgroups may include: Do patients display relatively stable or unstable genomes? Are there dominant clonal populations present in cancer samples? Which phase (stepwise or punctuated) summarizes the majority of the cancer cell population?

For a cell population within the stepwise phase, the main strategy would be to slow down progression and avoid triggering the chaotic phase. Approaches to achieve this rather than aiming to eliminate the population may include changing cellular types, epigenetic statuses, or by decreasing the speed of proliferation. For highly dynamic and unstable populations, these could ideally be induced to establish clonal population by adjusting constraint.

(b) Rather than directly killing cancer cells, more attention should be paid to improve system constraints above the individual cell level. Immuno-therapy, for instance, should have better outcomes. In particular, immuno-therapies should focus more on overall immuno-capability (including its dynamics) and are not just specific to cancer gene mutations. Similarly, focus should shift to cancer cell constraint. Furthermore, other supporting aspects of patients need to be included into the treatment strategy equation, including nutrition, physical activity levels, psychological status, family/group support, and other social environmental factors.

(c) Low-dose approaches should be tested with different types of drugs, specific or non-specific alike. These could be combined with the promising adaptive therapy approach, where cancer response profiles are used and dosage is adjusted to avoid overtreatment.

(d) For most diverse sporadic cancers, developing more target specific drugs will not be the ideal approach. In contrast, less specific and broader coverage should be the focus, as molecular specificity can be easily changed. Developing more general agents to cover the overall evolutionary potential should be more effective.

(e) As the key for most treatments is to slow down the cancer evolutionary process, especially following the surgical removal of cancer, many approaches beyond drug treatments should be studied. If different drugs ultimately function as system stresses, and certain stresses can change the system status, it is possible that, some other physical stress (e.g. heating) could be used for the same purpose, either for maintaining or altering the evolutionary pattern of cancer. A similar idea has been suggested to explain the treatment of some cancer cases [Coffey *et al.*, 2006]. In fact, many physicians know this well. For many patients who respond to a specific treatment well, it is likely that somehow the treatment has resettled the system rather than a specific molecular mechanism that worked as designed. For example, some cases of metronomic chemotherapy resulted in the activation of anti-tumor immunity [Kareva *et al.*, 2015].

8.6 The Dilemma of Pushing for Early Diagnosis and Treatment and the Role of Luck in Cancer

After decades of promoting the idea that "catching cancer earlier saves lives", early diagnosis has become a top clinical priority. As discussed in previous chapters, if cancer is a stepwise progress, then specific biomarkers of earlier stages are important as they could be used to identify early cases, and if targetable, could be exploited in order to eliminate cancer early on and save lives. With increased investments in instrumentations and biomarker identification, especially following various screening projects, many otherwise undetected cases have been identified. While these earlier diagnoses have increased the overall survival time for specific cancers, it has also become an issue of concern for some, as many early detections have led to overdiagnosis and overtreatment. Based on the highly heterogeneous trajectory of cancer cells and the fact that cancer cells can enter dormancy, a special stage where cancer cells cease dividing but survive in a quiescent state with proliferation potential, it is easy to predict that treatment could promote cancer cells to harm patients. There are clinical observations that support this view. For example, some individuals never exhibit cancer symptoms despite having the cancer status, some patients can become "cured" without any treatment, and some exceptional later stage patients can still survive much longer after treatment has failed. In contrast, in various *in vivo* models, while successful treatment can slow down tumor growth in general, there are often exceptions where the same treatment can speed up tumors. In some cases, surgical removal of one tumor could trigger unexpected accelerated growth of other tumors from different locations in the same animal. It is now becoming clear that surgery itself contributes to both local recurrences and distant metastases [van der Bij *et al.*, 2009]. With these experimental and clinical observations plus the fact that there are no known mechanisms to predict cancer trajectory, there should be more cautions and careful studies on earlier diagnosis and treatment, as there are both trade-offs and luck involved.

8.6.1 *Overdiagnosis and overtreatment*

Overdiagnosis refers to when a disease condition is diagnosed that would otherwise not go on to cause symptoms or death [Welch and Black, 2010]. In the case of cancer, this can be explained by two clinical observations. First, without medical intervention, the cancer never progresses or sometimes regresses, and second, the cancer progresses slowly enough that patient dies of other causes prior to the cancer becoming symptomatic. Significantly, overdiagnosis is a serious problem in the cancer clinic, as the estimated magnitude of overdiagnosis is high in various major cancers (e.g. approximately 25% of mammographically detected breast cancer, 50% of chest X-ray and/or sputum-detected lung cancer, 60% of prostate specific antigen-detected prostate cancer).

The highly heterogeneous nature of cancer explains the unpredictability of most cancer cases. It also questions the rationale of using specific gene mutation signatures for earlier diagnosis and treatment. The higher the genome heterogeneity, the more cases that will be overdiagnosed, as the high heterogeneity of cancer progression contributes to the clinical fact that some early stage cancers will not progress to later stages. It is thus useful to know how much benefit and harm early screening will bring to patients. The "trade-off" of screening mammography in 50-year-old women has been analyzed by Welch and

Black based on information from US Preventive Service Task Force and other publications, and the important message was summarized into a simple draft balance sheet. Among 1000 women (50-year-old) undergoing annual mammography for 10 years, the benefit is that one woman will avoid a breast cancer death. In contrast, the harms are surprising: 2–10 women will be overdiagnosed and treated needlessly; 5–15 women will be informed that they have breast cancer earlier than they would, however, this will have no effect on their prognosis; 200–500 women will have "false alarms" or false-positives (50–200 will be biopsied). Similarly, a prostate cancer screening trial in the United States (reported in 2009 and again in 2012) did not find evidence of any benefits, while an European study found some benefit [Concato, 2013]. In addition to the disappointments from these contradictory studies, and the low benefit ratio even from positive reports plus the costs on the quality of life (we have asked many researchers and physicians about the same question, and the expected benefits are much higher), more of an issue is the subsequent unnecessary treatment that will bring risk to patients, including pushing some slow cancers to become more aggressive. The treatment will likely change the evolutionary status of cancer patients in different degrees and in different directions (including beneficial or harmful). Two questions need to be asked from the societal point of view: first, how much should the health system spend for early diagnosis and overtreatment? Second, how do we balance the benefit for lucky individuals against the harm for unlucky individuals as they co-exist within the screened population?

It is thus important to investigate whether patients of certain cancer types would benefit more from earlier screening, so the medical community can specifically push for it. More importantly, we need quantitative data of the benefits and harms of screening for major cancer types and promote the early diagnosis accordingly. With these hard data, we can have meaningful conversations among researchers, physicians, as well as patients. How we direct these conversations is key. Suppose we only emphasize the one life saved during the annual screening of 1000 individuals in 10 years, people will likely see this is worth the cost. However, if up to 20 individuals will actually be harmed (some very seriously), and hundreds of patients will undergo unnecessary biopsies before being cleared, would the reactions to these discussions be the same? The challenge lies in the treatment, as the complexity of cancer limits the prediction of the treatment. Different from eliminating a bacterial infection with antibiotics, cancer treatment will most likely change the evolutionary phase of the disease, thus having a negative impact even when on the surface there is no significant response (i.e. minimal cell death). It is interesting to note that prolonged use of common antibiotics actually could induce mitochondrial dysfunction and oxidative damage in mammalian cells [Kalghatgi *et al.*, 2013]. Together, these support an ancient Chinese viewpoint where all drugs are poisons. The prevention of overtreatment should thus be an important subject.

8.6.2 *How does luck play a major role in cancer?*

The dilemma for cancer diagnosis, treatment, prevention and even research is dealing with trade-offs, where the known and unknown represent challenges. We know a large number of gene mutations will contribute to cancer, but we do not know which one(s) will be a key contributing factor for a given individual; we know some viral infections lead to cancer, but we do not know which infected patient will be the one or when this will happen; we know a variety of probabilities that exist at the population level, from lung cancer odds of tobacco smokers, to 5-year survival rates for a specific treatment option, but we do not know how to predict the outcome of a specific individual. We know that any diagnosis and/or treatment will be a benefit to some, result in no improvement or be detrimental to others, but we do not know who will fall into which group. In summary, we know there is a relationship that is difficult to understand between the population average and exceptions, but we do not know how to sort out the exceptions due to the highly dynamic nature of genome/environmental interactions.

However, this difficult situation is not only unique to cancer. It is a universal truth that is impossible to separate the good from bad. The list is extensive: young and old, certainty and uncertainty, yin and yang, two sides of the same coin. It is a package with trade-offs that reflect at multiple levels (e.g. molecular/pathway, cellular, individual, and society levels). In other words, everything represents a paradox. We should not forget this when pushing our research, diagnosis, and treatment.

As we discussed in Chapters 5 and 6, cancer represents a high evolutionary price we pay. Specifically, while fuzzy inheritance provides the most possible adaptive capability, it also creates the opportunities for cancer evolution. Various stress conditions and adaptation can generate new genetic variations. Whether or not these variations will be eliminated or result in cancer onset often involves many incidences of good or bad luck. For example, many stages of cancer formation, from initiation to progression, are associated with bad luck (or good luck in the perspective of cancer), as multiple system constraints make it hard for cancer to be successful. A given factor could be good or bad for a specific stage of cancer evolution, or individual, or environment. Since there are so many factors, the formation of a good or bad combination really boils down to an issue of probability or luck. This viewpoint has become more visible as the field has failed to identify major genetic and environmental causative factors. It further explains the "why me" question, especially for those individuals who do not have a family history of cancer, watch their lifestyle carefully, and do everything "right" (exercise regularly, abstain from smoking, etc.), but are still hit by cancers.

Recently, the variation in cancer risk among different tissues has been correlated to the number of stem cell divisions. Briefly, the number of stem cell divisions was determined by using the number of cells in an organ and the percentage of the long-lived stem cells. The reasoning behind this is if for every cell division, there is a risk of a cancer causing mutation in a daughter cell, then the tissues hosting the greatest number of stem cell divisions are the most vulnerable to cancer. The number of stem cell divisions of major organs and actual cancer statistics were compared. As the number of stem cell divisions in a given tissue rises, so does the chance of cancer striking that tissue/organ site, leading to the conclusion that random gene mutations, or molecular "bad luck," serve as the main cause of cancer [Tomasetti and Vogelstein, 2015]. The authors have further concluded that this mathematical model explains two-thirds of all cancers [Counzin-Frankel, 2015]. It even suggests that most cancer cases are not preventable, as environmental or inherited contributors are less important. This conclusion contradicts the efforts of promoting cancer prevention and has generated heated debate.

Emphasizing stochasticity in cancer is long overdue. Years ago, by associating NCCAs to stochastic cancer progression, we promoted the role of randomness in cancer formation for most sporadic cancer types, as genetic heterogeneity, and especially stochastic genome alterations, are crucial for cancer evolution (Chapter 5). This departs from focusing on specific genetic targets. Based on this viewpoint, the traditionally ignored "genetic noise" (i.e. stochastic genetic alterations) has been appreciated as a key to cellular adaptation (good luck) but unfortunately also leads to increased cancer risk (bad luck) as a trade-off [Horne *et al.*, 2014]. Unfortunately, the concept that stochasticity is essential to adaptability as well as cancer evolution is still ignored by many.

While stochasticity is most important for cancer evolution, one must not interpret it as a reason to disregard prevention. Just because it is difficult to link diverse environmental factors to cancer (except for dominant factors such as viral infections or smoking), it does not mean they are less important. Cancer formation requires the "perfect storm," and many factors contribute to increased good or bad luck. Maintaining a healthy lifestyle should reduce the probability of cancer formation by lowering stress-induced stochasticity. While some tissues have a higher cancer risk due to a higher baseline of stochasticity, it cannot explain why different individuals have good or bad luck when they have similar baselines (the similar status of the stem cells), aside from those individuals with genetic predispositions. Thus, this mathematical model should be explained very differently to support cancer prevention. In addition, while cellular division represents an important opportunity for generating variation, the environmental impact on the genome (which can be adjusted by cancer prevention strategies) serves as a key constraint of cancer evolution.

Nevertheless, the issue of good and bad luck needs to be seriously discussed in the context of cancer prevention, diagnostics and treatment, especially in the framework of how fuzzy inheritance and evolution works. Good and bad luck are both necessary for a bio-system and are entirely dependent on the environment. Since environmental changes are difficult to predict, the bad fortune of cancer and its treatment response will also be hard to predict.

To illustrate the issue of bad luck in cancer, different levels that involve cancer should be discussed. First, as the entire human population needs variance, some individuals will pay the price to ensure the survival of the human race. Since environments are highly dynamic, the population has to have individuals ready to provide needed variations. Due to unforeseeable future events, the most effective way is to generate a certain degree of change is through the fuzzy inheritance mechanism rather than fixed specific changes. However, many variations will be neutral or display disadvantages compared with wild type following evolutionary selection. As illustrated in Chapter 5, many outliers can save the population in a special situation even though it means that these outliers represent disadvantages to population growth under normal conditions. As soon as these outliers pass their genomes and save the cellular population, fuzzy inheritance will repopulate the genetic landscape of survived cells similar to the parental population.

Let us discuss a few extreme circumstances: a p53 mutation is well known for its harmful features, as it causes genome instability and cancer. However: (1) It is essential for survival in mice with a BRCA mutation. The deletion of a BRCA gene is embryonically lethal and the p53 mutation can rescue it [Deng and Brodie, 2001]. Sure some mutations can increase the chance of cancer later in life, but they can also balance out or compensate for other mutations, impacting the individual's survival and odds to pass the genome. We anticipate a p53 gene mutation can rescue many other genetic alterations as well; (2) p53-/- mice display a survival advantage when irradiated in utero [Norimura *et al.*, 1996], representing an example of a normally deleterious gene becoming advantageous in a changed environment. It is not hard to speculate if the environment became highly radioactive, some individuals with a p53 mutation would have better a chance to survive and pass the genome to future generations; (3) Furthermore, under certain high stress events, cellular adaptation should be higher in cells with a p53 mutation. It would be interesting to investigate whether an individual with a p53 mutation would have a less stringent sexual filter (see Chapter 7). Even more remarkable, if the species is no longer survivable (e.g. during a potential extinction event), an individual with a p53 mutation may be more prone to genome chaos induction at the germline level, which could increase the odds for new species formation. All together, these examples emphasize the importance of the "bad for now, but good for future potential" of many variations. Unfortunately, most of the potential advantages of variation are invisible under current conditions. Nevertheless, certain variations could be good for the population as an insurance policy, but bad for some "unlucky" individuals who pay the price.

Second, at the somatic cell level, not all genetic variation will lead to cancer (not all gene mutations will lead to genome changes; not all genome alterations will form dominant genome systems), and not even the same powerful environmental factors can have 100% penetration among the population (not all infections lead to cancer; most heavy smokers do not get lung cancer despite that smoking contributes to most lung cancers). While generated variation is essential for cellular function, not all will pay the same high price.

Considering the stochasticity that exists at the cellular and individual levels, the reality of cancer is not very surprising: some are luckier than others under the same screening and/or treatments. Oncologists know this very well. They have luckier patients that benefit from a certain treatment more than other patients. However, predicting who will be the luckiest is often mission impossible. There are often paradoxical benefits and harms. For example, Barrett's esophagus can become esophageal adenocarcinoma. It seems like a no brainer to remove it at an earlier stage. However, the operation procedure can pose a higher death risk than the frequency of adenocarcinoma formation. A balanced treatment option is thus important. Health care conditions, surgical experience, and oncologists all serve as additional layers of luck for patients, in addition to the social support patients can have.

Perhaps we need to take the advantage of the much-improved certainty at the population level and push the benefit of prevention, diagnosis and treatment at a maximal average level. In this case, we need to separate average populations (majority) from two extremely lucky and unlucky populations. Using smoking as an example, for the average individual, a non-smoking lifestyle will drastically reduce the risk of cancer. However, those who smoke like a chimney and never get cancer and those who never smoke and still have lung cancer represent a "lottery" group in the context of a single contributing factor. Their profile will interfere with the understanding of the average profile. By separating these cases, the cancer prevention effect argument with respect to a specific environmental factor will be much stronger.

Different patient subpopulations (characterized by genetic and environmental profiles) will require different diagnostic and treatment strategies, as different approaches will offer different benefits and harms. The next question is how to distinguish average and outliers, as the benefit/harms relationship could be very different among them. If a solid genetic profile can be used to achieve this goal, then the mutation profile is useful. However, since most cancer types display diverse and dynamic genetic profiles, a new method to classify patients may be the incorporation of overall genome instability, as cell populations with stable genomes favor average behavior while the unstable genome favors outlier behavior.

It is important to use cellular population heterogeneity to measure the average and outlier-represented cancer potential, improve overall lifestyle, reduce environmental chemical contamination, and decrease other high stress conditions of human life in order to reduce cancer incidence within the population. One of the often ignored aspects is to reduce medical-related stress, especially during cancer diagnosis and treatment. Another key factor is patient psychological status. A positive cancer diagnosis could destroy someone. Knowing a patient's attitude towards cancer and their drug response potential are important before offering the diagnosis and treatment to them.

More research is needed to study how stochasticity is selected under different degrees of stress during cancer treatment. For example, are the major benefits and harms estimated prior to diagnosis and treatment? Should overall genome instability be used to separate patients based on different degrees of stochasticity? How do we take the current treatment effect into account to ensure longer term benefit? Is it possible to monitor the evolutionary potential in the effort to limit evolutionary phase transition?

Lastly, patient education is important. By and large, individuals set their own odds to have cancer or not, as good lifestyle could reduce two-thirds of overall cancer incidence. For those who have cancer, quantitative data are needed to help these patients make very difficult medical decisions. Perhaps due to optimistic human nature, many patients may think of themselves as being the lucky ones or the exceptions, and that they will beat the odds when facing this disease. While it is important to be positive, it is also important to face the medical reality. Therefore, both the positive and negative aspects as well as truly average outcomes should be seriously presented to patients.

8.7 Why Most Cancer Animal Models are Not Working

The increased heterogeneity observed in cancer has promoted researchers to search for a system that better reflects the reality of cancer rather than "cell culture artifacts". "*In vitro* systems are not reliable for cancer research and must be confirmed by *in vivo* mouse models" some have claimed, as if mouse models will finally deliver the missing answers. In fact, the mouse model already serves as the standard prior to many clinical trials. It has also become a precondition for publishing in some high impact journals. Unfortunately, most cancer data obtained from mouse models have limitations when applied to the clinic, as these models have provided poor prediction power for understanding real human cancer.

Traditional explanations of this disparity have focused on the obvious biological differences between mice and humans as well as statistical concerns. With new findings demonstrating that the true disparity is overwhelming as observed among different patients with the same type of cancer, within the same tumor up to the single cell level, and within the same tumor at different time windows of examination, the question asked needs to be changed. The key might not be the difference between *in vitro* and *in vivo* or mouse and human, but between real cancer with high heterogeneity and models with eliminated or largely reduced heterogeneity or noise (see Chapter 6).

8.7.1 *Various animal models in cancer research and their limitations*

Mouse models have been used to illustrate various molecular mechanisms of cancer by linking cancer to specific gene mutations and pathways, to monitor drug response and in particular to serve as pre-clinical trial models.

The rationale of using mouse models seems to be straightforward and logical, as mouse models allow for experimentation that would otherwise be impractical and unethical. Mouse models are significantly different than *in vitro* models, allowing for the study of the whole organismal system. For example, it is nearly impossible to study tumor invasiveness into adjacent tissue using *Petri dishes.* After all, we are all mammals and share most major organ and physiological systems and, in particular, share conserved molecular pathways as well as most important genes [Pennacchio, 2003]. Additional advantages include their small size, ease in breeding and handling, and an intermediate lifespan that is longer than worms and flies but shorter than primates.

There are several major types of mouse models and most of them are based on completely inbred strains. For environmentally-induced carcinogenic models (by radiation, chemicals, viral carcinogens and extreme physiological/pathological conditions), they result in a restricted subset of tumor types with incomplete penetrance and variable latency, limiting their application to understanding human cancers [Frese and Tuveson, 2007]. In addition, the induced condition is often much more extreme than actual patient conditions.

For genetically engineered models (GEM), these include both transgenic GEM and endogenous GEM [Frese and Tuveson, 2007; Heyer *et al.*, 2010]. GEM mice allow for the study of individual genes in analyzing genetic, environmental and pharmaceutical aspects of cancer. However, the specificity of the manipulated gene is comprised due to off-target passenger mutations, and the effects of cancer gene mutations are artificially increased in these animal models (the creation of highly clonal cell populations representing an entire organ or animal does not replicate the natural situation that occurs during sporadic cancer but only represents the conditions of limited familial cancers). As a result, in models, the process of tumorigenesis is very robust, particularly in leukemia-genesis, resulting in a high incidence of disease occurring within weeks to a few months [McCormick and Kavet, 2004]. Various types of non-germline genetically engineered mouse models are now established to mimic genes in a more natural setting in terms of temporal, spatial, dosage and heterogenic conditions [Heyer *et al.*, 2010]. The combination of various engineered models (by experimental breeding) has been used to illustrate the interaction among genes to enhance or repress cancer progression. However, less complex, more homogeneous models are significantly different from real patients. For example, combinations of well-known cancer gene mutations are rarely detected from one individual tumor, as illustrated by cancer genome sequencing projects (Chapter 4).

With regards to xenograft models, compromised immunological systems provide a convenient way to study tumor cell lines or explants; however such an artificially created condition does not translate to actual cancer. This is underscored by the ineffectiveness of these studies' results in the clinical setting [Johnson *et al.*, 2001]. It has been noted that the dynamic genome behavior of cancer cells is significantly different in mice with and without functioning immunological systems. Without immunological stress, there is much less genetic heterogeneity in tumors, leading to better replication of results and validation.

In vivo imaging mouse models have been recently developed, which might provide needed dynamic information applicable to patient diagnosis [Grimm *et al.*, 2005; Sahai, 2007]. The dynamics of cancer progression can now be directly visualized or measured, and this approach may provide a step in the right direction by attempting to accurately mimic human population responses.

Unfortunately, all types of mouse models display high disparities when compared with cancer patients, and such disparities are simply blamed by obvious differences between humans and small animals. These include an exhaustive list of physiological and genetic conditions (such as metabolic type, immune system, body size/weight, generation time and possible genome architecture, including gene structure, regulatory elements, repetitive sequences and telomeres, and genome context/genomic topology and instability) [Frese and Tuveson, 2007;

Gawrylewski, 2007; Ye *et al.*, 2007]. Even at the basic level of transcription regulation, the highly conserved transcription factor binding sites differ between the livers of humans and mice in 41% to 89% of the examined events [Odom *et al*, 2007]. Not surprisingly, there is a call to develop large animal models to replace mouse models.

Along with biological differences, further explanations of these disparities have been attributed to statistical limitations (randomization and bias), methodological flaws, lack of exact replication, and problems in measurement and evaluation of the effects of treatment. In addition, negative results are often omitted, resulting in significant publication bias [Bebarta *et al.*, 2003; Perel *et al.*, 2007; Sena *et al.*, 2010; van der Worp *et al.*, 2010]. These technical limitations should be correctable, yet even in carefully designed experiments, the unsolvable disparity between animal models and patient reality persists. This not only creates perplexing problems, preventing the successful translation of years of research into therapies, but also wastes billions of dollars.

The limitations of using mouse models to study cancer become even more puzzling. Many cancer gene mutations and molecular pathways are highly conserved between mouse and human. It is especially hard to explain some of these well-designed animal experiments when the results lead to huge disappointments after clinical application.

As we have mentioned, Judah Folkman's breakthrough angiogenesis research, in particular results from animal models, led to predictions that cancer would be cured within a few years. This hype led to more than 2,500 laboratories and 375 companies engaging in angiogenesis research and drug development, with many dozens of clinical trials. However, like many other "breakthroughs", the clinical trials were disappointing.

This story is not an exception but represents a general trend in cancer research. In the past several decades of the War on Cancer, there have been many short-lived breakthrough stories sharing the following common pattern: A new approach targets a different molecular pathway; the concept seems to make sense as judged by the dominant gene-based cancer theory and most importantly, the animal studies are convincing. Unfortunately, the initial promise leads to disappointment as new therapies that rapidly and reproducibly eliminated tumors in animal models often cannot duplicate this success in the clinic. Clearly, the dominant pathways established in animal models do not fit the reality of patients where most "driver mutations" occur in less than 1% of patient populations and there are no commonly shared pathways (see Chapter 4). According to Robert Weinberg, "A fundamental problem which remains to be solved in the whole cancer research effort, in terms of therapies, is that the preclinical models of human cancer, in large part, stink." [Leaf, 2004].

8.7.2 *Why is the current mouse model effective for basic research when focused on a specific hypothesis, but not very useful for clinical implications?*

The most important issue that limits the mouse model in cancer research is the artificial reduction of system heterogeneity. The following discussions illustrate this point.

8.7.2.1 *Individual mouse models can effectively test a specific hypothesis by reducing system heterogeneity, which also changes the key feature of the subject under study*

Since genome heterogeneity mediated somatic cell evolution is a key feature of cancer progression and drug resistance, a realistic mouse model should preserve such key feature. However, in order to illustrate a specific molecular mechanism, the effective way would be to create an elegant model to focus on a single factor. While this much simpler model is useful for molecular understanding of a specific feature of cancer, it is paradoxically less relevant to real cancer. In other words, by reducing the heterogeneity, the linear model can effectively illustrate a given hypothesis. Unfortunately, the evolutionary mechanism of cancer equals to the collection of all of the individually identified molecular mechanisms, and we thus need an almost unlimited number of models to encompass the highly diverse aspects of cancer.

There are numerous ways to ignore heterogeneity, making the conclusions of these studies less applicable to real cancer.

First, gene knockout mouse models are designed to dissect and "prove" individual gene function in the "causation" of cancer. However, knockout models only represent systems without specific genes, and they do not exactly test the function of the knocked out genes. Importantly, as we discussed in Chapter 5, molecular manipulation at the gene level can lead to increased system heterogeneity. The increased cancer potential often should not be associated with the initial function of the knockout gene.

Second, in most mouse models, though simplified, a certain degree of heterogeneity still exists. However, this heterogeneity is obscured in the quest to identify gene based patterns. Typically, reports only highlight recurrent patterns while stochastic events are disregarded as insignificant noise. Though an increasing number of reports show the biological relevance of stochasticity, methods of molecular profiling often average out heterogeneity (Chapter 6). On top of this, there is incomplete phenotype penetration in the vast majority of models [Heyer *et al.*, 2010], not to mention the cell-to-cell variation observed in activated pathways. At the individual animal level, for example, when tumors are analyzed from multiple GEM carrying a cancer driver gene, a few individual animals' tumors may share recurrent molecular signatures such as c-Myc over-expression. This recurrent pattern then becomes the key message in the report. This is despite the fact that the majority of animals display diverse pathway involvement. Even in GEM models, penetrance is rarely 100%, underscoring the presence of heterogeneity at multiple levels in animal models.

In addition, heterogeneity in animal models (compared to natural populations) can be effectively obscured by various statistical analyses. Due to genetic and environmental homogeneity in animal models, many experiments show statistical significance of an isolated gene or pathway in the presence of limited heterogeneity. However, as soon as further heterogeneity is introduced (when natural populations are used), statistical significance drops dramatically. Further, simulation experiments show that even minor genetic alteration can significantly change the outcome of significance and/or predictability. For example, the failure to replicate genetic association studies has been linked to differing genetic architectures, as polymorphisms can fail to replicate interactions with other polymorphisms caused by distinct ethnic backgrounds or different geographic regions [Greene *et al.*, 2009]. In a similar vein, in inbred mouse strains, CNVs have been fixed by inbreeding, but they are highly variable among different strains [Liang *et al.*, 2008]. Thus, our analyses provide further explanation as to why many approaches proven to work in homogeneous animal models (as judged by statistical standards) fail in heterogeneous patient populations (no longer statistically significant).

Third, another important effect is environmental homogeneity. Similar to simplifying the genetics of the model, the approach of simplifying the environment also leads to significant disparities. It has been suggested that there is a covariant interaction between the environment and quantitative trait loci (QTLs) that account for genetic effects. Surprisingly, these interaction effects are more frequent and larger than the "main" effects [Valdar *et al.*, 2006]. This further poses a challenging question, by creating a homogeneous environment, we can reduce most of the environmentally associated noise (more correctly referred to as the signature of genome-environmental interactions). This increases our ability to get replicable data, but these data likely will not truly reflect real patients who display the diverse signatures of genome-environmental interactions.

Fourth, there is a strong caveat for the many manipulations on specific genes or environments in animal models. The immediate result of artificial selection and natural selection is drastically different from long term consequences. Within defined experimental conditions, there is no real competition (they are supplied with sufficient food, shielded from pathogens, and no competition exists from "normal" animals without genetic manipulation) and these altered animals can retain some features artificially, such as longer lifespan or maintaining the mutation status. However, in the natural world, they likely would not survive past a young age, lose these competitive advantages, or quickly reduce the "importance" of a specific gene mutation.

For example, mutation of the insulin receptor in *C. elegans* can double the lifespan, and it can increase lifespan in *Drosophila* closer to 50% [Clancy *et al.*, 2001; Kenyon *et al.*, 1993]. Various animals have also shown an increased lifespan with caloric reduction [Bioship and Guarente, 2007]. This dietary restriction approach has been linked to the insulin signaling pathway, the rapamycin-sensitive TOR signaling pathway and the Sir2 longevity pathway. Recent animal studies have demonstrated the possibility of using drugs to achieve longevity in mice. When rapamycin (an inhibitor of the mTOR pathway) was fed to mice late in life, the lifespan of mice was extended, suggesting a potential medical intervention of the natural aging process.

The use of rapamycin in mice (or similar approaches) represents artificially achieved conditions which might be difficult to extrapolate to humans. Rapamycin is an immunosuppressive agent which can compromise an individual's defense systems [Cox and Mattison, 2009]. In the mouse study, mice were isolated from infectious agents. Thus the lifespan increase caused by the drug appears to be beneficial within the artificial laboratory environment only when potential harmful effects of nature have been eliminated. Within the natural environment exposures to pathogens commonly occur, and many of the mice with reduced immune function would not survive and therefore would not live longer lifespans. Recent studies of the mTOR pathway support this point. Reduction of this pathway's activity might extend life under certain conditions; however, rapamycin treated individuals have a greater risk of dying from a bacterial infection, even when they are young. Similarly, this is applicable to the recent results of calorie restriction in rhesus monkeys [Colman *et al.*, 2009]. When the real evolutionary trade-off is involved, the picture will be quite different. That is the exact reason why mouse model-derived findings fail to translate to real cancer, where evolutionary selection rules.

8.7.2.2 *Reduced heterogeneity leads to limited prediction power of animal models*

By controlling both genetic and environmental circumstances with the use of highly homogeneous experimental models, the ability to replicate real systems that have high levels of heterogeneity is lost. This is strongly supported by the limited prediction power of many animal models.

While there is a general acceptance that animal models work, there is only a 10% prediction value using traditional toxicology tests in rats

[Sankar, 2005], a 7% prediction value in carcinogenic studies (with 79.3% false positive results) [Knight *et al*, 2006], and prediction value in xenograft models based on dozens of cancer drugs is low as well [Johnson *et al.*, 2001]. At the molecular front of research, using experimental systems one can artificially create a stepwise phase with a dominant pathway to illustrate the function of selected key cancer genes within animal models. However, in the real world of somatic cell evolution, these favored genes have a low probability of becoming dominant.

8.7.2.3 *Key considerations regarding the application of animal models to human disease*

Despite the continuous debate over the value of animal models with regards to human diseases [Shanks *et al.*, 2009; van der Worp *et al.*, 2010], there is reluctance to seriously debate some basic questions. Based on the above analyses, we must address the following questions:

8.7.2.3.1 *Is the current way of using animal models in cancer research truly useful to the clinic?*

This is not just a trivial academic issue as billions of dollars and the value of generations of research are on the line. One often used defensive argument to support the mouse model is that mice are not human and thus the expectation of consistency between mouse models and people is not reasonable. However, the reason why they are still used is based on the rationale that they hold great relevance to cancer and current disparities can eventually be reduced by improving our experimental designs!

From a purely academic point of view, all models are not true to nature yet many of them are scientifically useful. Simplification through the use of models occurs in many scientific fields and has worked well in classic mathematics, population genetics, and in many infectious disease studies. By increasing layers of complexity, a simple model could work, in theory, for cancer. In contrast, current mouse models are created to drastically reduce heterogeneity rather than incorporate the layers of complexity involved in the natural evolutionary selection process. In other words, by design, levels of complexity are drastically reduced in the model to determine which genes or pathways are responsible for the disease. Paradoxically, such simplification also creates disparities between these models and reality, as in these simplified models, the "causative relationship" is readily apparent, while in complex natural populations, the causative relationship is often cryptic.

We thus argue that any simplification must not fundamentally change the system that is under analysis. Key features that define a system cannot be simplified.

8.7.2.3.2 *If animal models are necessary to study or predict cancer, how do we improve these model systems?*

There are many suggestions to improve mouse models to ensure a more accurate representation of human cancer. The key should be addressing the issue of genetic and environmental heterogeneity rather than using homogeneous conditions, even when primates are used. As different strains of the same species have highly variable sensitivity to specific treatment, we could face the same challenge for standard human disease models even when using humans, as each patient is unique. However, it is possible that the degree of heterogeneity could be compared and monitored using different animal models. This way, an animal model could respond more like a patient population in terms of similar degrees of system heterogeneity and complexity.

For example, to mimic the system heterogeneity of cancer, the following options need to be considered when designing mouse models:

(1) Avoidance of using limited varieties of inbred mice when performing experiments directly related to cancer. If inbred mice are used, experiments should include as many different strains as possible. The establishment of various panels of mice with diverse genetic backgrounds might be necessary. Research is needed to establish such panels using heterogeneous stocks of mice, perhaps even wild mice. By doing so, it is possible that the reliability of animal model experiments will decrease, but the disparity between animal models and

patient populations will also decrease. Statistical and simulation-based experiments are needed to evaluate such panels or platforms to ensure the animal model displays heterogeneous responses toward certain treatments and that these models have relevant predictive ability.

(2) As multiple diverse molecular mechanisms can lead to cancer, in order to create a clinically useful animal model, a large number of animal models each based on one molecular mechanism would be required. For example, all key biological events need to be monitored using animal models including a model of DNA repair deficiency, cell cycle regulation errors, chromosomal segregation mistakes, mitochondrion dysfunction, cell death disorders, and abnormal transcription regulation. Each of these main types will involve hundreds of gene mutations and mutation combinations. Interestingly, it has been gradually realized that most common diseases are caused by arrays of genetic alteration and environmental stresses. Therefore large numbers of diverse animal models would be required for each type of common human disease study. Only within this type of seemingly unrealistic situation could animal models come close to replicating the diversity and complexity seen in patients.

(3) Variable animal housing conditions are needed to mimic the heterogeneous environments found in nature. For this purpose, the age of mice, sex of mice, health status and co-morbidity (weight, general health status, presence of infections), type of food, and quality of environments (the presence of stressful conditions) need to be considered [van der Worp *et al.*, 2010]. In fact it has been recently shown that the environment plays a large role in cancer evolution [Cao *et al.*, 2010]. Even better, evolutionary selection pressure should be considered.

(4) To achieve both genetic and environmental heterogeneity, animal model experiments performed by individual investigators become unrealistic. To solve this issue, major animal model centers are needed for systematic evaluation. On the surface, it sounds very expensive to have this type of comprehensive animal model center. However, when compared to the money spent on small scale experiments and their lack of medical applicability, such a collective effort would be justified. In addition, an international database is needed to deposit all animal model data including all negative data. Strong statistical and simulation teams

are also required by these centers to dynamically monitor and apply the data generated from animal models.

8.7.2.3.3 *Cautions are needed when interpreting current mouse model results*

The following questions should be frequently asked when interpreting the results of mouse models:

(1) Does a specific result derived from a model system represent a general rule or exception?

Biological model systems have been extensively used in genetic research since its inception. From early efforts on *D. melanogaster* to the modern molecular genetic era where *E. coli*, *S. cerevisiae* and *C. elegans* are extensively utilized, each model system has greatly contributed to the accumulation of our biological knowledge. Almost everything we know about the fundamental properties of living cells has come from the study of model organisms. So why have promising results derived from current mouse models failed in the cancer clinic? Is it because the knowledge we obtained from mouse models does not represent the general rule?

It turns out the key difference is that, historically, most bio-research has focused on "parts characterization," and current cancer research needs to face a cellular evolutionary process where the heterogeneity is the main feature (Chapters 5 and 6). In addition, previous studies did not incorporate genome level alterations, which are commonly detected in pathological conditions rather than in normal physiological conditions. Finally, expectations greatly differ between traditional basic research and mouse model work in cancer research. In fact, few are upset when some conclusions from fly research do not fit human conditions. However, there is always huge disappointment when the message from a preclinical mouse model fails in the clinic. The reasoning is rather simple, most mouse models are rationalized to be relevant to cancer, and whether or not the derived data fit the clinical understanding becomes a reality-check. In this sense, it is not unreasonable to question how much of our knowledge from mouse models (one type of artificial knowledge) could be applied to real-world, natural systems. Such conclusion also is applicable to other model systems. For example, approximately 80% of the 6000 genes of *S. cerevisiae* can be deleted under experimental conditions and are therefore defined as nonessential genes. It is hard to imagine that in natural populations the vast majority of genes could be deleted and the species would survive over eons. Surely, under natural conditions, these "nonessential" genes become essential. How applicable to the natural population is any genetic interaction network based on such a model system? No wonder a synthetic lethal pair in *S. cerevisiae* is difficult to validate in *D. melanogaster*. The truth is most well-defined models with reduced heterogeneity represent exceptional cases at best.

(2) Changing the expectations for animal models in human disease

It is therefore important to change expectations when using animal models. First, the approach of single factor analysis might not be ultimately useful for clinical prediction when there are so many factors involved with similar degrees of contribution. In order to illustrate each factor, non-natural conditions are created that result in inconsistent but reproducible results. In doing so, we ignore the importance of the package of life, as evolutionary selection is focused on the total package of life rather than each individual part or pathway. Often, recombining dissected parts cannot explain a complex system, as it is difficult to predict emergent properties based only on known mechanisms of individual parts. Different experimental models afford opportunities to study individual genes in order to investigate genetic, environmental and pathogenic aspects of specific diseases [Taneja and David, 2001]. But the real challenge here is to integrate diverse data into a coherent framework to understand emergent properties of natural populations and to improve clinical predictions based on large numbers of diverse factors.

Second, the conceptual framework must be correct prior to application of any model. Models do not precisely replicate actual patients, but good models should provide relevant information or at least be capable of closely approximating reality. The fundamental problem is the use of correct assumptions, even though many current basic assumptions have in fact led to faulty model designs and interpretations. For example, many have assumed that cancer is a disease of commonly shared gene mutations, and the creation of animal models with the same mutation should mimic the human condition. Tumor responses to certain treatments within homogenous animal systems are assumed to have similar effects on patients. These assumptions are not correct at all in light of the nature of complex systems, where genome level heterogeneity is a key feature.

Third, is it possible to predict system behavior without precisely monitoring the parts within a given system? To solve the problem of predicting natural populations with models, we might need to focus on the overall system behavior rather than specific responses of particular parts [Heng, 2008]. By monitoring the dynamics of phenotypes, it might be possible to determine overall probabilities in individuals (even though it would be challenging to predict an individual's responses based on population profiles, as the monitored factors contribute only a small percentage of probability). For example, it would be useful to know if certain medical interventions would have a broad positive or negative impact on diverse populations using improved animal models. Similarly, studies designed to predict whether interventions would have an effect on limited populations or if certain beneficial effects only apply to certain combinations of genetic and environmental conditions would be of value. This information would be useful in identifying subpopulations of humans for precision medicine. The concept of non-dissectible system heterogeneity in patient populations would also change the way we study and treat human diseases.

Fourth, based on the hypothesis that most common diseases are not caused by common genetic or environmental factors [Heng, 2010], and that there is a high level of inconsistency between lab animals and patient populations caused by the artificially reduced heterogeneity in animal models, negative animal data also need to be re-evaluated. For example, it is possible that certain treatments do not benefit a pure population of animals but can display real benefits in diverse patient populations, as different strains of mice can respond to the same treatment differently. This explains why positive results in patients can also result in negative results in animal models. Furthermore, understanding the issue of heterogeneity is of utmost importance in medicine. Debate on how much risk society is willing to tolerate in the development of new drugs is needed, as the same drug can be beneficial for some but harmful to others. There are examples of drugs being withdrawn from the market due to approximately 1% of users displaying drastic side effects [Topol, 2004]. New approaches and policies are needed to address this issue as more drugs will display heterogeneous responses to patient populations.

In closing, more in depth discussion is now needed to address both new strategies and profound philosophical questions. Knowing that the rationale of using simplified model systems is disadvantageous when applying that knowledge back to the real world of heterogeneity and complexity, what should we do next? The answers to this question will have a profound impact on biological research including how animal models are applied to human diseases and in precision medicine.

8.8 The Beauty of Molecular Specificity or the Mess of High Level Landscape Heterogeneity?

While cancer is a disease with overwhelming heterogeneity, current concepts and research tools are best suitable for linear model analysis with drastically reduced "noise". This fact not only explains many clinical difficulties, but also calls for a comprehensive review of the cancer theories, main technical strategies, and limitations of our previous knowledge.

Unfortunately, when these truly surprising data are presented, the syntheses are rather limited. There is a huge disconnection between new data and the rationale/theory to generate these data, and few want to seriously discuss this issue. For most reports, there are two strongly conflicting messages: On the one hand, authors acknowledge that high heterogeneity is a big surprise (and against original goals); on the other hand, authors can often ignore the massive heterogeneity they just reported and focus on their "success" (identification/targeting specific pathways or business as usual), as if the heterogeneity issue will not impact their pathway analysis. Even for many evolutionary biologists, despite acknowledging mind-bogglingly high levels of genetic diversity, they still like to tell a nice evolutionary story characterized by common genetic changes.

Then, what is the key conceptual difference prior to and after these discoveries? The urgently needed crucial synthesis is to answer the following key questions: does overwhelming heterogeneity fundamentally reduce the significance of a specific pathway? Why have some gene expression profiling studies successfully pinpointed key genes whereas the majority of studies have failed? How do we separate and estimate the contributions of macro- and microcellular evolution in cancer? Can the current gene mutation framework handle these surprises? If a well-accepted theory cannot be used in making correct predictions (as the increase of surprises has demonstrated), what should we, as scientists, do? Until such discussions become routine, the field is not going to change its attitude towards a new paradigm [Heng, 2007a; 2013c; Heng *et al.*, 2015].

Illustrated by previous chapters, searching for a specific molecular cure has been center stage of cancer research for decades, with limited success. It is now time to deal with the less specificity/high heterogeneity defined clinical reality. The myth has been that magic gene-based molecular targets can be eventually be found despite massive heterogeneity, as long as we keep working hard to uncover them. The fact is, high heterogeneity makes the mission impossible for the majority of cases [Heng *et al.*, 2010a; 2010b; 2013a; 2013b]. This requires a new way of thinking, strategies and even new expectations, as when the level of heterogeneity is high in the face of evolutionary selection, the specificity of individual molecular contributions will become less significant (these molecular understandings are useful for publishing papers but less useful in the clinic). The explanation is simple: when researchers focus on a specific pathway in a well-designed linear system or cherry pick favored pathways from massive amounts of data, other possibilities resulting from disease process dynamics are ignored (with the help of statistical tools). Nature does not favor fixed specific pathways under crisis, especially when these dominant pathways observed by researchers are the result of average profiling, as evolution selects the winner from the dynamic process with high unpredictability (and often relies on outliers during transition events, which are not easily detected by current methodologies) (see Chapter 6).

8.8.1 *Genome heterogeneity, transcriptome dynamics and evolutionary potential*

We have recently established the linkage between genome heterogeneity and transcriptome dynamics by comparing cell populations representing both macro- and microcellular evolutionary phases [Heng *et al.*, 2008; Stevens *et al.*, 2013a; 2014]. In this experiment, the transcriptome of the cancer cell population is highly dynamic and less predictable, even though it is possible to identify some gene signatures at certain time points where the population is highly homogenous (within the microcellular evolutionary phase, for example). This case study offers a very good example to explain the conflict between molecular specificity (which can be achieved at a certain phase of cancer evolution, at the individual cell level or even within a given homogenous cell population) and the non-specificity of heterogeneity-mediated evolutionary potential (where in reality, winners can be stochastically selected from a large number of candidates during cancer evolution).

It is well known that it is challenging to identify common expression signatures in cancer samples. Often, variation in expression signatures has been attributed to technological problems. It was hoped that if all technical issues were improved, in a perfect world, the common expression signatures would be detected. However, despite technical improvements made over the last decade, uncertainty still persists.

According to the genome theory, such uncertainty is normal for cancer evolution, and the detected transcriptome dynamics/uncertainty are not simply caused by technical factors, but reflect the internal key features of cancer. Much has been said in previous chapters when discussing fuzzy inheritance, heterogeneity-mediated evolutionary potential and landscape models. To test these ideas, we have systematically compared the pattern of genome alteration and expression dynamics using a model of immortalization to watch the transcriptome in action (see Chapter 5).

RNA was isolated from cell populations representative to prior to and after immortalization. Specifically, passages 7, 17, 25, 54 and 109 were used (see **Fig. 5-4**). According to the previous studies, passages 17 and 25 are within the punctuated phase, and the karyotypes are highly unstable and heterogeneous, while passages 54 and 109 are within the stepwise stage, and the karyotypes are stable and homogenous.

Co-measurement of karyotypic alteration and global gene expression throughout the immortalization process revealed the following:

(1) Karyotype alteration drastically alters gene expression, supporting the concept that genome context defines the network structure reflected by the drastically altered transcriptome (see **Fig. 8-1**).

(2) Karyotypic and gene expression change during immortalization is highly dynamic, as each stage of progression has a unique expression pattern. Across the entire immortalization process, there is no gradual accumulation of the gene expression pattern (such as oncogene expression becoming higher and higher during progression, a prediction of gene mutation theory). In contrast, the expression of a gene fluctuates, especially within the punctuated phase of evolution. This finding demonstrates that the karyotype, transcriptome, and transcriptome-determined pathways are in constant flux during somatic cellular evolution (particularly during the macro-evolutionary phase where karyotype replacement dominates).

(3) Due to the high dynamics observed from the punctuated phase, the expression pattern differs in the stepwise phase as well. This is consistent with the observation that repeating the same experiments results in different expression profiles during both phases. Thus, different runs of evolution display different expression profiles.

(4) Replicate samples from stages with stable genomes (within stepwise phase) are more similar to each other than are replicate samples displaying karyotype heterogeneity (punctuated phase), suggesting that averaging profiling of unstable cancer cell populations yields inaccurate results (or the average cannot be used as a measure in the macro-evolutionary phase), as there is no average cancer cell population when heterogeneity is high (see Chapter 6).

(5) When cell populations enter into the stable stepwise phase (from p54 to p109), dominant pathways become much clearer compared to during the punctuated phase (from p17 to p25). However, different runs of experiments lead to different short lists of genes.

(6) Drug treatment can drastically alter karyotypes leading to new expression profiles. This is especially true for harsh drug treatments, as high stress will push genomes into the chaotic phase, which will lead to rapid drug resistance (see Genome Chaos).

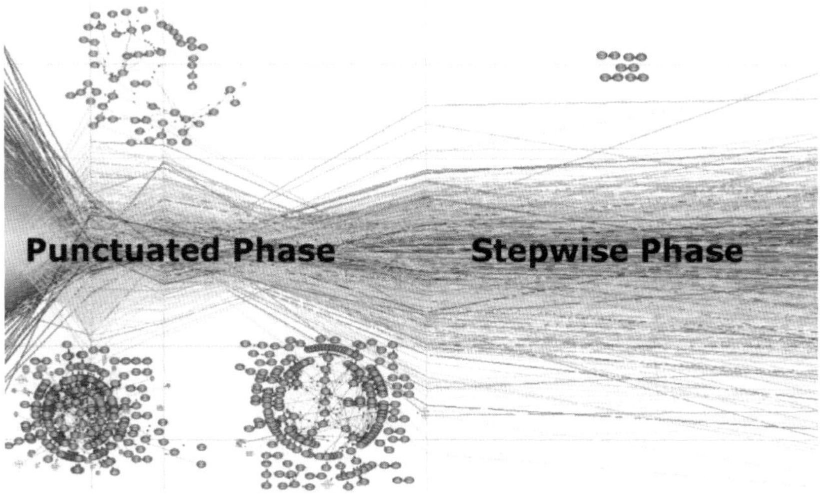

Fig. 8-1. Cell populations with unstable karyotypes display unstable expression.

Line graph of normalized intensity values throughout passages of each expressed transcript. Here each line represents each of the 22,185 transcripts profiled throughout the time course of immortalization. Large changes in expression patterns are evident within the punctuated phase, where extensive karyotypic change has occurred. Within the stepwise phase, the pattern is much more stable when little karyotypic change takes place. Interestingly, all three stages examined within the punctuated phase display complicated network structures (represented by the involvement of large numbers of pathways). In contrast, the network structure is much simpler and does not show much change during the stepwise phase (for more details, please see Stevens *et al.*, 2014). The four pasted neato layout images (three for punctuated phase and one for stepwise phase) of Natural Language Processing pathways were created in Genespring using genes found to be significantly expressed with unconnected nodes removed in nearest neighbor comparisons at each stage (Reproduced with permission from [Stevens *et al.*, 2014]).

8.8.2 Implications

Interestingly, this experiment can explain many confusing issues regarding the roles of genetic specificity and genome heterogeneity. For example, under a highly specific condition, such as a single cell, a relatively pure cell population, a linear model, or a specific stage of investigation (such as the micro-evolutionary phase in an immortalization model), the molecular specificity can be highly significant. That is the reason why the expression profile is somewhat reliable for most stable cell lines or cancer cases displaying more linear evolution. However, when a high level of heterogeneity is involved (especially genome level heterogeneity), such as in a mixed cell population, the selection process under high stress conditions, or during the punctuated phase, "parts" specificity is less important than heterogeneity-mediated evolutionary selection. Under dynamic conditions, it is the evolutionary potential and not a specific pathway that holds the winning ticket. Yes, pathway specificity is still important for a given cell or some sub-populations, but if evolution does not favor these specific cells, they will be eliminated along with the significance of these pathways. Furthermore, most current molecular characterizations are achieved using "average profiling". In reality, however, the macro-evolutionary process often selects outliers as winners, which is beyond our isolated understanding of molecular specificity. That is the reason why a specific pathway holds less importance in the evolutionary process when genome alteration is dominant.

As long as genomes can be effectively changed, transcriptome dynamics will provide increased evolutionary potential for cancer evolution. This result fully supports the evolutionary mechanism of cancer as well as the multiple level landscape model. The significance of specific pathways is diminished by genome heterogeneity. As illustrated in **Fig. 8-2**, in a typical tumor, different cells display highly diverse genomes, which define different genetic networks with different transcriptomes. When genome heterogeneity is high, there are many molecular packages that can be chosen, and each of these packages becomes less significant among large numbers of cells and patients in an evolutionary sense. Clearly, the average genetic signature profile will provide limited information of clinical value for improving long-term treatment benefit, as the winners or surviving tumor cells will be difficult to predict under treatment and certainly will not be of currently dominant cell populations.

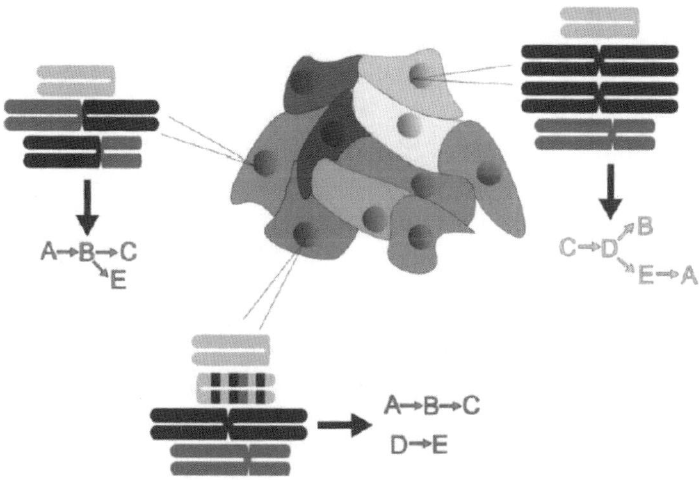

Fig. 8-2. Intratumoral genome and transcriptome diversity.
The mechanism behind tumor heterogeneity can be illustrated by the genome and transcriptome relationship. When the majority of cells display different karyotypes, there is a huge number of diverse molecular specificities available (agents of simplified networks are represented by letters A–E and interactions are represented by arrows). Each of the networks is real but less significant within the game of evolution. Phenotypic heterogeneity derived from karyotypic or genomic heterogeneity increases the probability that one or more cells can escape key selection pressures and acquire phenotypes such as transformation, invasiveness and drug resistance. Since most patients undergo different runs of evolution, molecular specificity is very low amongst the patient population and within the entire process of cancer (Reproduced with permission from [Stevens *et al.*, 2013a]).

Epilogue

Two years have passed since I started this writing project. During this process, I was often advised by colleagues to "Be careful of what you write, as you cannot be too critical of so many popular beliefs..." Swimming against the tide is never a good idea. I still remember when a famous professor told me that being critical to others could be very costly, since you never know who will review your next paper or grant application.

This is very true, and I learned it the hard way. Ever since we realized the importance of stochastic genome alteration in cancer evolution in the early 2000s and began promoting the idea, it has never been easy to publish our findings or secure NIH funding. In times of tight budgets and limited funding for the research community, who would not want to have more friendly reviewers with whom you see eye to eye?

However, I have to be honest to myself and to this extremely important field where people's lives are on the line. Knowing that we had discovered both novel genetic and evolutionary concepts that are crucial to solving many paradoxes of cancer research, I personally believed that I had no choice but to share our discoveries with other researchers to unify the field and to be candid and critical to each issue that deserves debate, especially fundamental topics that prevent the progression of cancer research. This decision came from my scientific conscience. Long before I became a scientist, I dreamed of following the footsteps of so many who searched for the truth no matter how unpopular it was. Despite my nervousness, it is now my opportunity to do just that.

To synthesize common themes from the diverse discussions in this book, I would like to highlight a few take-home messages. In particular, I would like to clarify some of our positions and further explain our rationale to promote them.

First, debating scientific legitimacy and/or major limitations of current popular concepts and strategies is neither quitting nor being pessimistic towards science. Quite the opposite! Searching for a better framework to solve key paradoxes represents having the ultimate optimism in science. The future is full of hope, but identifying the right path is crucial. Rather than continuing current data accumulation under less relevant theories, such as searching for the next molecular pathway-defined magic bullet or molecular profiling massive numbers of samples, the establishment of a new conceptual framework based on bio-complexity and evolution should relieve confusion in the research field and provide better answers to clinical reality.

In the past two decades, various –omics technologies have surprisingly revealed the previously unthinkable genetic and non-genetic complexity to the fields of genetics and genomics [Ball P, 2013; Chakravarti, 2011; Heng, 2009; Heng *et al.*, 2011; Joyner and Prendergast, 2014; Page *et al.*, 2011]. New facts generate confusion and demand new theories of genetics and evolution. For example, the contribution of the DNA sequence to genetic information is drastically reducing, and missing inheritability has increased the knowledge gaps between gene-mediated molecular mechanisms and phenotypes. In fact, these confusions reflect as anomalies in current genetic and evolutionary theory based on the gene-centric view. As discussed in previous chapters, this fact-theory conflict is so significant that we predict a new bio-revolution is right around the corner! Applying this new thinking into cancer research is not just necessary, but the only way out of this puzzlement, as genetic and evolutionary theories are the basis of molecular cancer research.

It is interesting to point out that a major paradigm shift is not an event only for the past or future. We have witnessed paradigm emergence in science, and we should be ready to push this new one. Just in 1998, the discovery that the expansion of the universe is accelerating rather than decelerating revolutionized cosmology. The equally disruptive observation regarding gene and genome in biology in general, and in cancer evolution in particular, should revolutionize cancer research.

The long waited revolution in biology was called upon following the initial completion of the human genome sequencing project [Strohman, 1999]. In recent years, many genetic concepts are finally changing with the movement of celebrating the unknowns and embracing the genome based new paradigm [Heng, 2009; Heng *et al.*, 2011a; Ball, 2013]. Have we now reached a point of transition?

According to Thomas Kuhn, a scientific revolution involves replacing a prevailing dominant paradigm (which defines a scientific world view, together with research strategies and their utility) with a contending paradigm [Kuhn, 1962; Strohman, 1997; 1999]. Such a revolution has several signatures. (1) Incommensurability: The scientists on either side of the paradigmatic divide have ultimate difficulty in understanding the other's way of thinking or reasons for adopting it, as they see the same facts differently. (2) The accumulation of anomalies: The mainstream science of the current paradigm unintentionally produces an increasing number of observations that do not support that paradigm and in fact question technological methods and theoretical framework. The paradigm, with the accumulated anomalies, loses the confidence of scientists in a minority of the field who are then ready and willing for change. (3) The paradigm shift encounters resistance from the old guard: Such resistance to change is based not only on a scientific incommensurability but also on traditional ways of teaching and training new generations in the (old) ways of thinking and research. (4) There is enormous inertia: If there is no fully competent paradigm ready to take its place, the scientific enterprise will continue business as usual. After reading this book, I hope you would agree with us that it is now time for a paradigm shift in cancer research.

Second, the main goal of this book is to initiate debate by asking many important questions, a necessary step in the search for new direction and strategies. It was said that the single most important habit for innovative thinkers is asking questions. Many credit Albert Einstein for saying that given 20 days to solve a problem, he would spend 19 days defining it. Unfortunately in cancer research, the opposite is true. The majority of researchers spend most of their time collecting data and testing specific hypotheses without much effort devoted to defining the conceptual questions in the first place. Asking many key questions forcefully illustrates that the current paradigm of cancer research is less relevant. This is the precondition to searching for a new paradigm and not the other way around.

During a phone conversation with the editor-in-chief of *JAMA*, I was informed that the journal was interested in publishing my commentary titled "The conflict between complex systems and reductionism." In this piece I was critical to the reductionist approach in medicine. The editor mentioned that others had concerns over my message, as even though my points were correct, I did not present specific solutions. I countered, "What if this piece represents the original discovery that the earth is not the center of the universe?" She laughed, and my piece was soon printed in *JAMA* [Heng, 2008]. This short publication has received much attention from the medical community. Even today, we have no idea where the center of the universe is, but departing from the geocentric view was surely fundamental! As science can ultimately self-correct, better strategies will be found with efforts focused on searching for new paradigms while departing from the antiquated.

Third, by outlining the recently accumulated anomalies of cancer research, we are NOT disregarding the significance of studying specific genes and pathways for pure basic research, as the knowledge of defined molecular mechanisms is very interesting from a purely academic point of view, and parts characterization has played an important role in the realization of the knowledge gap between understanding parts and whole system emergence. However, since cancer evolution represents a highly dynamic process, the selection process does not favor fixed, specific pathways, as there are so many and most are not highly penetrant within the patient population. It is thus hard to predict winners among a large number of mutations/pathways and their possible combinations. Each run of somatic cell evolution is different, and all involve multiple cycles of NCCAs and CCAs. This fact reduces the clinical significance of a given molecular mechanism, especially under treatment. The evolutionary mechanism of cancer powerfully explains this point, which is articulated by the multiple level genetic and non-genetic landscape models.

It is necessary to emphasize that the field of cancer research is not just an academic field for the purpose of satisfying scientific curiosity and accumulating knowledge. Ultimately, the field exists to solve the riddles of this terrible disease and produce practical solutions that translate to actual clinical situations. This realization is not a criticism of basic research, but puts basic research into the context of clinical relevance. If some lab approaches look elegant but are totally impractical, we need to reevaluate and have alternative options to solve problems so that they will have clinical utility. For example, we must not drastically reduce the heterogeneous nature of cancer in favor of studying specific molecular mechanisms. Accordingly, researchers should propose fewer artificial hypotheses that obviously hold less clinical implications. With the conception of the evolutionary mechanism of cancer, we predicted an almost unlimited number of individual molecular mechanisms; should we test all of them for the sake of testing? Certainly not! Equally important, it is known that the knowledge gap between the normal physiological process and the pathological process is drastically large, which challenges the traditional viewpoint that one can study the abnormal to understand the normal, or vice versa. It also explains why the knowledge obtained from the developmental process (with high predictability) often fails to explain cancer (with high stochasticity), as well as expected results from an experimental setting frequently differs from the clinical situation. Furthermore, a large portion of our current molecular understanding of cancer is based on exceptions rather than the general rule, as they have been derived from highly selected cases or well controlled experimental conditions. We do appreciate their value and elegance under exceptional conditions, but we understand their fundamental limitations in the vast majority of cancer cases, where stochastic cancer evolution dominates.

Fourth, one challenge is to re-clarify many concepts and scientific language, as many terminologies mean very different things under gene and genome perspectives. For example, the majority of genome researchers do not perform karyotype analysis. Furthermore, there is no specific framework to evaluate or explain genome level changes if there is no linkage to specific genes. Under the status quo, "genetic alteration" mainly refers to gene mutations and copy number variations, and the karyotype level is still ignored by many molecular researchers. When we mention "genome alteration", we refer to the large scale changes of karyotypes, either structurally or numerically, which are identifiable by molecular cytogenetic analysis.

Fifth, despite the diverse nature of the various subjects discussed, we did spell out some specific strategies throughout the book. For example, we need to make the following transitions in cancer evolutionary studies: (1) emphasize the previously ignored genome level rather than concentrate solely on cancer genes/pathways; (2) focus on stochastic genetic events during cancer progression rather than studying only commonly shared fixed patterns; (3) switch from a stepwise linear model to a stochastic punctuated model of evolution; and (4) establish a general evolutionary mechanism of cancer rather than continue categorizing the burgeoning list of individual molecular mechanisms. By accepting the evolutionary mechanism of cancer and understanding the key role environmental stresses play in cancer evolution, the cancer-environment interaction and cancer prevention become priorities. More action needs to be on reducing environmental contributions than just studying genetics.

As for the clinical implication, a systematic analysis is now needed to demonstrate which types of genetic or non-genetic profiles display the highest power of clinical prediction. In particular, the following issues need to be considered: (1) how do we classify patients according to overall genome instability rather than gene mutation profiles? (2) How do we test the new concept of applying higher level system constraint to reduce cancer evolution without inducing genome chaos? (3) Is it possible to develop new therapeutic strategies that involve lower and less specific drug treatment rather than targeting cancer cells with deadly force? (4) How do we balance the benefits and harms of diagnosis and treatment?

At the public health front, the environmental impact on cancer, including potential harms from various commercial products, needs much attention. By and large, the environmental impact on cancer is much more dominant than any specific genetic factor, and many environmental factors can be avoided by societal intervention. Despite the difficulty to identify causative mechanisms of specific chemical and commercial products with regard to cancer, their strong associations with cancer are serious enough for a call to action.

To achieve these goals, new methodologies are needed to compare the individual cell and cell populations during different phases of evolution. One important issue is the drastically different response between individual cells and cell populations under normal physiological and extreme pathological conditions, especially when genome instability is involved. According to the genome theory, the average molecular profile might be a good representation of normal developmental or physiological conditions. However, during pathological processes where there are many stochastic genome alterations involved, the outliers become more meaningful. This is the case in drug resistant clones, as the average is less useful in this situation.

Another challenge is to develop a large scale experimental platform for karyotype analysis. Current SKY analysis requires mitotically active cells that are difficult to access directly from clinical samples without tedious culture procedures. Systems are currently being developed to induce SKY-ready mitotic figures directly from clinical samples within hours. Alternatively, interphase FISH has also been compared with SKY to investigate whether the overall genome heterogeneity could be measured without depending on mitotic figures.

Sixth, we are convinced that establishing the genome theory would be an important step in departing from the current gene mutation theory of cancer. That is the rationale behind making the comparisons of alternative theories in Chapter 3. One important consideration of focusing on the genome level is that the genome defines system inheritance; since it represents the highest level of genetic organization, the genome must be involved in the somatic cell evolutionary process. However, if readers prefer other levels of genetic or non-genetic organization, we encourage them to re-examine all major current concepts of cancer in light of the new data of inheritance and evolution. Such re-examination will likely improve the genome theory, as much more synthesis is needed to mature, or perhaps bring new paradigms to replace the genome theory. If better theories emerge as the result of such research, we are willing to give up ours, as the ultimate goal is to establish the best available theory to advance cancer research.

Finally, we need action NOW! The newly identified inheritance (system inheritance and fuzzy inheritance) and evolutionary mechanisms of cancer have provided a new foundation for future frameworks of cancer research. Since cancer is not just an accident, but likely the price we pay for cellular adaptation, it is less likely that we will eliminate cancer altogether. In contrast, we need to find better methods for cancer prevention and management through constraining strategies without triggering increased unpredictability. For example, many known unknowns in the field of cancer research can be re-interpreted by cancer evolutionary concepts. Despite current efforts, little is known about the basis of cancer drug resistance. How can drug resistance be overcome? How are healthy tissues and genes involved? How does cancer spread (Bourzac, 2014)? As we discussed in Chapters 5–8, as soon as cancer is considered as a genome defined emergent evolutionary system, the answers to these questions become much more obvious.

Likewise, we need to resolve increasing confusion within the research community. For example, authors of many high-profile publications are admitting that cancer is a highly stochastic process, characterized by punctuated evolution where genome chaos is a common phenomenon. However, most of these authors often revert to traditional reasoning and support the stepwise evolutionary pattern in discussions and conclusions. This practice leads to a new paradox: on one hand, they discovered unmanageable genetic diversity, but on the other hand, they cherry-picked a few genes to illustrate their points and claim success, which is against the real meaning of the data. It is hard for readers to understand their true messages. Evidently, such paradoxes represent the need for new frameworks to explain these data. Further accumulation of data will not automatically result in a new theory, however searching for a new framework will.

More researchers (especially future generations) are needed to change the status quo. This will require courage, sacrifice, out-of-the-box thinking, and persistence. The mindset to "just do it" (produce more data) is not enough and could even be wasteful and/or misleading, especially when resources are limited and a new paradigm is urgently needed. For example, cancer genome sequencing has identified many "drivers" and massive "passengers" mutations. Targeting the drivers has become a new buzzword. Unfortunately, the dynamic relationship between "drivers" and "passengers" will once again disappoint us, as without drivers, passengers will just do the job by becoming drivers, or perhaps even better under many unique situations such as treatment induced macrocellular evolution. By the way, many normal tissues also display high levels of typical driver mutations. Clearly, we need a new way of thinking that is not just based in average bio-specificity and part characterization, but a system thinking based on the evolutionary selection among different genome defined packages. Controlling cancer and benefitting patients will require constructive responses and well thought out actions in cancer research. We urge you to act now and keep these debates and changes going!

Bibliography

Abecasis, G.R., Auton, A., Brooks, L.D., DePristo, M.A., Durbin, R.M., Handsaker, R.E., Kang, H.M., Marth, G.T. and McVean, G.A. (2012). An integrated map of genetic variation from 1,092 human genomes, *Nature*, 491, pp. 56–65.

Abdallah, B.Y., Horne, S.D., Stevens, J.B., Liu, G., Ying, A.Y., Vanderhyden, B., Krawetz, S.A., Gorelick, R. and Heng, H.H. (2013). Single cell heterogeneity: Why unstable genomes are incompatible with average profiles, *Cell Cycle*, 12, pp. 3640–3649.

Abdallah, B.Y., Horne, S.D., Kurkinen, M., Stevens, J.B., Liu, G., Ye, C.J., Barbat, J., Bremer, S.W. and Heng, H.H. (2014). Ovarian cancer evolution through stochastic genome alterations: Defining the genomic role in ovarian cancer, *Syst Biol Reprod Med*, 60, pp. 2–13.

Acar, M., Mettetal, J.T. and van Oudenaarden, A. (2008). Stochastic switching as a survival strategy in fluctuating environments, *Nat Genet*, 40, pp. 471–475.

Akagi, T., Sasai, K. and Hanafusa, H. (2003). Refractory nature of normal human diploid fibroblasts with respect to oncogene-mediated transformation, *Proc Natl Acad Sci USA*, 100, pp. 13567–13572.

Aktipis, C.A., Kwan, V.S., Johnson, K.A., Neuberg, S.L. and Maley, C.C. (2011). Overlooking evolution: A systematic analysis of cancer relapse and therapeutic resistance research, *PLoS One*, 6, e26100.

Aktipis, C.A., Boddy, A.M., Gatenby, R.A., Brown, J.S. and Maley, C.C. (2013). Life history trade-offs in cancer evolution, *Nat Rev Cancer*, 13, pp. 883–892.

Albertson, T.M., Ogawa, M., Bugni, J.M., Hays, L.E., Chen, Y., Wang, Y., Treuting, P.M., Heddle, J.A., Goldsby, R.E. and Preston, B.D. (2009).

DNA polymerase epsilon and delta proofreading suppress discrete mutator and cancer phenotypes in mice, *Proc Natl Acad Sci USA*, 106, pp. 17101–17104.

Almendro, V., Cheng, Y.K., Randles, A., Itzkovitz, S., Marusyk, A., Ametller, E., Gonzalez-Farre, X., Muñoz, M., Russnes, H.G., Helland, A., Rye, I.H., Borresen-Dale, A.L., Maruyama, R., van Oudenaarden, A., *et al.*, (2014). Inference of tumor evolution during chemotherapy by computational modeling and in situ analysis of genetic and phenotypic cellular diversity, *Cell Rep*, 6, pp. 514–527.

Andersen, B.L., Yang, H.C., Farrar, W.B., Golden-Kreutz, D.M., Emery, C.F., Thornton, L.M., Young, D.C. and Carson, W.E. (2008). Psychological intervention improves survival for breast cancer patients, *Cancer*, 113, pp. 3450–3458.

Anderson, A.R., Weaver, A.M., Cummings, P.T. and Quarant, V. (2006). Tumor morphology and phenotypic evolution driven by selective pressure from the microenvironment, *Cell*, 127, pp. 905–915.

Annual report of president's cancer panel. (2008–2009). Reducing environmental cancer risk: What we can do now. http://deainfo.nci.nih.gov/advisory/pcp/annualReports/pcp08-09rpt/PCP_Report_08-09_508.pdf

Annual report of the president's panel. (2010–2011). The future of cancer research: Accelerating scientific innovation. http://deainfo.nci.nih.gov/advisory/pcp/annualReports/index.htm

Ao, P., Galas, D., Hood, L., Yin, L. and Zhu, X.M. (2010). Towards predictive stochastic dynamical modeling of cancer genesis and progression, *Interdiscip Sci*, 2, pp. 140–144.

Armitage, P. and Doll, R. (1954). The age distribution of cancer and a multi-stage theory of carcinogenesis, *Br J Cancer*, 8, pp. 1–12.

Armitage, P. and Doll, R. (1957). A two-stage theory of carcinogenesis in relation to the age distribution of human cancer, *Br J Cancer*, 11, pp. 161–169.

Attardi, L.D. (2005). The role of p53-mediated apoptosis as a crucial anti-tumor response to genomic instability: Lessons from mouse models, *Mutat Res*, 569, pp. 145–157.

Aubele, M., Mattis, A., Zitzelsberger, H., Walch, A., Kremer, M., Hutzler, P., Höfler, H. and Werner, M. (1999). Intratumoral heterogeneity in breast carcinoma revealed by laser-microdissection and comparative genomic hybridization, *Cancer Genet Cytogenet*, 110, pp. 94–102.

Baca, S.C., Prandi, D., Lawrence, M.S., Mosquera, J.M., Romanel, A., Drier, Y., Park, K., Kitabayashi, N., Macdonald, T.Y., Ghandi, M., Van Allen, E., Kryukov, G.V., Sboner, A., Theurillat, J.P., Soong, T.D., *et al.*, (2013). Punctuated evolution of prostate cancer genomes, *Cell*, 153, pp. 666–677.

Baker, G.S. (2012/2013). Paradox-driven cancer research, *Disrupt Sci Technol*, 1, pp. 143–148.

Balaban, N.Q., Merrin, J., Chait, R., Kowalik, L. and Leibler, S. (2004). Bacterial persistence as a phenotypic switch, *Science*, 305, pp. 1622–1625.

Baldo, B.A. and Pham, N.H. (2013). Adverse reactions to targeted and non-targeted chemotherapeutic drugs with emphasis on hypersensitivity response and the invasive metastatic switch, *Cancer Metastasis Rev*, 32, pp. 723–761.

Ball, P. (2013). DNA: Celebrate the unknowns, *Nature*, 496, pp. 419–420.

Banfield, W.G., Woke, P.A., Mackay, C.M. and Cooper, H.L. (1965). Mosquito transmission of a reticulum cell sarcoma of hamsters, *Science*, 148, pp. 1239–1240.

Barbieri, C.E., Baca, S.C., Lawrence, M.S., Demichelis, F., Blattner, M., Theurillat, J.P., White, T.A., Stojanov, P., Van Allen, E., Stransky, N., Nickerson, E., Chae, S.S., Boysen, G., Auclair, D., Onofrio, R.C. *et al.*, (2012). Exome sequencing identifies recurrent SPOP, FOXA1 and MED12 mutations in prostate cancer, *Nat Genet*, 44, pp. 685–689.

Bartos, J.D., Gaile, D.P., McQuaid, D.E., Conroy, J.M., Darbary, H., Nowak, N.J., Block, A., Petrelli, N.J., Mittelman, A., Stoler, D.L. and Anderson, G.R. (2007). aCGH local copy number aberrations associated with overall copy number genomic instability in colorectal cancer: Coordinate involvement of the regions including BCR and ABL, *Mutat Res*, 615, pp. 1–11.

Basanta, D., Gatenby, R.A. and Anderson, A.R. (2012). Exploiting evolution to treat drug resistance: Combination therapy and the double bind, *Mol. Pharm*, 9, pp. 914–921.

Bauer, K.H. (1928). *Die Mutationstheorie der Geschwulstentstehung. Übergang von Körperzellen in Geschwulstzellen durch Gen-Änderung*, (Springer, Berlin).

Bauer, K.H. (1949). *Das krebspproblem* (The Cancer Problem), (Springer, Berlin, Gottingen. Heidelberg).

Bayani, J., Selvarajah, S., Maire, G., Vukovic, B., Al-Romaih, K., Zielenska, M. and Squire, J.A. (2006). Genomic mechanisms and measurement of structural and numerical instability in cancer cells, *Semin Cancer Biol*, 17, pp. 5–18.

Bebarta, V., Luyten, D. and Heard, K. (2003). Emergency medicine animal research: Does use of randomization and blinding affect the results? *Acad Emerg Med*, 10, pp. 684–687.

Beerenwinkel, N., Antal, T., Dingli, D., Traulsen, A., Kinzler, K.W., Velculescu, V.E., Vogelstein, B. and Nowak, M.A. (2007). Genetic progression and the waiting time to cancer, *PLoS Comput Biol*, 3, e225.

Begley, C.G. and Ellis, L.M. (2012). Drug development: Raise standards for preclinical cancer research, *Nature*, 483, pp. 531–533.

Belov, K. (2012). Contagious cancer: Lessons from the devil and the dog, *Bioessays*, 34, pp. 285–292.

Bentivegna, S., Zheng, J., Namsaraev, E., Carlton, V.E., Pavlicek, A., Moorhead, M., Siddiqui, F., Wang, Z., Lee. L., Ireland, J.S., Suyenaga, K., Willis, T.D., Faham, M. and Seymour, A.B. (2008). Rapid identification of somatic mutations in colorectal and breast cancer tissues using mismatch repair detection (MRD), *Hum Mutat*, 29, pp. 441–450.

Berenblum, I. and Shubik, P. (1947). A new quantitative approach to the study of the stages of chemical carcinogenesis in the mouse skin, *Br J Cancer*, 1, pp. 383–391.

Berenblum, I. and Shubik, P. (1949). An experimental study of the initiating state of carcinogenesis, and a re-examination of the somatic cell mutation theory of cancer, *Br J Cancer*, 3, pp. 109–118.

Berger, M.F., Lawrence, M.S., Demichelis, F., Drier, Y., Cibulskis, K., Sivachenko, A.Y., Sboner, A., Esgueva, R., Pfluege, D., Sougnez, C., Onofrio, R., Carter, S.L., Park, K., Habegger, L., Ambrogio, L. *et al.*, (2011). The genomic complexity of primary human prostate cancer, *Nature*, 470, pp. 214–220.

Berger, M.F., Hodis, E., Heffernan, T.P., Deribe, Y.L., Lawrence, M.S., Protopopov, A., Ivanova, E., Watson, I.R., Nickerson, E., Ghosh, P., Zhang, H., Zeid, R., Ren, X., Cibulskis, K., Sivachenko, A.Y., *et al.*, (2012). Melanoma genome sequencing reveals frequent PREX2 mutations, *Nature*, 485, pp. 502–506.

Bernard, I., Weinstein, I.B. and Case, K. (2008). The history of cancer research: Introducing an AACR centennial series, *Cancer Res*, 68, pp. 6861–6862.

Bielas, J.H. and Loeb, L.A. (2005). Quantification of random genomic mutations, *Nat Meth*, 2, pp. 285–290.

Bielas, J.H., Loeb, K.R., Rubin, B.P., True, L.D. and Loeb, L.A. (2006). Human cancers express a mutator phenotype, *Proc Natl Acad Sci USA*, 103, pp. 18238–18242.

Biesterfeld, S., Gerres, K., Fischer-Wein, G. and Bocking, A. (1994). Polyploidy in non-neoplastic tissues, *J Clin Pathol*, 471, pp. 38–42.

Bigger, J.W. (1944a). Treatment of staphylococcal infections with penicillin by intermittent sterilisation, *Lancet,* 244, pp. 497–500.

Bigger, J.W. (1944b). The bactericidal action of penicillin on staphylococcus pyogenes, *Irish J Med Sci*, 19, pp. 585–595.

Bishop, N.A. and Guarente, L. (2007). Genetic links between diet and lifespan: Shared mechanisms from yeast to humans, *Nat Rev Genet*, 8, pp. 835–844.

Blelloch, R.H., Hochedlinger, K., Yamada, Y., Brennan, C., Kim, M., Mintz, B., Chin, L. and Jaenisch, R. (2004). Nuclear cloning of embryonal carcinoma cells, *Proc Natl Acad Sci USA*, 101, pp. 13985–13990.

Blount, Z.D., Barrick, J.E., Davidson, C.J. and Lenski, R.E. (2012). Genomic analysis of a key innovation in an experimental Escherichia coli population, *Nature*, 489, pp. 513–518.

Bodmer, W., Bielas, J.H. and Beckman, R.A. (2008). Genetic instability is not a requirement for tumor development, *Cancer Res*, 68, pp. 3558–3560.

Bourzac, K. (2014). Biology: Three known unknowns, *Nature*, 509, pp. S69–71.

Boveri, T. (1914) *Zur Frage der Entstehung maligner Tumoren*, (Gustav Fischer Verlag, Jena, Germany).

Boveri, T. (2008). Concerning the origin of malignant tumours. *J Cell Sci*, 121 (Supplement 1), pp. 1–84.

Brash, D. and Cairns J. (2009). The mysterious steps in carcinogenesis, *Br J Cancer*, 101, pp. 379–380.

Breivik, J. (2005). The evolutionary origin of genetic instability in cancer development, *Semin Cancer Biol*, 15, pp. 51–60.

Butler, D. (2008). Crossing the valley of death (News feature), *Nature*, 453, pp. 840–842.

Cahill, D.P., Kinzler, K.W., Vogelstein, B. and Lengauer, C. (1999). Genetic instability and darwinian selection in tumours, *Trends Cell Biol*, 9, pp. M57–60.

Cairns, J. (1975). Mutation selection and the natural history of cancer, *Nature*, 255, pp. 197–200.

Calabrese, P., Tavaré, S. and Shibata, D. (2004). Pretumor progression: Clonal evolution of human stem cell populations, *Am J Pathol*, 164, pp. 1337–1346.

Callaway, E. (2013). Deal done over HeLa cell line, *Nature*, 500, pp. 132–133.

Campbell, L.L. and Polyak, K. (2007). Breast tumor heterogeneity: Cancer stem cells or clonal evolution? *Cell Cycle*, 6, pp. 2332–2338.

Cancer Genome Atlas Network (2012). Comprehensive molecular characterization of human colon and rectal cancer, *Nature*, 487, pp. 330–337.

Cancer Genome Atlas Research Network (2012). Comprehensive genomic characterization of squamous cell lung cancers, *Nature*, 489, pp. 519–525.

Cao, L., Liu, X., Lin, E.J., Wang, C., Choi, E.Y., Riban, V., Lin, B. and During, M.J. (2010). Environmental and genetic activation of a brain-adipocyte BDNF/leptin axis causes cancer remission and inhibition, *Cell*, 142, pp. 52–64.

Cao, L. and During, M.J. (2012). What is the brain-cancer connection? *Annu Rev Neurosci*, 35, pp. 331–345.

Caulin, A.F. and Maley, C.C. (2011). Peto's Paradox: Evolution's prescription for cancer prevention, *Trends Ecol Evol*, 26, pp. 175.82.

Celton-Morizur, S. and Desdouets, C. (2010). Polyploidization of liver cells, *Adv. Exp. Med. Biol*, 676, pp. 123–135.

Chaffer, C.L., Brueckmann, I., Scheel, C., Kaestli, A.J., Wiggins, P.A., Rodrigues, L.O., Brooks, M., Reinhardt, F., Su, Y., Polyak, K., Arendt, L.M., Kuperwasser, C., Bierie, B. and Weinberg, R.A. (2011). Normal and neoplastic nonstem cells can spontaneously convert to a stem-like state, *Proc Natl Acad Sci USA*, 108, pp. 7950–7955.

Chakravarti, A. (2011). Genomics is not enough, *Science*, 334, pp. 15.

Chambwe, N., Kormaksson, M., Geng, H., De, S., Michor, F., Johnson, N.A., Morin, R.D., Scott, D.W., Godley, L.A., Gascoyne, R.D., Melnick, A., Campagne, F. and Shaknovich, R. (2014). Variability in DNA methylation defines novel epigenetic subgroups of DLBCL associated with different clinical outcomes, *Blood*, 123, pp. 1699–1708.

Chandrakasan, S., Ye, C.J., Chitlur, M., Mohamed, A.N., Rabah, R., Konski, A., Heng, H.H. and Savaşan, S. (2011). Malignant fibrous histiocytoma two years after autologous stem cell transplant for Hodgkin lymphoma: Evidence for genomic instability, *Pediatr Blood Cancer*, 56, pp. 1143–1145.

Chang, S.L., Lai, H.Y., Tung, S.Y. and Leu, J.Y. (2013). Dynamic large-scale chromosomal rearrangements fuel rapid adaptation in yeast populations, *PLoS Genet*, 9, e1003232.

Chen, C.L., Rappailles, A., Duquenne, L., Huvet, M., Guilbaud, G., Farinelli, L., Audit, B., d'Aubenton-Carafa, Y., Arneodo, A., Hyrien, O. and Thermes, C. (2010). Impact of replication timing on non-CpG and CpG substitution rates in mammalian genomes, *Genome Res*, 20, pp. 447–457.

Chen, K.W. and Pienta, K.J. (2011). Modeling invasion of metastasizing cancer cells to bone marrow utilizing ecological principles, *Theor Biol Med Model*, 8, e36.

Chen, Y., Chi, P., Rockowitz, S., Iaquinta, P.J., Shamu, T., Shukla, S., Gao, D., Sirota, I., Carver, B.S., Wongvipat, J., Scher, H.I., Zheng, D. and Sawyers, C.L. (2013). ETS factors reprogram the androgen receptor cistrome and prime prostate tumorigenesis in response to PTEN loss, *Nat Med*, 19, pp. 1023–1029.

Chida, Y., Hamer, M., Wardle, J. and Steptoe, A. (2008). Do stress-related psychological factors contribute to cancer incidence and survival? *Nat Clin Pract Oncol*, 5, pp. 466–475.

Choi, P.S., van Riggelen, J., Gentles, A.J., Bachireddy, P., Rakhra, K., Adam, S.J., Plevritis, S.K and Felsher, D.W. (2011). Lymphomas that recur after MYC suppression continue to exhibit oncogene addiction, *Proc Natl Acad Sci USA*, 108, pp. 17432–17437.

Ciriello, G., Miller, M.L., Aksoy, B.A., Senbabaoglu, Y., Schultz, N. and Sander, C. (2013). Emerging landscape of oncogenic signatures across human cancers, *Nat Genet*, 45, pp. 1127–1133.

Clancy, D.J., Gems, D., Harshman, L.G., Oldham, S., Stocker, H., Hafen, E., Leevers, S.J. and Partridge, L. (2001). Extension of life-span by loss of CHICO, a Drosophila insulin receptor substrate protein, *Science*, 292, pp. 104–106.

Cleary, A.S., Leonard, T.L., Gestl, S.A. and Gunther, E.J. (2014). Tumour cell heterogeneity maintained by cooperating subclones in Wnt-driven mammary cancers, *Nature*, 508, pp. 113–117.

Coccé, M.F., Lubieniecki, U., Kordes, D., Alderete, and Gallego, M. (2011). A complex karyotype in an atypical teratoid/rhabdoid tumor: Case report and review of the literature, *J Neurooncol*, 104, pp. 375–380.

Coffey, D.S., Getzenberg, R.H. and DeWeese, T.L. (2006). Hyperthermic biology and cancer therapies: A hypothesis for the "Lance Armstrong effect", *JAMA*, 296, pp. 445–448.

Cohan, F.M. (2002). What are bacterial species? *Ann Rev Microbiol*, 56, pp. 457–487.

Cohen, A.A., Geva-Zatorsky, N., Eden, E., Frenkel-Morgenstern, M., Issaeva, I., Sigal, A., Milo, R., Cohen-Saidon, C., Liron, Y., Kam, Z., Cohen, L., Danon, T., Perzov, N. and Alon, U. (2008). Dynamic proteomics of individual cancer cells in response to a drug, *Science*, 322, pp. 1511–1516.

Collins, F. and Barker, A.D. (2007). Mapping the cancer genome. Pinpointing the genes involved in cancer will help chart a new course across the complex landscape of human malignancies, *Sci Am*, 296, pp. 50–57.

Collins, F. and Varmus, H. (2015). A new initiative on precision medicine, *N Engl J Med*, 372, pp. 793–795.

Colman, R.J., Anderson, R.M., Johnson, S.C., Kastman, E.K., Kosmatka, K.J., Beasley, T.M., Allison, D.B., Cruzen, C., Simmons, H.A., Kemnitz, J.W. and Weindruch, R. (2009). Caloric restriction delays disease onset and mortality in rhesus monkeys, *Science*, 325, pp. 201–204.

Concato, J. (2013). Probability, uncertainty, and prostate cancer, *Ann Intern Med*, 158, pp. 211–212.

Conrad, D.F., Keebler, J.E., DePristo, M.A., Lindsay, S.J., Zhang, Y., Casals, F., Idaghdour, Y., Hartl, C.L., Torroja, C., Garimella, K. V., Zilversmit, M., Cartwright, R., Rouleau, G.A., Daly, M., Stone, E.A., Hurles, M.E. and Awadalla, P. (2011). Variation in genome-wide mutation rates within and between human families, *Nat Genet*, 43, pp. 712–714.

Copley, S.D. (2012). Moonlighting is mainstream: Paradigm adjustment required, *Bioessays*, 34, pp. 578–588.

Counzin-Frankel, J. (2015). The bad luck of cancer. Analysis suggests most cases can't be prevented, *Science*, 347, pp. 12.

Cox, L.S., Mattison, J.A. (2009). Increasing longevity through caloric restriction or rapamycin feeding in mammals: Common mechanisms for common outcomes? *Aging Cell*, 8, pp. 607–613.

Crasta, K., Ganem, N.J., Dagher, R., Lantermann, A.B., Ivanova, E.V., Pan, Y., Nezi, L., Protopopov, A., Chowdhury, D. and Pellman, D. (2012). DNA breaks and chromosome pulverization from errors in mitosis, *Nature*, 482, pp. 53–58.

Creekmore, A.L., Silkworth, W.T., Cimini, D., Jensen, R.V., Roberts, P.C. and Schmelz, E.M. (2011). Changes in gene expression and cellular architecture in an ovarian cancer progression model, *PLoS One*, 63, pp. e17676.

Crespi, B. and Summers, K. (2005). Evolutionary biology of cancer, *Trends Ecol Evol*, 20, pp. 545–552.

Crespi, B.J. and Summers, K. (2006). Positive selection in the evolution of cancer, *Biol Rev Camb Philos Soc*, 81, pp. 407–424.

Crespi, B., Foster, K. and Francisco Úbeda, F. (2014). First principles of Hamiltonian medicine, *Phil Trans R Soc B*, 369, 20130366.

Dang, C.V. (2013). MYC, metabolism, cell growth, and tumorigenesis, *Cold Spring Harb Perspect Med*, 3, e8.

Davies, P. (2011). Cancer: The beat of an ancient drum? Rather than rogue cells gone berserk, cancers may be the foot soldiers of ages-old atavisms, The Guardian, Monday 25 April 2011.

Davoli, T. and de Lange, T. (2011). The causes and consequences of polyploidy in normal development and cancer, *Annu Rev Cell Dev Biol*, 27, pp. 585–610.

de Pagter, M.S., van Roosmalen, M.J., Baas, A.F., Renkens, I., Duran, K.J., van Binsbergen, E., Tavakoli-Yaraki, M., Hochstenbach, R., van der Veken, L.T., Cuppen, E. and Kloosterman, W.P. (2015). Chromothripsis in healthy individuals affects multiple protein-coding genes and can result in severe congenital abnormalities in offspring, *Am J Hum Genet*, 96, pp. 651–656.

de The, H. and Chen, Z. (2010). Acute promyelocytic leukaemia: Novel insights into the mechanisms of cure, *Nat Rev Cancer*, 10, pp. 775–783.

Deng, C.X. and Brodie, S.G. (2001). Knockout mouse models and mammary tumorigenesis, *Semin Cancer Biol*, 11, pp. 387–394.

Dewhurst, S.M., McGranahan, N., Burrell, R.A., Rowan, A.J., Grönroos, E., Endesfelder, D., Joshi, T., Mouradov, D., Gibbs, P., Ward, R.L., Hawkins, N.J., Szallasi, Z., Sieber, O.M. and Swanton, C. (2014). Tolerance of whole-genome doubling propagates chromosomal instability and accelerates cancer genome evolution, *Cancer Discov*, 4, pp. 175–185.

Dexter, D.L., Kowalski, H.M., Blazar, B.A., Fligiel, Z., Vogel, R. and Heppner, G.H. (1978). Heterogeneity of tumor cells from a single mouse mammary tumor, *Cancer Res*, 38, pp. 3174–3181.

Dolberg, D.S. and Bissell, M.J. (1984). Inability of Rous sarcoma virus to cause sarcomas in the avian embryo, *Nature*, 309, pp. 552–556.

Druker, B.J., Tamura, S., Buchdunger, E., Ohno, S., Segal, G.M., Fanning, S., Zimmermann, J. and Lydon, N.B. (1996). Effects of a selective inhibitor of the Abl tyrosine kinase on the growth of Bcr-Abl positive cells, *Nat Med*, 2, pp. 561–566.

Duesberg, P.H. and Vogt, P.K. (1970). Differences between the ribonucleic acids of transforming and nontransforming avian tumor viruses, *Proc Natl Acad Sci USA*, 67, pp. 1673–1680.

Duesberg, P., Rausch, C., Rasnick, D. and Hehlmann, R. (1998). Genetic instability of cancer cells is proportional to their degree of aneuploidy, *Proc Natl Acad Sci USA*, 95, pp. 13692–13697.

Duesberg, P. and Rasnick, D. (2000). Aneuploidy, the somatic mutation that makes cancer a species of its own, *Cell Motil Cytoskeleton*, 47, pp. 81–107.

Duesberg, P., Mandrioli, D., McCormack, A. and Nicholson, J.M. (2011). Is carcinogenesis a form of speciation? *Cell Cycle*, 10, pp. 2100–2114.

Durfee, L.A., Lyon, N., Seo, K. and Huibregtse, J.M. (2010). The ISG15 conjugation system broadly targets newly synthesized proteins: Implications for the antiviral function of ISG15, *Mol Cell*, 38, pp. 722–732.

Eldredge, N. and Gould, S.J. (1972). *Models in Paleobiology*, ed. Schopf, T.J.M. "Punctuated equilibria: An alternative to phyletic gradualism," (Freeman Cooper, San Francisco) pp. 82–115.

Elenbaas, B., Spirio, L., Koerner, F., Fleming, M.D., Zimonjic, D.B., Donaher, J.L., Popescu N.C., Hahn, W.C. and Weinberg, R.A. (2001). Human breast cancer cells generated by oncogenic transformation of primary mammary epithelial cells, *Genes Dev*, 15, pp. 50–65.

Elowitz, M.B., Levine, A.J., Siggia, E.D. and Swain, P.S. (2002). Stochastic gene expression in a single cell, *Science*, 297, pp. 1183–1186.

Esteller, M. (2008). Epigenetics in cancer, *N Engl J Med*, 358, pp. 1148–1159.

Ewald, P.W. (2009). An evolutionary perspective on parasitism as a cause of cancer, *Adv Parasitol*, 68, pp. 21–43.

Faggioli, F., Vijg, J. and Montagna, C. (2011). Chromosomal aneuploidy in the aging brain, *Mech Ageing Dev*, 132, pp. 429–436.

Fallows, J. (2014). When Will Genomics Cure Cancer? A conversation with the biogeneticist Eric S. Lander about how genetic advances are transforming medical treatment. The Atlantic. January/ February Issue.

Fearon, E.R. and Vogelstein, B. (1990). A genetic model for colorectal tumorigenesis, *Cell*, 61, pp. 759–767.

Feinberg, A.P., Ohlsson, R. and Henikoff, S. (2006). The epigenetic progenitor origin of human cancer, *Nat Rev Genet*, 7, pp. 21–33.

Feinberg, A.P. and Irizarry, R.A. (2010). Evolution in health and medicine Sackler colloquium: Stochastic epigenetic variation as a driving force of development, evolutionary adaptation, and disease, *Proc Natl Acad Sci USA*, 107, Suppl 1, pp. 1757–1764.

Feinberg, A.P. (2014). Epigenetic stochasticity, nuclear structure and cancer: The implications for medicine, *J Intern Med*, 276, pp. 5–11.

Feller, S.M. and Lewitzky, M. (2012). Very 'sticky' proteins – not too sticky after all? *Cell Commun Signal*, 10, pp. 15.

Feng, Z., Hu, W., Rajagopal, G. and Levine, A.J. (2008). The tumor suppressor p53: Cancer and aging, *Cell Cycle*, 7, pp. 842–847.

Ferguson-Smith, M.A. and Trifonov, V. (2007). Mammalian karyotype evolution, *Nat Rev Genet*, 8, pp. 950–962.

Fernandez, A.A. and Morris, M.R. (2008). Mate choice for more melanin as a mechanism to maintain a functional oncogene, *Proc Natl Acad Sci USA*, 105, pp. 13503–13507.

Fidler, I.J. and Hart, I.R. (1981). Biological and experimental consequences of the zonal composition of solid tumors, *Cancer Res*, 41, pp. 3266–3267.

Fillmore, C. and Kuperwasser, C. (2007). Human breast cancer stem cell markers CD44 and CD24: Enriching for cells with functional properties in mice or in man? *Breast Cancer Res*, 9, e303.

Fisher, J.C. and Hollomon, J.H. (1951). A hypothesis for the origin of cancer foci, *Cancer*, 4, pp. 916–918.

Fisher, J.C. (1958). Multiple-mutation theory of carcinogenesis, *Nature*, 181, pp. 651–652.

Fojo, T. and Grady, C. (2009). How much is life worth: Cetuximab, non-small cell lung cancer, and the $440 billion question, *J Natl Cancer Inst*, 101, pp. 1044–1048.

Forment, J.V., Kaidi, A. and Jackson, S.P. (2012). Chromothripsis and cancer: Causes and consequences of chromosome shattering, *Nat Rev Cancer*, 12, pp. 663–670.

Foulds, L. (1954). The experimental study of tumor progression: A review, *Cancer Res*, 14, pp. 327–339.

Foulds, L. (1975). *Neoplastic Development*, Vol. 1. (Academic Press, New York).

Fox, E.J. and Loeb, L.A. (2014). Cancer: One cell at a time, *Nature*, 512, pp. 143–144.

Fox, E.J., Salk, J.J. and Loeb, L.A. (2009). Cancer genome sequencing—an interim analysis, *Cancer Res*, 69, pp. 4948–4950.

Fragouli, E. and Wells, D. (2011). Aneuploidy in the human blastocyst, *Cytogenet. Genome Res*, 133, pp. 149–159.

Frank, S.A. and Nowak, M.A. (2004). Problems of somatic mutation and cancer, *Bioessays*, 26, pp. 291–299.

Frei, E., 3rd and Canellos, G. P. (1980). Dose: A critical factor in cancer chemotherapy, *Am J Med*, 69, pp. 585–594.

Frei, E., 3rd, Elias, A., Wheeler, C., Richardson, P. and Hryniuk, W. (1998). The relationship between high-dose treatment and combination chemotherapy: The concept of summation dose intensity, *Clin Cancer Res*, 4, 2027–2037.

Frese, K.K. and Tuveson, D.A. (2007). Maximizing mouse cancer models, *Nat Rev Cancer*, 7, pp. 645–658.

Friedewald, W.F. and Rous, P. (1944). The initiating and promoting elements in tumor production: An analysis of the effects of tar, benzpyrene, and methylcholanthrene on rabbit skin, *J Exp Med*, 80, pp. 101–126.

Fujii, H., Szumel, R., Marsh, C., Zhou, W. and Gabrielson, E. (1996). Genetic progression, histological grade, and allelic loss in ductal carcinoma in situ of the breast, *Cancer Res*, 56, pp. 5260–5265.

Fukasawa, K., Wiener, F., VandeWoude, G.F. and Mai, S.B. (1997). Genomic instability and apoptosis are frequent in p53 deficient young mice, *Oncogene*, 15, pp. 1295–1302.

Galipeau, P.C., Li, X., Blount, P.L., Maley, C.C., Sanchez, C.A., Odze, R.D., Ayub, K., Rabinovitch, P.S., Vaughan, T.L. and Reid, B.J. (2007). NSAIDs modulate CDKN2A, TP53, and DNA content risk for progression to esophageal adenocarcinoma, *PLoS Med.*, 4, e67.

Garber, K. (2005). Human cancer genome project moving forward despite some doubts in community, *J Natl Cancer Inst*, 97, pp. 1322–1324.

Garitaonandia, I., Amir, H., Boscolo, F.S., Wambua, G.K., Schultheisz, H.L., Sabatini, K., Morey, R., Waltz, S., Wang, Y.C., Tran, H., Leonardo, T.R., Nazor, K., Slavin, I., Lynch, C., Li, Y., Coleman, R., *et al.*, (2015). Increased risk of genetic and epigenetic instability in human embryonic stem cells associated with specific culture conditions, *PLoS One*, 10, e0118307.

Garraway, L.A. and Lander, E.S. (2013). Lessons from the Cancer Genome, *Cell*, 153, pp. 17–37.

Garsed, D.W., Marshall, O.J., Corbin, V.D., Hsu, A., Di Stefano, L., Schröder, J., Li, J., Feng, Z.P., Kim, B.W., Kowarsky, M., Lansdell, B., Brookwell, R., Myklebost, O., Meza-Zepeda, L., Holloway, A.J., *et al.*, (2014). The architecture and evolution of cancer neochromosomes, *Cancer Cell*, 26, pp. 653–667.

Gärtner, H.V., Seidl, C., Luckenbach, C., Schumm, G., Seifried, E., Ritter, H. and Bültmann, B. (1996). Genetic analysis of a sarcoma accidently transplanted from a patient to a surgeon, *N Engl J Med*, 335, pp. 1494–1496.

Gatenby, R.A. (1996). Application of competition theory to tumour growth: Implications for tumour biology and treatment, *Eur J Cancer*, 32A, pp, 722–726.

Gatenby, R.A. (2009). A change of strategy in the war on cancer, *Nature*, 459, pp. 508–509.

Gatenby, R.A., Silva, A.S., Gillies, R.J. and Frieden, B.R. (2009). Adaptive therapy, *Cancer Res*, 69, pp. 4894–4903.

Gawrylewski, A. (2007). The trouble with animal models, *The Scientist*, 21, July 1, 2007.

Gentric, G., Desdouets, C. and Celton-Morizur, S. (2012). Hepatocytes polyploidization and cell cycle control in liver physiopathology, *Int J Hepatol* 2012, e282430.

Germain, P.L. (2012). Cancer cells and adaptive explanations, *Biol Philos*, 27, pp. 785–810.

Gibbs, W.W. (2003). Untangling the roots of cancer, *Sci Am*, 289, pp. 56–65.

Gillies, R.J., Verduzco, D. and Gatenby, R.A. (2012). Evolutionary dynamics of carcinogenesis and why targeted therapy does not work, *Nat Rev Cancer*, 12, pp. 487–493.

Giorgi, E.E. (2015). From many, one, *The Scientist Magazine*, 4, pp. 50–55.

Gisselsson, D. (2003). Chromosome instability in cancer: How, when, and why? *Adv Cancer Res*, 87, pp. 1–29.

Gluckman, P.D., Beedle, A.S., Hanson, M.A. and Low, F.M. (2013). Human growth: Evolutionary and life history perspectives, *Nestle Nutr Inst Workshop Ser*, 71, pp. 89–102.

Gluckman, P.D., Hanson, M.A. and Low, F.M. (2011). The role of developmental plasticity and epigenetics in human health, *Birth Defects Res C Embryo Today*, 93, pp. 12–18.

Godefroy, N., Lemaire, C., Renaud, F., Rincheval, V., Perez, S., Parvu-Ferecatu, I., Mignotte, B. and Vayssière, J.L. (2004). P53 can promote mitochondria- and caspase-independent apoptosis, *Cell Death Differ*, 11, pp. 785–787.

Gorelick, R. and Carpinone, J. (2009). Origin and maintenance of sex: The evolutionary joys of self sex, *Biol J Linn Soc*, 98, pp. 707–728.

Gorelick, R. and Heng, H.H. (2011). Sex reduces genetic variation: A multidisciplinary review, *Evolution*, 65, pp. 1088–1098.

Gould, S.J. and Eldredge, N. (1993). Punctuated equilibrium comes of age, *Nature*, 366, pp. 223–227.

Grant, A. (2009). The big idea that might beat cancer and cut health-care costs by 80 percent, *Discover Magazine*, Sept 30, 2009.

Graphodatsky, A.S., Trifonov, V.A. and Stanyon, R. (2011). The genome diversity and karyotype evolution of mammals, *Mol Cytogenet*, 4, e22.

Grasso, C.S., Wu, Y.M., Robinson, D.R., Cao, X., Dhanasekaran, S.M., Khan, A.P., Quist, M.J., Jing, X., Lonigro, R.J., Brenner, J.C., Asangani, I.A., Ateeq, B., Chun, S.Y., Siddiqui, J., Sam, L., *et al.*, (2012). The mutational landscape of lethal castration-resistant prostate cancer, *Nature*, 487, pp. 239–243.

Greene, C.S., Penrod, N.M., Williams, S.M. and Moore, J.H. (2009). Failure to replicate a genetic association may provide important clues about genetic architecture, *PLoS One*, 4, e5639.

Grimm, J., Kirsch, D.G., Windsor, S.D., Kim, C.F., Santiago, P.M., Ntziachristos, V., Jacks, T. and Weissleder, R. (2005). Use of gene expression profiling to direct in vivo molecular imaging of lung cancer, *Proc Natl Acad Sci USA*, 102, pp. 14404–14409.

Gupta, P.B., Fillmore, C.M., Jiang, G., Shapira, S.D., Tao, K., Kuperwasser, C. and Lander, E.S. (2011). Stochastic state transitions give rise to phenotypic equilibrium in populations of cancer cells, *Cell*, 146, pp. 633–644.

Hahn, W.C., Counter, C.M., Lundberg, A.S., Beijersbergen, R.L., Brooks, M.W. and Weinberg, R.A. (1999). Creation of human tumour cells with defined genetic elements, *Nature*, 400, pp. 464–468.

Hanahan, D. and Weinberg, R.A. (2000). The hallmarks of cancer, *Cell*, 100, pp. 57–70.

Hanahan, D. and Weinberg, R.A. (2011). Hallmarks of cancer: The next generation, *Cell*, 144, pp. 646–674.

Hansemann, D.P. (1897) *Die mikroskopische Diagnose der boesartigen Geschwuelste*, (Berlin: August Hirschwald).

Harewood, L., Schutz, F., Boyle, S., Perry, P., Delorenzi, M., Bickmore, W.A. and Reymond, A. (2010). The effect of translocation-induced nuclear reorganization on gene expression, *Genome Res*, 20, pp. 554–564.

Harris H. (2005). A long view of fashions in cancer research, *Bioessays*, 27, pp. 833–838.

Hayden, E.C. (2009). Cutting off cancer's supply lines, *Nature*, 458, pp. 686–687.

Hayden, E.C. (2010a). Human genome at ten: Life is complicated, *Nature*, 464, pp. 664–667.

Hayden, E.C. (2010b). Sex bias blights drug studies, *Nature*, 464, pp. 332–333.

Henderson, S., Chakravarthy, A., Su, X., Boshoff, C. and Fenton, T.R. (2014). APOBEC-mediated cytosine deamination links PIK3CA helical domain mutations to human papillomavirus-driven tumor development, *Cell Rep*, 7, pp. 1833–1841.

Heng, H.H. (2007a). Cancer genome sequencing: The challenges ahead, *Bioessays*, 29, pp. 783–794.

Heng, H.H. (2007b). Karyotypic chaos, a form of non-clonal chromosome aberrations, plays a key role for cancer progression and drug resistance, *FASEB: Nuclear Structure and Cancer*. Vermont Academy, Saxtons River, Vermont.

Heng, H.H. (2007c). Elimination of altered karyotypes by sexual reproduction preserves species identity, *Genome*, 50, pp. 517–524.

Heng, H.H. (2008). The conflict between complex systems and reductionism. *JAMA*, 300, pp. 1580–1581.

Heng, H.H. (2009). The genome-centric concept: Resynthesis of evolutionary theory, *Bioessays*, 31, pp. 512–525.

Heng, H.H. (2010). Missing heritability and stochastic genome alterations, *Nat Rev Genet*, 11, pp. 813.

Heng, H.H. (2013a). Genomics: HeLa genome versus donor's genome, *Nature*, 501, pp. 167.

Heng, H.H. (2013b). Henrietta Lacks (HeLa) cell genome: Cell line identity and the personal privacy, in Molecular Cytogenetics Blog. (Biomed Central, 2013). http://blogs.biomedcentral.com/on-biology/2013/09/17/debating-the-hela-cell-controversy-karyotype-defined-cell-line-identity-and-the-privacy-of-patients-2/

Heng, H.H. (2013c). *Handbook on Systems and Complexity in Health*, eds. Sturmberg, J.P. and Martin, C.M., Chapter 12 "Bio-complexity: Challenging reductionism," (Springer, New York) pp. 193–208.

Heng, H.H. (2014). Distinguishing constitutional and acquired nonclonal aneuploidy, *Proc Natl Acad Sci USA*, 111, E972.

Heng, H.Q. and Chen, W.Y. (1985). The study of the chromatin and the chromosome structure for *Bufo gargarizans* by the light microscope, *J Sichuan Univ Nat Sci*, 2, pp 105–109.

Heng, H.Q., Chen, W.Y. and Yosida, T.H. (1987). Studies on amphibian chromosome by the high resolution banding, *Proc Japan Acad*, 62, pp. 53–55.

Heng, H.Q., Chen, W.Y. and Wang, Y.C. (1988). Effects of pingyanymycin on chromosomes: A possible structural basis for chromosome aberration, *Mutat Res*, 199, pp. 199–205.

Heng, H.H., Tsui, L.C. and Moens, P.B. (1994). Organization of heterologous DNA inserts on the mouse meiotic chromosome core *Chromosoma*, 103, pp. 401–407.

Heng, H.H., Chamberlain, J.W., Shi, X.M., Spyropoulos, B., Tsui, L.C. and Moens, P.B. (1996). Regulation of meiotic chromatin loop size by chromosomal position, *Proc Natl Acad Sci USA*, 93, pp. 2795–2800.

Heng, H.H., Liu, G., Lu, W., Bremer, S., Ye, C.J., Hughes, M. and Moens, P. (2001a). Spectral karyotyping (SKY) of mouse meiotic chromosomes, *Genome*, 44, pp. 293–298.

Heng, H.H., Krawetz, S.A., Lu, W., Bremer, S., Liu, G. and Ye, C.J. (2001b). Re-defining the chromatin loop domain, *Cytogenet Cell Genet*, 93, pp. 155–161.

Heng, H.H., Ye, C.J., Yang, F., Ebrahim, S., Liu, G., Bremer, S.W., Thomas, C.M., Ye, J., Chen, T.J., Tuck-Muller, C., Yu, J.W., Krawetz, S.A. and Johnson, A. (2003). Analysis of marker or complex chromosomal rearrangements present in pre- and post-natal karyotypes utilizing a combination of G-banding, spectral karyotyping and fluorescence in situ hybridization, *Clin Genet*, 63, pp. 358–367.

Heng, H.H., Stevens, J.B., Liu, G., Bremer, S.W., Ye, C.J. (2004a). Imaging genome abnormalities in cancer research, *Cell Chromosome*, 3:e1.

Heng, H.H., Goetze, S., Ye, C.J., Liu, G., Stevens, J.B., Bremer, S.W., Wykes, S.M., Bode, J. and Krawetz, S.A. (2004b). Chromatin loops are selectively anchored using scaffold/matrix-attachment regions, *J Cell Sci*, 117, pp. 999–1008.

Heng, H.H., Stevens, J.B., Liu, G., Bremer, S.W., Ye, K.J., Reddy, P.V., Wu, G.S., Wang, Y.A., Tainsky, M.A. and Ye, C.J. (2006a). Stochastic cancer progression driven by non-clonal chromosome aberrations, *J Cell Physiol*, 208, pp. 461–472.

Heng, H.H., Bremer, S.W., Stevens, J., Ye, K.J., Miller, F., Liu, G. and Ye, C.J. (2006b). Cancer progression by non-clonal chromosome aberrations, *J Cell Biochem*, 98, pp. 1424–1435

Heng, H.H., Liu, G., Bremer, S., Ye, K.J., Stevens, J., Ye, C.J. (2006c). Clonal and non-clonal chromosome aberrations and genome variation and aberration, *Genome*, 49, pp. 195–204.

Heng, H.H., Stevens, J.B., Lawrenson, L., Liu, G., Ye, K.J., Bremer, S.W. and Ye, C.J. (2008). Patterns of genome dynamics and cancer evolution, *Cell. Oncol*, 30, pp. 513–514.

Heng, H.H., Bremer, S.W., Stevens, J.B., Ye, K.J., Liu, G. and Ye, C.J. (2009). Genetic and epigenetic heterogeneity in cancer: A genome-centric perspective, *J Cell Physiol*, 220, pp. 538–547.

Heng, H.H., Stevens, J.B., Bremer, S.W., Ye, K.J., Liu, G. and Ye, C.J. (2010a). The evolutionary mechanism of cancer, *J Cell Biochem*, 220, pp. 538–547.

Heng, H.H., Liu, G., Stevens, J.B., Bremer, S.W., Ye, K.J. and Ye, C.J. (2010b). Genetic and epigenetic heterogeneity in cancer: The ultimate challenge for drug therapy, *Curr Drug Targets*, 11, pp. 1304–1316.

Heng, H.H., Liu, G., Stevens, J.B., Bremer, S.W., Ye, K.J., Abdallah, B.Y., Horne, S.D. and Ye, C.J. (2011a). Decoding the genome beyond sequencing: The new phase of genomic research, *Genomics*, 98, pp. 242–252.

Heng, H.H., Stevens, J.B., Bremer, S.W., Liu, G., Abdallah, B.Y. and Ye, C.J. (2011b). Evolutionary mechanisms and diversity in cancer, *Adv Cancer Res*, 112, pp. 217–253.

Heng, H.H., Liu, G., Stevens, J.B., Abdallah, B.Y., Horne, S.D., Ye, K.J., Bremer, S.W., Chowdhury, S.K. and Ye, C.J. (2013a). Karyotype heterogeneity and unclassified chromosomal abnormalities, *Cytogenet Genome Res*, 139, pp. 144–157.

Heng, H.H., Bremer, S.W., Stevens, J.B., Horne, S.D., Liu, G., Abdallah, B.Y., Ye, K.J. and Ye, C.J. (2013b). Chromosomal instability (CIN): what it is and why it is crucial to cancer evolution, *Cancer Metastasis Rev*, 32, pp. 325–340.

Heng, H.H., Horne, S.D., Stevens, J.B., Abdallah, B.Y., Liu, G., Chowdhury, S.K., Bremer, S.B., Zhang, K. and Ye, C.J. (2015). Heterogeneity mediated system complexity: The ultimate challenge for studying common and complex diseases, *Proceeding of The First International*

Conference on Systems and Complexity in Health, (Springer, New York) (in press).

Heppner, H.G. (1984). Tumor heterogeneity, *Cancer Res*, 44, pp. 2259–2265.

Heppner, G.H. and Miller, F.R. (1998). The cellular basis of tumor progression, *Int Rev Cytol*, 177, pp. 1–56.

Heppner, G.H. (1989). Tumor cell societies, *J Natl Cancer Inst*, 81, pp. 648–649.

Heppner, G.H. and Miller, B.E. (1989). Therapeutic implications of tumor heterogeneity, *Semin Oncol*, 16, pp. 91–105.

Heyer, J., Kwong, L.N., Lowe, S.W. and Chin, L. (2010). Non-germline genetically engineered mouse models for translational cancer research, *Nat Rev Cancer*, 10, pp. 470–480.

Hochedlinger, K., Blelloch, R., Brennan, C., Yamada, Y., Kim, M., Chin, L. and Jaenisch, R. (2004). Reprogramming of a melanoma genome by nuclear transplantation, *Genes Dev*, 18, pp. 1875–1885.

Holland, A.J. and Cleveland, D.W. (2012). Chromoanagenesis and cancer: Mechanisms and consequences of localized, complex chromosomal rearrangements, *Nat Med*, 18, pp. 1630–1638.

Hood, L. (2014). Statement for Institute for System Biology. http://www.systemsbiology.org/isbresearch#sthash.7IlZVKqN.dpuf

Horne, S.D., Stevens, J.B., Abdallah, B.Y., Liu, G., Bremer, S.W., Ye, C.J. and Heng, H.H. (2013a). Why imatinib remains an exception of cancer research, *J Cell Physiology*, 228, pp. 665–670.

Horne, S. D., Abdallah, B.Y., Stevens, J. B., Liu, G., Ye, K. J., Bremer, S. W. and Heng, H. H. (2013b). Genome constraint through sexual reproduction: Application of 4D-Genomics in reproductive biology, *Syst Biol Reprod Med*, 59, pp. 124–130.

Horne, S.D., Chowdhury, S.K. and Heng, H.H. (2014). Stress, genomic adaptation, and the evolutionary trade-off, *Front Genet*, 5, e92.

Horne, S. and Heng, H. (2014). Genome chaos, chromothripsis and cancer evolution, *J Cancer Stud Ther*, 2014 Sep 22.

Horne, S.D., Pollick, S.A. and Heng, H.H. (2015a). Evolutionary mechanism unifies the hallmarks of cancer, *Int J Cancer*, 136, pp. 2012–2021.

Horne, S.D., Ye, C.J., Abdallah, B.Y., Liu, G. and Heng, H.H. (2015b). Cancer genome evolution, *Transl Cancer Res*, 4, pp. 303–313.

Huang, S., Ernberg, I. and Kauffman, S. (2009). Cancer attractors: A systems view of tumors from a gene network dynamics and developmental perspective, *Semin Cell Dev Biol*, 20, pp. 869–876.

Huang, S. (2013). Genetic and non-genetic instability in tumor progression: Link between the fitness landscape and the epigenetic landscape of cancer cells, *Cancer Metastasis Rev*, 32, pp. 423–448.

Hudson, K.L. and Collins, F.S. (2013). Biospecimen policy: Family matters, *Nature*, 500, pp. 141–142.

Hudson, T.J. Anderson, W. Artez, A. Barker, A.D. Bell, C. Bernabé, R.R. Bhan, M.K. Calvo, F. Eerola, I. Gerhard, D.S. Guttmacher, A. Guyer, M. Hemsley, F.M. Jennings, J. L. Kerr, D., *et al.*, (2010). International network of cancer genome projects, *Nature*, 464, pp. 993–998.

Hultén, M.A., Jonasson, J., Iwarsson, E., Uppal, P., Vorsanova, S.G., Yurov, Y.B. and Iourov, I.Y. (2013). Trisomy 21 mosaicism: We may all have a touch of Down syndrome, *Cytogenet Genome Res*, 139, pp. 189–192.

Huxley, J. (1956). Cancer biology: Comparative and genetic, *Biol Rev*, 31, pp. 474–514.

Illmensee, K. and Mintz, B. (1976). Totipotency and normal differentiation of single teratocarcinoma cells cloned by injection into blastocysts, *Proc Natl Acad Sci USA*, 73, pp. 549–553.

Inaki, K. and Liu, E.T. (2012). Structural mutations in cancer: Mechanistic and functional insights, *Trends Genet*, 28, pp. 550–559.

International Cancer Genome Consortium, 2010. https://icgc.org/

International Human Genome Sequencing Consortium. (2001). Initial sequencing and analysis of the human genome, *Nature*, 409, pp. 860–921

International Human Genome Sequencing Consortium. (2004). Finishing the euchromatic sequence of the human genome, *Nature*, 431, 931–945.

Iourov, I.Y., Vorsanova, S.G. and Yurov, Y.B. (2008). Chromosomal mosaicism goes global, *Mol Cytogenet*, 1, e26.

Iourov, I.Y., Vorsanova, S.G. and Yurov, Y.B. (2012). Single cell genomics of the brain: focus on neuronal diversity and neuropsychiatric diseases, *Curr Genomics*, 13, pp. 477–488.

Isalan, M., Lemerle, C., Michalodimitrakis, K., Horn, C., Beltrao, P., Raineri, E., Garriga-Canut, M. and Serrano, L. (2008). Evolvability and hierarchy in rewired bacterial gene networks, *Nature*, 452, pp. 840–845.

ISCN (2009). A*n International System for Human Cytogenetic Nomenclature*, Shaffer LG, Slovak ML, Campbell LJ (eds), (S Karger, Basel 2009).

Jaffe, L.F. (2005). A calcium-based theory of carcinogenesis, *Adv Cancer Res*, 94, pp. 231–263.

Jiao, X., Wood, L.D., Lindman, M., Jones, S., Buckhaults, P., Polyak, K., Sukumar, S., Carter, H., Kim, D., Karchin, R. and Sjöblom, T. (2012).

Somatic mutations in the notch, NF-KB, PIK3CA, and hedgehog pathways in human breast cancers, *Gene Chromosomes Cancer*, 51, pp. 480–489.

Johnson, B.E., Mazor, T., Hong, C., Barnes, M., Aihara, K., McLean, C.Y., Fouse, S.D., Yamamoto, S., Ueda, H., Tatsuno, K., Asthana, S., Jalbert, L.E., Nelson, S.J., *et al.*, (2014). Mutational analysis reveals the origin and therapy-driven evolution of recurrent glioma, *Science*, 343, pp. 189–193.

Johnson, J.I., Decker, S., Zaharevitz, D., Rubinstein, L.V., Venditti, J.M., Schepartz, S., Kalyandrug, S., Christian, M., Arbuck, S., Hollingshead, M. and Sausville, E.A. (2001). Relationships between drug activity in NCI preclinical in vitro and in vivo models and early clinical trials, *Br J Cancer*, 84, pp. 1424–1431.

Johnson, J.L. (1973). Use of nucleic acid homologies in the taxonomy of anaerobic bacteria, *Int J Syst Bacteriol*, 23, pp. 308–315.

Jones, M.J. and Jallepalli, P.V. (2012). Chromothripsis: Chromosomes in crisis, *Dev Cell*, 23, pp. 908–917.

Jones, P.A. and Baylin, S.B. (2002). The fundamental role of epigenetic events in cancer, *Nat Rev Genet*, 3, pp. 415–428.

Jones, S., Chen, W.D., Parmigiani, G., Diehl, F., Beerenwinkel, N., Antal, T., Traulsen, A., Nowak, M.A., Siegel, C., Velculescu, V.E., Kinzler, K.W., Vogelstein, B., Willis, J. and Markowitz, S.D. (2008). Comparative lesion sequencing provides insights into tumor evolution, *Proc Natl Acad Sci USA*, 105, pp. 4283–4288.

Jonkers, J. and Berns, A. (2004). Oncogene addiction: sometimes a temporary slavery, *Cancer Cell*, 6, pp. 535–538.

Joyner, M.J. and Prendergast, F.G. (2014.) Chasing Mendel: Five questions for personalized medicine, *J Physiol*, 592, pp. 2381–2388.

Joyner, M.J. (2015). 'Moonshot' medicine will let us down, The New York Time, Jan 29, 2015.

Kaiser, J. (2010). UPDATED: A Skeptic Questions Cancer Genome Projects, ScienceInsider (Breaking news and analysis from the world of science policy), 23 April 2010.
http://news.sciencemag.org/2010/04/updated-skeptic-questions-cancer-genome-projects#disqus_thread

Kalghatgi, S., Spina, C.S., Costello, J.C., Liesa, M., Morones-Ramirez, J.R., Slomovic, S., Molina, A., Shirihai, O.S. and Collins, J.J. (2013).

Bactericidal antibiotics induce mitochondrial dysfunction and oxidative damage in Mammalian cells, *Sci Transl Med*, 5, 192ra85.

Kareva, I., Waxman, D.J. and Lakka Klement, G. (2015). Metronomic chemotherapy: An attractive alternative to maximum tolerated dose therapy that can activate anti-tumor immunity and minimize therapeutic resistance, *Cancer Lett*, 358, pp. 100–106.

Kenyon, C., Chang, J., Gensch, E., Rudner, A. and Tabtiang, R. (1993). A C. elegans mutant that lives twice as long as wild type, *Nature*, 366, pp. 461–464.

Kerbel, R.S. and Kamen, B.A. (2004). The anti-angiogenic basis of metronomic chemotherapy, *Nat Rev Cancer*, 4, pp. 423–436.

King, M. (1993). *Species Evolution. The Role of Chromosome Change,* (Cambridge University Press, Cambridge).

Klein, C.A., Blankenstein, T.J., Schmidt-Kittler, O., Petronio M, Polzer, B., Stoecklein, N.H. and Riethmüller, G. (2002). Genetic heterogeneity of single disseminated tumour cells in minimal residual cancer, *Lancet*, 360, pp. 683–689.

Klein, C.A. (2006). Random mutations, selected mutations: A PIN opens the door to new genetic landscapes, *Proc Natl Acad Sci USA*, 103, pp. 18033–18034.

Knauss, S. and Klein, A. (2012). From aneuploidy to cancer: The evolution of a new species? *J. Biosci*, 37, pp. 211–220.

Knight, A., Bailey, J. and Balcombe, J. (2006). Animal carcinogenicity studies: 1. Poor human predictivity, *Altern Lab Anim*, 34, pp. 19–27.

Knudson, A.G. (1971). Mutation and cancer: Statistical study of retinoblastoma, *Proc Natl Acad Sci USA*, 68, pp. 820–823.

Koboldt, D. (2008). Harold Varmus' New Era of Cancer Research, Mass Genomics (June 13, 2008). http://massgenomics.org/2008/06/harold-varmus-new-era-of-cancer-research.html

Kohn, M., Högel, J., Vogel, W., Minich, P., Kehrer-Sawatzki, H., Graves, J.A. and Hameister, H. (2006). Reconstruction of a 450-My-old ancestral vertebrate protokaryotype, *Trends Genet*, 22, pp. 203–210.

Koren, A., Polak, P., Nemesh, J., Michaelson, J.J. Sebat, J. Sunyaev, S.R. and McCarroll, S.A. (2012). Differential relationship of DNA replication timing to different forms of human mutation and variation, *Am J Hum Genet*, 91, pp. 1033–1040.

Kozlov, A.P. (2010). The possible evolutionary role of tumors in the origin of new cell types, *Med Hypotheses*, 74, pp. 177–185.

Kreso, A. and Dick, J.E. (2014). Evolution of the cancer stem cell model, *Cell Stem Cell*, 14, pp. 275–291.

Kryvenko, O.N., Jankowski, M., Chitale, D.A., Tang, D., Rundle, A., Trudeau, S. and Rybicki, B.A. (2012). Inflammation and preneoplastic lesions in benign prostate as risk factors for prostate cancer, *Mod Pathol*, 25, pp. 1023–1032.

Kuhn, T.S. (1962). *The Structure of Scientific Revolutions*, (University of Chicago Press, Chicago).

Kulkarni, P., Shiraishi, T. and Kulkarni, R.V. (2013). Cancer: Tilting at windmills? *Mol Cancer*, 12, e108.

Kumar, A., White, T.A., MacKenzie, A.P., Clegg, N., Lee, C., Dumpit, R.F., Coleman, I., Ng, S.B., Salipante, S.J., Rieder, M.J., Nickerson, D.A., Corey, E. and Lange, P.H., *et al.*, (2011). Exome sequencing identifies a spectrum of mutation frequencies in advanced and lethal prostate cancers, *Proc Natl Acad Sci USA*, 108, pp. 17087–17092.

Kussell, E. and Leibler, S. (2005). Phenotypic diversity, population growth, and information in fluctuating environments, *Science*, 309, pp. 2075–2078.

Lamm, E. (2014). Inheritance Systems. *The Stanford Encyclopedia of Philosophy*. http://plato.stanford.edu/

Landan, G., Cohen, N.M., Mukamel, Z., Bar, A., Molchadsky, A., Brosh, R., Horn-Saban, S., Zalcenstein, D. A., Goldfinger, N., Zundelevich, A., Gal-Yam, E.N., Rotter, V. and Tanay, A. (2012). Epigenetic polymorphism and the stochastic formation of differentially methylated regions in normal and cancerous tissues, *Nat Genet*, 44, pp. 1207–1214.

Lawrence, M.S., Stojanov, P., Polak, P., Kryukov, G.V., Cibulskis, K., Sivachenko, A., Carter, S.L., Stewart, C., Mermel, C.H., Roberts, S.A., Kiezun, A., Hammerman, P.S., McKenna, A., Drier, Y., Zou, L., *et al.*, (2013). Mutational heterogeneity in cancer and the search for new cancer-associated genes, *Nature*, 499, pp. 214–218.

Lawrenson, L. (2010). Tracking profiles of genomic instability in spontaneous transformation and tumorigenesis. Wayne State University Press, Wayne State University, Detroit, USA.

Leaf, C. (2004). Why we're losing the war on cancer, and how to win it, FORTUNE Magazine, March 22, 2004.

Ledford, H. (2010). Big science: The cancer genome challenge, *Nature*, 464, pp. 972–974.

Lee, E., Iskow, R., Yang, L., Gokcumen, O., Haseley, P., Luquette, L.J., Lohr, J.G., Harris, C.C., Ding, L., Wilson, R.K., Wheeler, D.A., Gibbs, R.A., Kucherlapati, R., Lee, C., Kharchenko, P.V. and Park, P.J. (2012). Landscape of somatic retrotransposition in human cancers, *Science*, 337, pp. 967–971.

Lee, R.S., Stewart, C., Carter, S.L., Ambrogio, L., Cibulskis, K., Sougnez, C., Lawrence, M.S., Auclair, D., Mora, J., Golub, T.R., Biegel, J.A., Getz, G. and Roberts, C.W. (2012). A remarkably simple genome underlies highly malignant pediatric rhabdoid cancers, *J Clin Invest*, 122, pp. 2983–2988.

Leiserson, M.D.M., Blokh, D., Sharan, R. and Raphael, B.J. (2013). Simultaneous identification of multiple driver pathways in cancer, *PLoS Comput Biol*, 9, pp. e1003054.

Levine, B. and Abrams, J. (2008). p53: The Janus of autophagy? *Nat Cell Biol*, 10, pp. 637–639.

Liang, Q., Conte, N., Skarnes, W.C. and Bradley, A. (2008). Extensive genomic copy number variation in embryonic stem cells, *Proc Natl Acad Sci USA*, 105, pp. 17453–17456.

Lidstrom, M.E. and Konopka, M.C. (2010). The role of physiological heterogeneity in microbial population behavior, *Nat Chem Biol*, 6, pp. 705–712.

Liu, G., Stevens, J.B., Horne, S.D., Abdallah, B.Y., Ye, K.J., Bremer, S.W., Ye, C.J., Chen, D.J. and Heng, H.H. (2014). Genome chaos: Survival strategy during crisis, *Cell Cycle*, 13, pp. 528–537.

Liu, P., Erez, A., Nagamani, S.C., Dhar, S.U., Kołodziejska, K.E., Dharmadhikari, A.V., Cooper, M.L., Wiszniewska, J., Zhang, F., Withers, M.A., Bacino, C.A., Campos-Acevedo, L.D., Delgado, M.R., Freedenberg, D., Garnica, A., Grebe, T.A., Hernández-Almaguer, D., *et al.*, (2011). Chromosome catastrophes involve replication mechanisms generating complex genomic rearrangements, *Cell*, 146, pp. 889–903.

Loeb, L.A., Springgate, C.F. and Battula, N. (1974). Errors in DNA replication as a basis of malignant changes, *Cancer Res*, 34, pp. 2311–2321.

Loeb, L.A., Bielas, J.H. and Beckman, R.A. (2008). Cancers exhibit a mutator phenotype: Clinical implications, *Cancer Res*, 68, pp. 3551–3557.

Louro, R., Smirnova, A.S. and Verjovski-Almeida, S. (2009). Long intronic noncoding RNA transcription: Expression noise or expression choice? *Genomics*, 93, pp. 291–298.

Lu, P., Weaver, V.M. and Werb, Z. (2012). The extracellular matrix: A dynamic niche in cancer progression, *J Cell Biol*, 196, pp. 395–406.

Luo, J., Li, M., Tang, Y., Laszkowska, M., Roeder, R.G. and Gu, W. (2004). Acetylation of p53 augments its site-specific DNA binding both in vitro and in vivo, *Proc Natl Acad Sci*, 101, pp. 2259–2264.

Luo, J., Solimini, N.L. and Elledge, S.J. (2009). Principles of Cancer Therapy: Oncogene and Non-oncogene Addiction, *Cell*, 136, pp. 823–837.

Lupski, J.R. (2007). Genomic rearrangements and sporadic disease, *Nat Genet*, 39(7 Suppl), pp. S43–47.

Mack, S.C., Witt, H., Piro, R.M., Gu, L. and Zuyderduyn, S. (2014). Epigenomic alterations define lethal CIMP-positive ependymomas of infancy, *Nature*, 506, pp. 445–450.

Maley, C.C., Galipeau, P.C., Li, X., Sanchez, C.A., Paulson, T.G., Blount, P.L. and Reid, B.J. (2004a). The combination of genetic instability and clonal expansion predicts progression to esophageal adenocarcinoma, *Cancer Res*, 64, pp. 7629–7633.

Maley, C.C., Reid, B.J. and Forrest, S. (2004b). Cancer prevention strategies that address the evolutionary dynamics of neoplastic cells: Simulating benign cell boosters and selection for chemosensitivity, *Cancer Epidemiol Biomarkers Prev*, 13, pp. 1375–1384.

Maley, C.C., Galipeau, P.C., Finley, J.C., Wongsurawat, V.J., Li, X., Sanchez, C.A., Paulson, T.G., Blount, P.L., Risques, R.A., Rabinovitch, P.S. and Reid, B.J. (2006). Genetic clonal diversity predicts progression to esophageal adenocarcinoma, *Nat Genet*, 38, pp. 468–473.

Malhotra, A., Lindberg, M., Faust, G.G., Leibowitz, M.L., Clark, R.A., Layer, R.M., Quinlan, A.R. and Hall, I.M. (2013). Breakpoint profiling of 64 cancer genomes reveals numerous complex rearrangements spawned by homology-independent mechanisms, *Genome Res*, 23, pp. 762–776.

Martinez-Castro, P., Ramos, M.C., Rey, J.A., Benitez, J. and Sanchez Cascos, A. (1984). Homozygosity for a Robertsonian translocation (13q14q) in three offspring of heterozygous parents, *Cytogenet Cell Genet*, 38, pp. 310–312.

Martincorena, I., Roshan, A., Gerstung, M., Ellis, P., Van Loo, P., McLaren, S, .Wedge, D.C., Fullam, A., Alexandrov, L.B., Tubio, J.M., Stebbings, L., Menzies, A., Widaa, S., Stratton, M.R., Jones, P.H. and Campbell, P.J. (2015). Tumor evolution. High burden and pervasive positive selection of somatic mutations in normal human skin, *Science*, 348, pp. 880–886.

Marusyk, A., Almendro, V. and Polyak, K. (2012). Intra-tumour heterogeneity: A looking glass for cancer? *Nat Rev Cancer*, 12, pp. 323–334.

Marx, J. (2002). Debate surges over the origins of genetic defects in cancer, *Science*, 297, pp. 544–546.

Mayr, E. (1942). *Systematics and the Origin of Species*, (Columbia Univ. Press, New York).

McCann, J., Choi, E., Yamasaki, E., Ames, B.N. (1975). Detection of carcinogens as mutagens in the Salmonella/microsome test: Assay of 300 chemicals, *Proc Natl Acad Sci USA*, 72, pp. 5135–5139.

McCormick, D.L. and Kavet, R. (2004). Animal models for the study of childhood leukemia: Considerations for model identification and optimization to identify potential risk factors, *Int J Toxicol*, 23, pp. 149–161.

McDermott, D.H., Gao, J.L., Liu, Q., Siwicki, M., Martens, C., Jacobs, P., Velez, D., Yim, E., Bryke, C.R., Hsu, N., Dai, Z., Marquesen, M.M., Stregevsky, E., Kwatemaa, N., Theobald, N., *et al.*, (2015). Chromothriptic cure of WHIM syndrome, *Cell*, 160, pp. 686–699.

McFarland, C.D., Korolev, K.S., Kryukov, G.V., Sunyaev, S.R. and Mirny, L.A. (2013). Impact of deleterious passenger mutations on cancer progression, *Proc Natl Acad Sci USA*, 110, pp. 2910–2915.

McKenzie, G.J., Lee, P.L., Lombardo, M.J., Hastings, P.J. and Rosenberg, S.M. (2001). SOS mutator DNA polymerase IV functions in adaptive mutation and not adaptive amplification, *Mol Cell*, 7, pp. 571–579.

Meyerson, M. and Pellman, D. (2011). Cancer genomes evolve by pulverizing single chromosomes, *Cell*, 144, pp. 9–10.

Merlo, L.M., Pepper, J.W., Reid, B.J. and Maley, C.C. (2006). Cancer as an evolutionary and ecological process, *Nat Rev Cancer*, 6, pp. 924–935.

Merlo, L.M., Kosoff, R.E., Gardiner, K.L. and Maley, C.C. (2011). An in vitro co-culture model of esophageal cells identifies ascorbic acid as a modulator of cell competition, *BMC Cancer*, 11, e461.

Metzger, B.P., Yuan, D.C., Gruber, J.D., Duveau, F. and Wittkopp, P.J. (2015). Selection on noise constrains variation in a eukaryotic promoter, *Nature*, [Epub ahead of print].

Michor, F., Frank, S.A., May, R.M., Iwasa, Y. and Nowak, M.A. (2003). Somatic selection for and against cancer, *J Theor Biol*, 225, pp. 377–382.

Miklos, G. L. (2005). The Human Cancer Genome Project[mdash]one more misstep in the war on cancer, *Nat Biotech*, 23, pp. 535–537.

Miller, B.E., Miller, F.R. and Heppner, G.H. (1981). Interactions between tumor subpopulations affecting their sensitivity to the antineoplastic agents cyclophosphamide and methotrexate, *Cancer Res*, 41, pp. 4378–4381.

Mitelman, F. (2006). 50,000 tumors, 40,000 aberrations, and 300 fusion genes: How much remains? 50 years of 46 human chromosomes: Progress in cytogenetics, NCI, NIH. USA.

Mittral, I. (2007). The disconnection between tumor response and survival, *Nat Clin Pract Oncol*, 4, pp. 203.

Moreira, L.M.A., Santo, L.D.E. and Carvalho, A.F.L. (2013). Down syndrome due to rare inherited 15/21 Robertsonian translocation: Genetics and reproductive counseling, *Rev Med Int Sindr Down*, 17, pp. 36–38.

Moyed, H.S. and Broderick, S.H. (1986). Molecular cloning and expression of hipA, a gene of Escherichia coli K-12 that affects frequency of persistence after inhibition of murein synthesis, *J Bacteriol*, 166, pp. 399–403.

Mukherjee, S. (2010). *The Emperor of All Maladies, A Biography of Cancer.* (Scribner, New York).

Mulla, W., Zhu, J. and Li, R. (2010). Yeast: A simple model system to study complex phenomena of aneuploidy, *Nature*, 468, pp. 321–325.

Muller, H.J. (1951). *Science in Progress: Seventh Series*, ed. Baitsell, G. A., "Radiation damage to the genetic material," (Yale University Press, New Haven) pp. 93–165.

Muotri, A.R., Marchetto, M.C., Coufal, N.G. and Gage, F.H. (2007). The necessary junk: New functions for transposable elements, *Hum Mol Genet.*, 16, pp. R159–167.

Murphy, W.J., Larkin, D.M., Everts-van der Wind, A., Bourque, G., Tesler, G., Auvil, L., Beever, J.E., Chowdhary, B.P., Galibert, F., Gatzke, L., Hitte, C., Meyers, S.N., Milan, D., Ostrander, E.A., Pape, G., *et al.*, (2005). Dynamics of mammalian chromosome evolution inferred from multispecies comparative maps, *Science*, 309, pp. 613–617.

Navarro, A. and Barton, N.H. (2003). Chromosomal speciation and molecular divergence-accelerated evolution in rearranged chromosomes, *Science*, 300, 321–324.

Navin, N., Kendall, J., Troge, J., Andrews, P., Rodgers, L., McIndoo, J., Cook, K., Stepansky, A., Levy, D., Esposito, D., Muthuswamy, L., Krasnitz, A., McCombie, W.R., Hicks, J. and Wigler, M. (2011). Tumour evolution inferred by single-cell sequencing, *Nature*, 472, pp. 90–94.

Nelson, C.M. and Bissell, M.J. (2006). Of extracellular matrix, scaffolds, and signaling: Tissue architecture regulates development, homeostasis, and cancer, *Annu Rev Cell Dev Biol*, 22, pp. 287–309.

Nicholson, J.M. (2013). Will we cure cancer by sequencing thousands of genomes? *Mol Cytogenet*, 6, e57.

Nik-Zainal, S., Alexandrov, L.B., Wedge, D.C., Van Loo, P., Greenman, C.D., Raine, K., Jones, D., Hinton, J., Marshall, J., Stebbings, L.A., Menzies, A., Martin, S., Leung, K., *et al.*, (2012). Mutational processes molding the genomes of 21 breast cancers, *Cell*, 149, pp. 979–993.

Noble, D. (2015). Evolution beyond neo-Darwinism: A new conceptual framework, *J Exp Biol*, 218, pp. 7–13.

Noble, D. (2013). Physiology is rocking the foundations of evolutionary biology, *Exp Physiol*, 98, pp. 1235–1243.

Nordling, C.O. (1953). A new theory on the cancer-inducing mechanism. *Brit J Cancer*, 7, pp. 68–72.

Norimura, T., Nomoto, S., Katsuki, M., Gondo, Y. and Kondo, S. (1996). p53-dependent apoptosis suppresses radiation-induced teratogenesis, *Nat Med*, 2, pp. 577–580.

Nowak, M.A., Komarova, N.L., Sengupta, A., Jallepalli, P.V., Shih, I.M., Vogelstein, B. and Lengauer, C. (2002). The role of chromosomal instability in tumor initiation, *Proc Natl Acad Sci USA*, 99, pp. 16226–16231.

Nowak, M.A., Michor, F. and Iwasa, Y. (2006). Genetic instability and clonal expansion, *J Theor Biol*, 241, pp. 26–32.

Nowell, P.C. (1976). The clonal evolution of tumor cell populations, *Science*, 194, pp. 23–28.

Ober, C., Loisel, D.A. and Gilad, Y. (2008). Sex-specific genetic architecture of human disease, *Nat Rev Genet*, 9, pp. 911–922.

Odom, D.T., Dowell, R.D., Jacobsen, E.S., Gordon, W., Danford, T.W., MacIsaac, K.D., Rolfe, P.A., Conboy, C.M., Gifford, D.K. and Fraenkel, E. (2007). Tissue-specific transcriptional regulation has diverged significantly between human and mouse, *Nat Genet*, 39, pp. 730–732.

Ohshima, S. and Seyama, A. (2010). Cellular aging and centrosome aberrations, *Ann N Y Acad Sci*, 1197, 108–117.

Olivier, M., Petitjean, A., Marcel, V., Pétré, A., Mounawar, M., Plymoth, A., de Fromentel, C.C. and Hainaut, P. (2009). Recent advances in p53 research: An interdisciplinary perspective, *Cancer Gene Ther*, 16, pp. 1–12.

Oshimura, M., Sasaki, M. and Makino, S. (1973). Chromosomal banding patterns in primary and transplanted venereal tumors of the dog, *J Natl Cancer Inst*, 51, pp. 1197–1203.

Page, D., Cheung, V., Haussler, D. and Gibbs, R. (2011). Human genome: What's been most surprising? *Cell*, 147, pp. 9–10.

Park, S.Y., Gonen, M., Kim, H.J., Michor, F. and Polyak, K. (2010). Cellular and genetic diversity in the progression of in situ human breast carcinomas to an invasive phenotype, *J Clin Invest*, 120, pp. 636–644.

Parker, M., Mohankumar, K.M., Punchihewa, C., Weinlich, R., Dalton, J.D., Li, Y., Lee, R., Tatevossian, R.G., Phoenix, T.N., Thiruvenkatam, R., White, E., Tang, B., Orisme, W., *et al.*, (2014). C11orf95-RELA fusions drive oncogenic NF-[kgr]B signalling in ependymoma, *Nature*, 506, pp. 451–455.

Pasquier, E., Kavallaris, M. and André, N. (2010). Metronomic chemotherapy: New rationale for new directions, *Nat Rev Clin Oncol*, 7, pp. 455–465.

Patel, A.P., Tirosh, I., Trombetta, J.J., Shalek, A.K., Gillespie, S.M., Wakimoto, H., Cahill, D.P., Nahed, B.V., Curry, W.T., Martuza, R.L., Louis, D.N., Rozenblatt-Rosen, O., Suvà, M.L., Regev, A. and Bernstein, B.E. (2014). Single-cell RNA-seq highlights intratumoral heterogeneity in primary glioblastoma, *Science*, 344, pp. 1396–1401.

Pathak, S. (1990) *Trends in Chromosome Research*, ed. Sharma, T., Chromosome alterations in speciation and neoplastic transformation: A parallelism, (Springer Verlag, Narosa Publishing House, New Delhi), pp. 204–220.

Patil, A., Kinoshita, K. and Nakamura, H. (2010). Domain distribution and intrinsic disorder in hubs in the human protein-protein interaction network, *Protein Sci*, 19, pp. 1461–1468.

Patra, P. and Klumpp, S. (2013). Population dynamics of bacterial persistence, *PLoS One*, 8, e62814.

Paulson, T.G., Maley, C.C., Li, X., Li, H., Sanchez, C.A., Chao, D.L., Odze, R.D., Vaughan, T.L., Blount, P.L. and Reid, B.J. (2009). Chromosomal instability and copy number alterations in Barrett's esophagus and esophageal adenocarcinoma, *Clin Cancer Res*, 15, pp. 3305–3314.

Pavelka, N., Rancati, G., Zhu, J., Bradford, W.D., Saraf, A., Florens, L., Sanderson, B.W., Hattem, G.L. and Li, R. (2010). Aneuploidy confers quantitative proteome changes and phenotypic variation in budding yeast, *Nature*, 468, pp. 321–325.

Pellestor, F., Gatinois, V., Puechberty, J., Geneviève, D. and Lefort, G. (2014). Chromothripsis: Potential origin in gametogenesis and preimplantation cell divisions: A review, *Fertil Steril*, 102, pp. 1785–1796.

Pennacchio, L.A. (2003). Insights from human/mouse genome comparisons, *Mamm Genome*, 14, pp. 429–436.

Pepper, J.W. (2011). *Targeted Therapies: Mechanisms of Resistance*, ed. Gioeli, D., Somatic evolution of acquired drug resistance in cancer, (Humana Press, New York) pp. 127–134.

Pepper, J.W. (2012). Drugs that target pathogen public goods are robust against evolved drug resistance, *Evol Appl*, 5, pp. 757–761.

Perel, P., Roberts, I., Sena, E., Wheble, P., Briscoe, C., Sandercock, P., Macleod, M., Mignini, L.E., Jayaram, P. and Khan, K.S. (2007). Comparison of treatment effects between animal experiments and clinical trials: systematic review, *BMJ*, 334, e197.

Pfau, S.J. and Amon, A. (2012). Chromosomal instability and aneuploidy in cancer: From yeast to man, *EMBO Rep*, 13, pp. 515–527.

Pollock, P.M., Harper, U.L., Hansen, K.S., Yudt, L.M., Stark, M., Robbins, C.M., Moses, T.Y., Hostetter, G., Wagner, U., Kakareka, J., Salem, G., Pohida, T., Heenan, P., Duray, P., Kallioniemi, O., *et al.*, (2003). High frequency of BRAF mutations in nevi, *Nat Genet*, 33, pp. 19–20.

Ponder, R.G., Fonville, N.C. and Rosenberg, S.M. (2005). A switch from high-fidelity to error-prone DNA double-strand break repair underlies stress-induced mutation, *Mol Cell*, 19, pp. 791–804.

Pujadas, E. and Feinberg, A.P. (2012). Regulated noise in the epigenetic landscape of development and disease, *Cell*, 148, pp. 1123–1131.

Quintana, E., Shackleton, M., Sabel, M.S., Fullen, D.R., Johnson, T.M. and Morrison, S.J. (2008). Efficient tumour formation by single human melanoma cells, *Nature*, 456, pp. 593–598.

Raffaghello, L., Lee, C., Safdie, F.M., Wei, M., Madia, F., Bianchi, G. and Longo, V.D. (2008). Starvation-dependent differential stress resistance protects normal but not cancer cells against high-dose chemotherapy, *Proc Natl Acad Sci USA*, 105, pp. 8215–8220.

Rainey, P.B., Beaumont, H.J., Ferguson, G.C., Gallie, J., Kost, C., Libby, E. and Zhang, X.X. (2011). The evolutionary emergence of stochastic phenotype switching in bacteria, *Microb Cell Fact*, 10, Suppl 1, pp. S14.

Rancati, G., Pavelka, N., Fleharty, B., Noll, A., Trimble, R., Walton, K., Perera, A., Staehling-Hampton, K., Seidel, C.W. and Li, R. (2008). Aneuploidy underlies rapid adaptive evolution of yeast cells deprived of a conserved cytokinesis motor, *Cell*, 135, pp. 879–893.

Reynolds, A.R., Hart, I.R., Watson, A.R., Welti, J.C., Silva, R.G., Robinson, S.D., Da Violante, G., Gourlaouen, M., Salih, M., Jones, M.C., Jones,

D.T., Saunders, G., Kostourou, V., Perron-Sierra, F., Norman, J.C., *et al.*, (2009). Stimulation of tumor growth and angiogenesis by low concentrations of RGD-mimetic integrin inhibitors, *Nat Med*, 15, pp. 392–400.

Righolt, C. and Mai, S. (2012). Shattered and stitched chromosomes-chromothripsis and chromoanasynthesis-manifestations of a new chromosome crisis? *Genes Chromosomes Cancer*, 51, pp. 975–981.

Rockoff, J.D. (2005). Genome project tackles cancer, Baltimore Sun, December 14.

Romiti, A., Cox, M.C., Sarcina. I., Di Rocco, R., D'Antonio, C., Barucca, V. and Marchetti, P. (2013). Metronomic chemotherapy for cancer treatment: A decade of clinical studies, *Cancer Chemother Pharmacol*, 72, pp. 13–33.

Roschke, A.V., Lababidi, S., Tonon, G., Gehlhaus, K.S., Bussey, K., Weinstein, J.N. and Kirsch, I.R. (2005). Karyotypic "state" as a potential determinant for anticancer drug discovery, *Proc Natl Acad Sci USA*, 102, pp. 2964–2969.

Roschke, A.V. and Kirsch, I.R. (2010). Targeting karyotypic complexity and chromosomal instability of cancer cells, *Curr Drug Targets*, 11, pp. 1341–1350.

Rowley, J.D. (1998). The critical role of chromosome translocations inhuman leukemias, *Annu Rev Genet*, 32, pp. 495–519.

Sahai, E. (2007). Illuminating the metastatic process, *Nat Rev Cancer*, 7, pp. 737–749.

Sandberg, R. (2014). Entering the era of single-cell transcriptomics in biology and medicine, *Nat Meth*, 11, pp. 22–24.

Sanders, R. (2011). Are cancers newly evolved species? *Media Relations*, (UC Berkeley News Center, University of Berkeley, California, USA).

Sankar, U. (2005). The delicate toxicity balance in drug discovery, *The Scientist*, 19: 32.

Schröck, E., du Manoir, S., Veldman, T., Schoell, B., Wienberg, J., Ferguson-Smith, M.A., Ning, Y., Ledbetter, D.H., Bar-Am, I., Soenksen, D., Garini, Y. and Ried, T. (1996). Multicolor spectral karyotyping of human chromosomes, *Science*, 273, pp. 494–497.

Sena, E.S., van der Worp, H.B., Bath, P.M., Howells, D.W. and Macleod, M.R. (2010). Publication bias in reports of animal stroke studies leads to major overstatement of efficacy, *PLoS Biol*, 8, e1000344.

Sengupta, S. and Harris,C.C. (2005). p53: Traffic cop at the crossroads of DNA repair and recombination, *Nat Rev Mol Cell Biol*, 6, pp. 44–55.

Setlur, S.R. and Lee, C. (2012). Tumor archaeology reveals that mutations love company, *Cell*, 149, pp. 959–961.

Shanks, N., Greek, R. and Greek, J. (2009). Are animal models predictive for humans? *Philos Ethics Humanit Med*, 4, 2.

Sheltzer, J.M., Blank, H.M., Pfau, S.J., Tange, Y., George, B.M., Humpton, T.J., Brito, I.L., Hiraoka, Y., Niwa, O. and Amon, A. (2011). Aneuploidy drives genomic instability in yeast, *Science*, 333, pp. 1026–1030.

Shen, K.C., Heng, H., Wang, Y., Lu, S., Liu, G., Deng, C.X., Brooks, S.C. and Wang, Y.A. (2005). ATM and p21 cooperate to suppress aneuploidy and subsequent tumor development, *Cancer Res*, 65, pp. 8747–8753.

Shih, C., Shilo, B.Z., Goldfarb, M.P., Dannenberg, A. and Weinberg, R.A. (1979). Passage of phenotypes of chemically transformed cells via transfection of DNA and chromatin, *Proc Natl Acad Sci USA*, 76, pp. 5714–5718.

Sieber, O.M., Heinimann, K. and Tomlinson, I.P. (2003). Genomic instability—the engine of tumorigenesis? *Nat Rev Cancer*, 3, pp. 701–708.

Silk, A.D., Zasadil, L.M., Holland, A.J., Vitre, B., Cleveland, D.W. and Weaver, B.A. (2013). Chromosome missegregation rate predicts whether aneuploidy will promote or suppress tumors, *Proc Natl Acad Sci USA*, 110, pp. E4134–4141.

Silliker, M.E., Monroe, J.A. and Jorden, M.A. (1997). Evaluation of the efficiency of sexual reproduction in restoring Podospora anserina mitochondrial DNA to wild-type, *Curr Genet*, 32, pp. 281–286.

Silva, A.S., Kam, Y., Khin, Z.P., Minton, S.E., Gillies, R.J. and Gatenby, R.A. (2012). Evolutionary approaches to prolong progression-free survival in breast cancer, *Cancer Res*, 72, pp. 6362–6370.

Sjoblom, T., Jones, S., Wood, L.D., Parsons, D.W., Lin, J., Barber, T.D., Mandelker, D., Leary, R.J., Ptak, J., Silliman, N., Szabo, S., Buckhaults, P., Farrell, C., *et al.*, (2006). The consensus coding sequences of human breast and colorectal cancers, *Science*, 314, pp. 268–2274.

Solé, R.V., Valverde, S., Rodriguez-Caso, C. and Sardanyés, J. Can a minimal replicating construct be identified as the embodiment of cancer? *Bioessays*, 36, pp. 503–512.

Solimini, N.L., Luo, J. and Elledge, S.J. (2007). Non-oncogene addiction and the stress phenotype of cancer cells, *Cell*, 130, pp. 986–9888.

Sonnenschein, C. and Soto, A.M. (2008). Theories of carcinogenesis: An emerging perspective, *Semin Cancer Biol*, 18, pp. 372–377.

Sonnenschein, C., Soto, A.M., Rangarajan, A. and Kulkarni, P. (2014). Competing views on cancer, *J Biosci*, 39, pp. 281–302.

Soto, A.M. and Sonnenschein, C. (2011). The tissue organization field theory of cancer: A testable replacement for the somatic mutation theory, *Bioessays*, 33, pp. 332–340.

Soto, A.M. and Sonnenschein, C. (2014). One hundred years of somatic mutation theory of carcinogenesis: Is it time to switch? *Bioessays*, 36, pp. 118–120.

Sottoriva, A., Kang, H., Ma, Z., Graham, T.A., Salomon, M.P., Zhao, J., Marjoram, P., Siegmund, K., Press, MF., Shibata, D. and Curtis, C. (2015). A Big Bang model of human colorectal tumor growth, *Nat Genet*, 47, pp. 209–216.

Spans, L., Clinckemalie, L., Helsen, C., Vanderschueren, D., Boonen, S., Lerut, E., Joniau, S. and Claessens, F. (2013). The genomic landscape of prostate cancer, *Int J Mol Sci*, 14, pp. 10822–10851.

Speicher, M.R., Gwyn Ballard, S. and Ward, D.C. (1996). Karyotyping human chromosomes by combinatorial multi-fluor FISH, *Nat Genet*, 12, pp. 368–375.

Spencer, S.L., Gerety, R.A., Pienta, K.J. and Forrest, S. (2006). Modeling somatic evolution in tumorigenesis, *PLoS Comput Biol*, 2, e108.

Sprouffske, K., Aktipis, A. C, Radich, J.P., Carroll, M., Nedelcu, A.M. and Maley, C.C. (2013). An evolutionary explanation for the presence of cancer nonstem cells in neoplasms, *Evol Appl*, 6, pp. 92–101.

Spudich, J.L. and Koshland, D.E., Jr. (1976). Non-genetic individuality: Chance in the single cell, *Nature*, 262, pp. 467–471.

Stepanenko, A., Andreieva, S., Korets, K., Mykytenko, D., Huleyuk, N., Vassetzky, Y. and Kavsan, V. (2015). Step-wise and punctuated genome evolution drive phenotype changes of tumor cells, *Mutation Research*, 771, pp. 56–69.

Stephens, P.J., Tarpey, P.S., Davies, H., Van Loo, P., Greenman, C., Wedge, D.C., Nik-Zainal, S., Martin, S., Varela, I., Bignell, G.R., Yates, L.R., Papaemmanuil, E., *et al.*, (2012). The landscape of cancer genes and mutational processes in breast cancer, *Nature*, 486, pp. 400–404.

Stephens, P.J., Greenman, C.D., Fu, B., Yang, F., Bignell, G.R., Mudie, L.J., Pleasance, E.D., Lau, K.W., Beare, D., Stebbings, L.A., McLaren, S., Lin, M.L., McBride, D.J., Varela, I., Nik-Zainal, S., *et al.*, (2011). Massive genomic rearrangement acquired in a single catastrophic event during cancer development, *Cell*, 144, pp. 27–40.

Stevens, J.B., Liu, G., Bremer, S.W., Ye, K.J., Xu, W., Xu, J., Sun, Y., Wu, G.S., Savasan, S., Krawetz, S.A., Ye, C.J. and Heng, H.H. (2007). Mitotic cell death by chromosome fragmentation, *Cancer Res*, 67, pp. 7686–7694.

Stevens, J.B., Abdallah, B.Y., Regan, S.M., Liu, G., Bremer, S.W., Ye, C.J. and Heng, H.H. (2010). Comparison of mitotic cell death by chromosome fragmentation to premature chromosome condensation, *Mol Cytogenet*, 3, e20.

Stevens, J.B., Abdallah, B.Y., Liu, G., Ye, C.J., Horne, S.D., Wang, G., Savasan, S., Shekhar, M., Krawetz, S.A., Hüttemann, M., Tainsky, M.A., Wu, G.S., Xie, Y., Zhang, K. and Heng, H.H. (2011). Diverse system stresses: Common mechanisms of chromosome fragmentation, *Cell Death Dis*, 2:e178.

Stevens, J.B., Horne, S.D., Abdallah, B.Y., Ye, C.J. and Heng, H.H. (2013a). Chromosomal instability and transcriptome dynamics in cancer, *Cancer Metastasis Rev*, 32, pp. 391–402.

Stevens, J.B., Abdallah, B.Y., Liu, G., Horne, S.D., Bremer, S.W., Ye, K.J., Huang, J.Y., Kurkinen, M., Ye, C.J. and Heng, H.H. (2013b). Heterogeneity of cell death, *Cytogenet Genome Res*, 139, pp. 164–73.

Stevens, J.B. and Heng, H.H. (2013). *Human Interphase Chromosomes: The Biomedical Aspects, eds.* Yurov, Vorsanova, Iourov., Chapter 5, "Differentiating chromosome fragmentation and premature chromosome condensation," (Springer, New York) pp. 85–105.

Stevens, J.B., Liu, G., Abdallah, B.Y., Horne, S.D., Ye, K.J., Bremer, S.W., Ye, C.Y., Krawetz, S.K. and Heng, H.H. (2014). Unstable genome elevate transcriptome dynamics, *Int J Cancer*, 134, pp. 2074–2087.

Storchová, Z., Breneman, A., Cande, J., Dunn, J., Burbank, K., O'Toole, E. and Pellman, D. (2006). Genome-wide genetic analysis of polyploidy in yeast, *Nature*, 443, pp. 541–547.

Sturmberg, J.P. and Martin, C. (2013). *Handbook of Systems and Complexity in Health.* (Springer, New York).

Strauss, B. (2013). Mutation and cancer: A view from retirement, *DNA Repair (Amst)*, 12, pp. 875.

Strohman, R. (1997). The coming Kuhnian revolution in biology, *Nat Biotechnol*, 15, pp. 194–120.

Strohman, R. (1999). The upcoming biological revolution: An interview with Richard Strohman, Wild Duck Review Vol. V No. 2 on "Biotechnology".

Summers, K. and Crespi, B. (2008). The androgen receptor and prostate cancer: A role for sexual selection and sexual conflict? *Med Hypotheses*, 70, pp. 435–443.

Taneja, V. and David, C.S. (2001). Lessons from animal models for human autoimmune diseases, *Nat Immunol*, 2, pp. 781–784.

Templeton, A.R. (1989). *Speciation and its Consequences*, eds. Otte, D. and Endler, J.A., The meaning of species and speciation: A genetic perspective, (Sinauer Associates, Inc. Sunderland, MA) pp. 3–27.

Timp, W. and Feinberg, A.P. (2013). Cancer as a dysregulated epigenome allowing cellular growth advantage at the expense of the host, *Nat Rev Cancer*, 13, pp. 497–510.

Tomasetti, C. and Vogelstein, B. (2015). Variation in cancer risk among tissues can be explained by the number of stem cell divisions, *Science*, 347, pp. 78–81.

Tompa, P. and Csermely, P. (2004). The role of structural disorder in the function of RNA and protein chaperones, *FASEB J*, 18, pp. 1169–1175.

Tompa, P. (2012). Intrinsically disordered proteins: A 10-year recap, *Trends Biochem Sci*, 37, pp. 509–516.

Topol, E.J. (2004). Failing the public health—rofecoxib, Merck, and the FDA, *N Engl J Med*, 351, pp. 1707–1709.

Torkelson, J., Harris, R.S., Lombardo, M.J., Nagendran, J., Thulin, C. and Rosenberg, S.M. (1997). Genome-wide hypermutation in a subpopulation of stationary-phase cells underlies recombination-dependent adaptive mutation, *EMBO J*, 16, pp. 3303–3311.

Torres, E.M., Dephoure, N., Panneerselvam, A., Tucker, C.M., Whittaker, C.A., Gygi, S.P., Dunham, M.J. and Amon, A. (2010). Identification of aneuploidy-tolerating mutations, *Cell*, 143, pp. 71–83.

Torti, D. and Trusolino, L. (2011). Oncogene addiction as a foundational rationale for targeted anti-cancer therapy: Promises and perils, *EMBO Mol Med*, 3, pp. 623–636.

T.S. Institute, (2014), vol. 2014.

Tubio, J.M. and Estivill, X. (2011). Cancer: When catastrophe strikes a cell, *Nature*, 470, pp. 476–477.

Twombly, R. (2005). Cancer surpasses heart disease as leading cause of death for all but the very elderly, *J Natl Cancer Inst*, 97, pp. 330–331.

Upender, M.B., Habermann, J.K., McShane, L.M., Korn, E.L., Barrett, J.C., Difilippantonio, M.J. and Ried, T. (2004). Chromosome transfer induced aneuploidy results in complex dysregulation of the cellular transcriptome in immortalized and cancer cells, *Cancer Res*, 64, pp. 6941–6949.

Uversky, V.N. and Dunker, A.K. (2010). Understanding protein non-folding, *Biochim Biophys Acta*, 1804, pp. 1231–1264.

Valdar, W., Solberg, L.C., Gauguier, D., Cookson, W.O., Rawlins, J.N., Mott, R. and Flint, J. (2006). Genetic and environmental effects on complex traits in mice, *Genetics*, 174, pp. 959–984.

Valind, A., Jin, Y., Baldetorp, B. and Gisselsson, D. (2013). Whole chromosome gain does not in itself confer cancer-like chromosomal instability, *Proc Natl Acad Sci USA*, 110, pp. 21119–21123.

Vander Heiden, M.G., Cantley, L.C. and Thompson, C.B. (2009). Understanding the Warburg effect: The metabolic requirements of cell proliferation, *Science*, 324, pp. 1029–1033.

van der Bij, G.J., Oosterling, S.J., Beelen, R.H., Meijer, S., Coffey, J.C. and van Egmond, M. (2009). The perioperative period is an underutilized window of therapeutic opportunity in patients with colorectal cancer, *Ann Surg*, 249, pp. 727–734.

van Valen, L. and Mairorana, V.C. (1991). HeLa, a new microbial species, *Evol Theor*, 10, pp. 71–74.

van der Worp, H.B., Howells, D.W., Sena, E.S., Porritt, M.J., Rewell, S., O'Collins, V. and Macleod, M.R. (2010). Can animal models of disease reliably inform human studies? *PLoS Med*, 7, e1000245.

Vaux, D.L. (2011). In defense of the somatic mutation theory of cancer, *Bioessays,* 33, pp. 341–343.

Veltman, J.A. and Brunner, H.G. (2012). De novo mutations in human genetic disease, *Nat Rev Genet*, 13, pp. 565–575.

Venet, D., Dumont, J.E. and Detours, V. (2011). Most random gene expression signatures are significantly associated with breast cancer outcome, *PLoS Comput Biol*, 7, pp. e1002240.

Venter, J.C., Adams, M.D., Myers, E.W., Li, P.W., Mural, R.J., Sutton, G.G., Smith, H.O., Yandell, M., Evans, C.A., Holt, R.A., Gocayne, J.D., Amanatides, P., Ballew, R.M., Huson. D.H., Wortman, J.R., *et al.* (2001). The sequence of the human genome, *Science*, 291, pp. 1304–1351.

Vincent, M.D. (2010). The animal within: Carcinogenesis and the clonal evolution of cancer cells are speciation events sensu stricto, *Evolution*, 64, pp. 1173–1183.

Vincent, T.L. and Gatenby, R.A. (2008). An evolutionary model for initiation, promotion, and progression in carcinogenesis, *Int J Oncol*, 32, pp. 729–737.

Visscher, P.M., Brown, M.A., McCarthy, M.I. and Yang, J. (2012). Five years of GWAS discovery, *Am J Hum Genet*, 90, pp. 7–24.

Vittecoq, M., Roche, B., Daoust, P., Ducasse, H., Missé, D., Abadie, J., Labrut, S., Renaud, F., Gauthier-Clerc, M. and Thomas, F. (2013). Cancer: A missing link in ecosystem functioning? *Trends Ecol Evol*, 28, pp. 628–635.

Vogelstein, B. and Kinzler, K.W. (1993). The multistep nature of cancer, *Trends Genet*, 9, pp. 138–141.

Vogelstein, B., Papadopoulos, N., Velculescu, V.E., Zhou, S., Diaz, L.A. and Kinzler, K.W. (2013). Cancer genome landscapes, *Science*, 339, pp. 1546–1558.

Waddington, C.H. (1957). *The Strategy of Genes.* (Allen and Unwin, New York).

Wallace, D.C. (2012). Mitochondria and cancer, *Nat Rev Cancer*, 12, pp. 685–698.

Wallace, D.C. (2013). A mitochondrial bioenergetic etiology of disease, *J Clin Invest*, 123, pp. 1405–1412.

Wang, F., Durfee, L.A. and Huibregtse, J.M. (2013). A cotranslational ubiquitination pathway for quality control of misfolded proteins, *Mol Cell*, 50, pp. 368–378.

Wang, Y., Waters, J., Leung, M.L., Unruh, A., Roh, W., Shi, X., Chen, K., Scheet, P., Vattathil, S., Liang, H., Multani, A., Zhang, H., Zhao, R., Michor, F., Meric-Bernstam, F. and Navin, N.E. (2014). Clonal evolution in breast cancer revealed by single nucleus genome sequencing, *Nature*, 512, pp. 155–160.

Warburg, O. (1956). On respiratory impairment in cancer cells, *Science*, 124, pp. 269–270.

Wayne, L.G., Brenner, D.J., Colwell, R.R., Grimont, P.A.D., Kandler, O., Krichevsky, M.I., Moore, L.H., Moore, W.E.C., Murray, R.G.E., Stackebrandt, E., Starr, M.P. and Trüper, H.G. (1987). Report of the ad hoc committee on reconciliation of approaches to bacterial systematics, *Int J Syst Bacteriol*, 37, pp. 463–464.

Weaver, B.A., Silk, A.D., Montagna, C., Verdier-Pinard, P. and Cleveland, D.W. (2007). Aneuploidy acts both oncogenically and as a tumor suppressor, *Cancer Cell*, 11, pp. 25–36.

Weaver, V.M., Petersen, O.W., Wang, F., Larabell, C.A., Briand, P., Damsky, C. and Bissell, M.J. (1997). Reversion of the malignant phenotype of human breast cells in three-dimensional culture and in vivo by integrin blocking antibodies, *J Cell Biol*, 137, pp. 231–245.

Weinberg, R.A. (1982). Fewer and fewer oncogenes, *Cell*, 30, pp. 3–4.

Weinberg, R.A. (2010). Point: Hypotheses first, *Nature*, 464, pp. 678.

Weinberg, R.A. (2014). Coming full circle-from endless complexity to simplicity and back again, *Cell*, 157, pp. 267–271.

Weinstein, I.B. (2002). Cancer. Addiction to oncogenes—the Achilles heal of cancer, *Science*, 297, pp. 63–64.

Weinstein, I.B. and Joe, A.K. (2006). Mechanisms of disease: Oncogene addiction—a rationale for molecular targeting in cancer therapy, *Nat Clin Pract Oncol*, 3, pp. 448–457.

Weischenfeldt, J., Simon, R., Feuerbach, L., Schlangen, K., Weichenhan, D., Minner, S., Wuttig, D., Warnatz, H.J., Stehr, H., Rausch, T., Jäger, N., Gu, L., Bogatyrova, O., *et al.*, (2013). Integrative genomic analyses reveal an androgen-driven somatic alteration landscape in early-onset prostate cancer, *Cancer Cell*, 23, pp. 159–170.

Welch, D.R., Krizman, D.B. and Nicolson, G.L. (1984). Multiple phenotypic divergence of mammary adenocarcinoma cell clones. I. In vitro and in vivo properties, *Clin Exp Metastasis*, 2, pp. 333–355.

Welch, D.R. (1989). *Carcinogenesis and Dietary Fat*, ed. S. Abraham., Chapter 17 "Factors involved in the development and maintenance of tumor heterogeneity." (Kluwer Academic Publishers, Boston) pp. 279–301.

Welch, H.G. and Black, W.C. (2010). Overdiagnosis in cancer, *J Natl Cancer Inst*, 5, pp. 102, pp. 605–613.

Welsh, J.S. (2011). Contagious cancer, *Oncologist*, 16.1, pp. 1–4.

White, F.V., Dehner, L.P., Belchis, D.A., Conard. K., Davis. M.M., Stocker, J.T., Zuppan, C.W., Biegel, J.A. and Perlman, E.J. (1999). Congenital disseminated malignant rhabdoid tumor: A distinct clinicopathologic entity demonstrating abnormalities of chromosome 22q11, *Am J Surg Pathol*, 23, pp. 249–256.

WHO World Cancer Report, 2014. http://www.who.int/mediacentre/factsheets/fs297/en/

Wicha, M.S., Liu, S. and Dontu, G. (2006). Cancer stem cells: An old idea—a paradigm shift, *Cancer Res*, 66, pp. 1883–1890.

Wilkins, A.S. (2010). The enemy within: An epigenetic role of retrotransposons in cancer initiation, *Bioessays*, 32, pp. 856–865.

Wilkins, A.S. and Holliday, R. (2009). The evolution of meiosis from mitosis, *Genetics*, 181, pp. 3–12.

Williams, B.R., Prabhu, V.R., Hunter, K.E., Glazier, C.M., Whittaker, C.A., Housman, D.E. and Amon, A. (2008). Aneuploidy affects proliferation and spontaneous immortalization in mammalian cells, *Science*, 322, pp. 703–709.

Wood, L.D., Parsons, D.W., Jones, S., Lin, J., Sjöblom, T., Leary, R.J., Shen, D., Boca, S.M., Barber, T., Ptak, J., Silliman, N., Szabo, S., Dezso, Z., Ustyanksky, V., Nikolskaya, T., *et al.*, (2007). The genomic landscapes of human breast and colorectal cancers, *Science*, 318, pp. 1108–1113.

Woodruff, M.F. (1983). Cellular heterogeneity in tumours, *Br J Cancer*, 47, pp. 589–594.

Wright, B.E. (1953a). Specific suppression by glycine of the phenotypic expression of partially reverted glycineless E. coli mutants, *Nature*, 172, pp. 308–309.

Wright, B.E. (1953b). Auto-adaptation: A new phenomenon observed in a bacterial population, *J Bacteriol*, 66, pp. 407–420.

Wright, S. (1932). The roles of mutation, inbreeding, crossbreeding, and selection in evolution, *Proceedings of the Sixth International Congress on Genetics*, pp. 355–366.

Wundwelich, V. (2006). Early references to the mutational origin of cancer, *Int J Epidemiol*, 36, pp. 246–247.

Yang, F., O'Brien, P.C., Wienberg, J. and Ferguson-Smith, M.A. (1997). A reappraisal of the tandem fusion theory of karyotype evolution in Indian muntjac using chromosome painting, *Chromosome Res*, 5, pp. 109–117.

Yang, K.R., Mooney, S.M., Zarif, J.C., Coffey, D.S., Taichman, R.S. and Pienta, K.J. (2014). Niche inheritance: A cooperative pathway to enhance cancer cell fitness through ecosystem engineering, *J Cell Biochem*, 115, pp. 1478–1485.

Yang, X., Schadt, E.E., Wang, S., Wang, H., Arnold, A.P., Ingram-Drake, L., Drake, T.A. and Lusis, A.J. (2006). Tissue-specific expression and regulation of sexually dimorphic genes in mice, *Genome Res*, 16, pp. 995–1004.

Ye, C.J., Lu, W., Liu, G., Bremer, S.W., Wang, Y.A., Moens, P., Hughes, M., Krawetz, S.A. and Heng, H.H. (2001). The combination of SKY and specific loci detection with FISH or immunostaining, *Cytogenet Cell Genet*, 93, pp. 195–202.

Ye, C.J., Liu, G., Bremer, S.W. and Heng, H.H. (2007). Cancer Genomics: The dynamics of cancer chromosomes and genomes, *Cytogenet Genome Res.*, 118, pp. 237–246.

Ye, C.J., Stevens, J.B., Liu, G., Bremer, S.W., Jaiswal, A.S., Ye, K.J., Lin, M.F., Lawrenson, L., Lancaster, W.D., Kurkinen, M., Liao, J.D., Gairola, C.G., Shekhar, M.P., Narayan, S., Miller, FR. and Heng, H.H. (2009a). Genome based cell population heterogeneity promotes tumorigenicity: The evolutionary mechanism of cancer, *J Cell Physiol*, 219, pp. 288–300.

Ye, C.J., Lawrenson, L., Liu, G., Stevens, J.B., Ye, K.J., Bremer, S.W. and Heng, H.H. (2009b). *Fluorescence In Situ Hybridization (FISH)—Application*, ed. Liehr, T., Chapter 19, "Simultaneous fluorescence immunostaining and FISH," (Springer Publishing) pp. 193–216.

Yoo, M.H. and Hatfield, D.L. (2008). The cancer stem cell theory: Is it correct? *Mol Cells*, 26, pp. 514–516.

Yu, C. Yu, J. Yao, X. Wu, K.K. Lu, Y. Tang, S. Li, X. Bao, L. Li, X. Hou, Y. Wu, R. Jian, M. Chen, R. *et al.*, (2014) Discovery of biclonal origin and a novel oncogene SLC12A5 in colon cancer by single-cell sequencing, *Cell Res*, 24, pp. 701–712.

Yu, M.H., Huang, Y.C. and Chang, P.L. (2014). Rapid screening of the heterogeneity of DNA methylation by single-strand conformation polymorphism and CE-LIF in the presence of electro-osmotic flow, *Electrophoresis*, 35, pp. 2378–2385.

Zaidi, S.K., Van Wijnen, A.J., Lian, J.B., Stein, J.L. and Stein, G.S. (2013). Targeting deregulated epigenetic control in cancer. Targeting deregulated epigenetic control in cancer, *J Cell Physiol*, 228, pp. 2103–2108.

Zhang, Y. and Rowley, J.D. (2011). Chronic myeloid leukemia: Current perspectives, *Clin Lab Med*, 31, pp. 687–698.

Zheng, J. (2012). Energy metabolism of cancer: Glycolysis versus oxidative phosphorylation, *Oncol Lett*, 4, pp. 1151–1157.

Zhu, J., Pavelka, N., Bradford, W.D., Rancati, G. and Li, R. (2012). Karyotypic determinants of chromosome instability in aneuploid budding yeast, *PLoS Genet*, 8, e1002719.

Zong, C., Lu, S., Chapman, A.R. and Xie, X.S. (2012). Genome-wide detection of single-nucleotide and copy-number variations of a single human cell, *Science*, 338, pp. 1622–1626.

Index

Abbreviations

APL — acute promyelocytic leukemia
BMR — basal mutation rate
CCA — clonal chromosome aberration
C-Frag — chromosome fragmentation
CGS — cancer genome sequencing
CIN — chromosomal instability
CML — chronic myelogenous leukemia
CMV — cytomegalovirus
CNA — copy number alteration
COSMIC — catalogue of somatic mutations in cancer
CTCT — canine transmissible venereal tumor
DMF — defective mitotic figure
Dox — doxorubicin
HGP — human genome project
HPV — human papillomavirus
IDPs — intrinsically disordered proteins
MSI — micro-satellite instability
NCAB — national cancer advisory board
NCCA — non-clonal chromosome aberration
NCI — National Cancer Institute
PIN — point mutation instability
SKY — spectral karyotyping
TCGA — The Cancer Genome Atlas
TDFT — tasmanian devil facial tumor
TEs — transposable elements
TOFT — tissue organization field theory
WGS — whole genome sequencing